House of Learning

ENVIRONMENTAL RISK ASSESSMENT OF GENETICALLY MODIFIED ORGANISMS

Volume 4. Challenges and Opportunities with Bt Cotton in Vietnam

ENVIRONMENTAL RISK ASSESSMENT OF
GENETICALLY MODIFIED ORGANISMS SERIES

Titles available

Volume 1. A Case Study of Bt Maize in Kenya
Edited by A. Hilbeck and D.A. Andow

**Volume 2. Methodologies for Assessing
Bt Cotton in Brazil**
Edited by A. Hilbeck, D.A. Andow and E.M.G. Fontes

Volume 3. Methodologies for Transgenic Fish
Edited by A.R. Kapuscinski, S. Li, K.R. Hayes and G. Dana

**Volume 4. Challenges and Opportunities with Bt
Cotton in Vietnam**
Edited by D.A. Andow, A. Hilbeck and Nguyễn Vặn Tuất

ENVIRONMENTAL RISK ASSESSMENT OF GENETICALLY MODIFIED ORGANISMS

Volume 4. Challenges and Opportunities with Bt Cotton in Vietnam

Edited by

David A. Andow
Department of Entomology
University of Minnesota
Saint Paul, Minnesota, USA

Angelika Hilbeck
Institute for Integrative Biology
Swiss Federal Institute
  of Technology
Zurich, Switzerland

Nguyễn Vặn Tuất
National Institute for Plant
  Protection
Hanoi, Vietnam

Series Editors

Anne R. Kapuscinski
Department of Fisheries,
  Wildlife and Conservation
  Biology
University of Minnesota
St Paul, Minnesota, USA

and

Peter J. Schei
Fridtjof Nansens Institute
Lysaker, Norway

www.cabi.org

CABI is a trading name of CAB International

CABI Head Office
Nosworthy Way
Wallingford
Oxfordshire OX10 8DE
UK

Tel: +44 (0)1491 832111
Fax: +44 (0)1491 833508
E-mail: cabi@cabi.org
Website: www.cabi.org

CABI North American Office
875 Massachusetts Avenue
7th Floor
Cambridge, MA 02139
USA

Tel: +1 617 395 4056
Fax: +1 617 354 6875
E-mail: cabi-nao@cabi.org

A catalogue record for this book is available from the British Library, London, UK.

Library of Congress Cataloging-in-Publication Data

Environmental risk assessment of genetically modified organism / edited by A. Hilbeck and D. Andow.
 p. cm.
 Includes bibliographical references and index.
 ISBN 0-85199-861-5 (alk. paper)
1. Crops--Genetic engineering--Environmental aspects. 2. Transgenic plants--Risk assessment. 3. Corn--Genetic engineering--Kenya--Case studies.
I. Hilback, A. (Angelika) II. Andow, David Alan. III. Title.

SB123.57.E59 2004
631.5'23--dc22

2004007981

ISBN-13: 978 1 84593 390 6

Typeset by SPi, Pondicherry, India.
Printed and bound in the UK by Biddles Ltd, King's Lynn.

Contents

The colour plate section can be found following p. 42.

Contributors

Important note: *All of the Vietnamese names in this book are stated in the standard format in Vietnamese, with the family name first, followed by the given name. Vietnamese typically address each other by the given name, in part because there are relatively few family names. Following international citation standards, Vietnamese references are called using only the family name. However, to facilitate identification of the actual author the entire given name is included in the references for each chapter.*

Amugune, Nelson Onzere, *Lecturer, School of Biological Sciences, University of Nairobi, PO Box 30197, Riverside Drive, Chiromo Campus, Nairobi, Kenya. noamugune@yahoo.co.uk*

Andow, David A., *Professor of Insect Ecology, Department of Entomology, University of Minnesota, 219 Hodson Hall, 1980 Folwell Avenue, St Paul, Minnesota 55108, USA. dandow@umn.edu*

Anyango, Beatrice, *Senior Lecturer, School of Biological Sciences, University of Nairobi, PO Box 30197, Riverside Drive, Chironomo Campus, Nairobi, GPO 00100, Kenya. banyango@uonbi.ac.ke*

Arpaia, Salvatore, *Researcher, Department of Biotechnology/Health and Environment Protection, ENEA – Italian National Agency for New Technologies, Energy and Environment, Research Centre Trisaia, S.S. 106 Ionica km 419+500, Rotondella, Montana I-75026, Italy. salvatore.arpaia@ trisaia.enea.it*

Birch, A.N.E., *LEAF Coordinator, SCRI, Invergowrie, Dundee DD2 5DA, Scotland, UK. N.Birch@scri.ac.uk*

Bùi Cách Tuyến, *Rector, Ho Chi Minh Agriculture and Forestry University (HCM AFU), Thu Duc, Ho Chi Minh City, Vietnam. bctuyen@hcm.vnn.vn*

Capalbo, Deise, *Embrapa Environment (CNPMA), Rodovia SP 340, Km 127.5, CP 69, Tanquinho Velho, Jaguariúna, SP, CEP 13820-000, Brazil. deise@cnpma.embrapa.br*

Caprio, Michael, *Professor of Insect Genetics, Entomology and Plant Pathology, Mississippi State University, Box 9775, Mississippi 39762, USA. mcaprio@entomology.msstate.edu*

Chen, Yolanda H., *Entomologist, Crop and Environmental Sciences Division, International Rice Research Institute (IRRI), DAPA Box 7777, Metro Manila, Philippines. y.chen@cgiar.org*

Đàng Năng Bửu, *Nhaho Research Institute for Cotton and Agricultural Development, Nha Ho, Ninh Son, Ninh Thuan, Vietnam. ktxricfc@yahoo.com*

Depicker, Anna, *Professor, Plant Systems Biology, Plant Gene Regulation, Ghent University, VIB, Technologie Park 927, Gent B-9052, Belgium. Anna.Depicker@UGent.be*

Đinh Quyết Tâm, *Bee Research Centre, MARD, Hanoi, Vietnam. dinhqtam@ hn.vnn.vn*

Dương Xuân Diêu, *Nhaho Research Institute for Cotton and Agricultural Development, Nha Ho, Ninh Son, Ninh Thuan, Vietnam. dxdieu@yahoo.com*

Fitt, Gary P., *Deputy Chief, CSIRO Entomology, Long Pocket Laboratories, 120 Meiers Road, Indooroopilly, Brisbane, Queensland, QLD 4068, Australia. Gary.Fitt@csiro.au*

Fontes, Eliana, *Research Group Leader, Biosafety, Biological Control Unit, Embrapa Genetic Resources and Biotechnology, Parque Estacão Biológica, Caixa Postal 02372, Av. W5 Norte – Final, Brasília, DF 70770-900, Brazil. gmo-era@cenargen.embrapa.br*

Ghosh, Kakoli, *Agriculture Officer (Capacity Building), Plant Production and Protection Division (AGP), Seed and Plant Genetic Resources Service (AGPS), Food and Agriculture Organization (FAO), Viale Delle Terme di Caracalla, Rome, Lazio I-00100, Italy. kakoli.ghosh@fao.org*

Giband, Marc, *CIRAD, Embrapa Algodão, Rua Osvaldo Cruz 1143 – Centenário, Campina Grande, PB 58 107-720, Brazil. marc.giband@cirad.fr*

Guo, Jian-Ying, *State Key Laboratory for Biology of Plant Diseases and Insect Pests, Institute of Plant Protection, Chinese Academy of Agricultural Sciences (CAAS), 12, Zhong-Guan-Cun, Nan-Da-Jie, Beijing 100081, China (P.R.). guojy@cjac.org.cn*

Hilbeck, Angelika, *Institute of Integrative Biology, Swiss Federal Institute of Technology ETHZ, Universitätsstrasse 16, Zürich CH-8092, Switzerland. angelika.hilbeck@env.ethz.ch*

Hồ Văn Chiến, *Director, Southern Regional Department, Vietnam Plant Protection Department (PPD), Long Dinh, Chau Thanh, Tien Giang province, Vietnam. ttbvtvpn@hcm.vnn.vn*

Hoàng Anh Tuấn, *Nhaho Research Institute for Cotton and Agricultural Development, Nha Ho, Ninh Son, Ninh Thuan, Vietnam. hahtuan@yahoo.com*

Hoang Ngoc Binh, *Deputy Head of Planning and Investment Department, Vietnam Cotton Company (VCC), Lot 1 15–17 D2 Street, Northern Van Thanh Ward 25, Binh Thanh District, Ho Chi Minh City, Vietnam. bongvietnam@vnn.vn*

Hoàng Thanh Nhàn, *BCH National Focal Point, Nature Conservation Division, Vietnam Environment Protection Agency (VEPA), Vietnam Ministry of Natural Resources and the Environment (MONRE), 67 Nguyen Du, Hanoi, Vietnam. hnhan@nea.gov.vn*

Johnston West, Jill, *Utah Division of Wildlife Resources, 1594 W North Temple, Suite 2110, Salt Lake City, Utah 84114, USA. jillwest@utah.gov*

Just, David R., *Assistant Professor, Applied Economics and Management, Cornell University, 254 Warren Hall, Ithaca, New York 14853, USA. drj3@cornell.edu*

Lã Phạm Lân, *Plant Protection Division, Institute of Agricultural Sciences of Southern Vietnam (IAS), MARD, 121 Nguyen Binh Khiem, District 1, Ho Chi Minh City, Vietnam. lphlan@hcmc.netnam.vn*

Lang, Andreas, *Institute of Environmental Geosciences, University of Basel, Bernoullistr. 30, CH-4056 Basel, Switzerland. andreas.lang@unibas.ch*

Lê Minh Sắt, *Deputy Director, Department of Science and Technology, Vietnam Ministry of Science and Technology (MOST), 39 Tran Hung Dao, Hoan Kiem District, Hanoi, Vietnam. lmsat@most.gov.vn*

Lê Quang Quyển, *Director, Nhaho Research Institute for Cotton and Agricultural Development, Nha Ho, Ninh Son, Ninh Thuan, Vietnam. quyen@nhahocotton.org.vn*

Lê Thị Thu Hồng, *Head of Division of Science and International Cooperation, Southern Fruit Research Institute (SOFRI), Box 203, My Tho, Long Dinh, Chau Thanh, Tien Giang, Vietnam. ltthong@hcm.vnn.vn*

Lövei, Gabor L., *Senior Scientist, Faculty of Agriculture, Department of Integrated Pest Management, University of Arhus, Flakkebjerg Research Centre, Slagelse DK-4200, Denmark. gabor.lovei@agrsci.dk*

Lu, Bao-Rong, *Deputy Director, Institute of Biodiversity Science, Fudan University, Handan Road 220, Shanghai, P.R. 200433, China (P.R.). brlu@fudan.edu.cn*

Mai Văn Hào, *Nhaho Research Institute for Cotton and Agricultural Development, Nha Ho, Ninh Son, Ninh Thuan, Vietnam. maivanhao@yahoo.com*

Manachini, Barbara, *Professor Researcher, Department of Animal Biology, University of Palermo, Via Archirafi 18, 2, 90123 Palermo, Italy. b.manachini@unipa.it*

Nelson, Kristen C., *Associate Professor, Department of Forest Resources and Department of Fisheries, Wildlife and Conservation Biology, University of Minnesota, 115 Green Hall, 1530 Cleveland Avenue North, St Paul, Minnesota 55108, USA. kcn@umn.edu*

Nguyễn Hồng Sơn, *Deputy Director, Plant Protection Research Institute (PPRI), MARD, Dong Ngac, Tu Liem, Hanoi, Vietnam. hanhson@hn.vnn.vn*

Nguyễn Hữu Hổ, *Institute of Tropical Biology – Vietnamese Academy of Science and Technology, 1 Mac Dinh Chi Street, 1st District, Ho Chi Minh City, Vietnam. geneng@hcm.vnn.vn*

Nguyễn Hữu Huân, *Vice Director General, Vietnam Plant Protection Department (PPD), MARD, 28 Mac Dinh Chi Street, 1st District, Ho Chi Minh City, Vietnam. ppdsouth@hcm.fpt.vn*

Nguyễn Thị Hai, *Head of VCC Liaison Office in Ho Chi Minh City, Nhaho Research Institute for Cotton and Agricultural Development, Nha Ho, Ninh Son, Ninh Thuan, Vietnam. hainhaho@yahoo.com*

Nguyễn Thị Thanh Bình, *Nhaho Research Institute for Cotton and Agricultural Development, Nha Ho, Ninh Son, Ninh Thuan, Vietnam. binhcotton@yahoo.com*

Nguyễn Thị Thu Cúc, *Associate Professor, College of Agriculture, Plant Protection Department, Can Tho University, Campus II, 3 Thang 2 Street, Can Tho, Vietnam. nttcuc@ctu.edu.vn*

Nguyễn Thơ, *Consultant, Vietnam Institute of Agricultural Engineering & Postharvest Technology, 29/2G, D2 Ward, Van Thanh Bac, P25, Binh Thanh District, Ho Chi Minh City, Vietnam. nguyen_tho@hcm.fpt.vn*

Nguyễn Văn Bộ, *Director, Vietnamese Academy of Agricultural Sciences (VAAS), Vinh Quynh, Thanh Tri, Hanoi, Vietnam. nvbo@hn.vnn.vn*

Nguyễn Văn Đĩnh, *Dean of Faculty, Associate Professor, Faculty of Agronomy, Hanoi Agricultural University Number 1, Trau Quy, Gia Lam, Hanoi, Vietnam. agronomy@hau1.edu.vn*

Nguyễn Văn Huýnh, *Associate Professor, College of Agriculture, Department of Plant Protection, Can Tho University, Staff Housing, 30 Thang Tu Street, Can Tho, Vietnam. nvhuynh@ctu.edu.vn*

Nguyễn Văn Tuất, *Director General, Food Crops Research Institute (FCRI), MARD, Gia Loc, Hai Duong, Vietnam. vantuat55@vnn.vn*

Nguyễn Văn Uyển, *Professor, Yersin University, 27/25 Hau Giang Street, Ward No. 4, Tan Binh District, Ho Chi Minh City, Vietnam. nguyenvanuyen@hcm.vnn.vn*

Nguyễn Xuân Hồng, *President, Vietnam Labour Union, 6 Nguyen Cong Tru Street, Hai Ba Trung District, Hanoi, Vietnam. cdnn@hn.vnn.vn*

Omoto, Celso, *Associate Professor, Arthropod Resistance to Control Tactics Laboratory, Universidade de São Paulo (USP), Escola Superior de Agricultura 'Luiz de Queiroz' (ESALQ), Avenida Pádua Dias 11, C.P. 9, Piracicaba SP 13418-900, Brazil. celomoto@esalq.usp.br*

Phạm Văn Lầm, *Head of Pest Diagnostics & Identification Division, Plant Protection Research Institute (PPRI), MARD, Dong Ngac, Tu Liem, Hanoi, Vietnam. gdct_lam@yahoo.com*

Pham Van Toan, *Head of Biotechnology Office, Office of Agricultural Biotechnology Programme, Vietnam Ministry of Agriculture and Rural Development (MARD), 2 Ngoc Ha, Ba Dinh, Hanoi, Vietnam. pvtoan@hn.vnn.vn*

Phan Đình Pháp, *School of Biotechnology, International University, Vietnam National University – HCMC, Quarter 6, Linh Trung Ward, Thu Duc District, Ho Chi Minh City, Vietnam. dinhphap@yahoo.com*

Phan Văn Chi, *Deputy Director, Protein Biochemistry Laboratory, Vietnam Institute of Biotechnology (IBT), Vietnamese Academy of Science and Technology (VAST), 18 Hoang Quoc Viet Road, Cau Giay District, Hanoi, Vietnam. pvchi@ibt.ac.vn*

Pires, Carmen Silvia Soares, *Researcher, Laboratory of Bioecology and Semiochemicals of Insects, Embrapa Genetic Resources and Biotechnology, Parque Estacão Biológica, Av. W5 Norte – Final, Brasília DF 70770-900, Brazil. cpires@cenargen.embrapa.br*

Songa, Josephine M., *Crop Protection Coordinator, Kenya Agricultural Research Institute (KARI), PO Box 14733, Nairobi 00800, Kenya. jmsonga@africaonline.co.ke*

Sujii, Edison Ryoiti, *Researcher Biological Control, Embrapa Genetic Resources and Biotechnology, Parque Estacão Biológica, Av. W5 Norte – Final, Brasília DF 70770-900, Brazil. sujii@cenargen.embrapa.br*

Trác Khương Lai, *Head of Division of Biological Control and IPM on Fruit and Vegetables, Southern Fruit Research Institute (SOFRI), PO Box 203, My Tho, Tien Giang, Vietnam. tklai_vn@yahoo.com*

Trần Anh Hào, *Deputy Director General, Vietnam Cotton Company (VCC), Lot 1 15–17, D2 Street, Ward 25, Binh Thanh District, Ho Chi Minh City, Vietnam. bongvietnam@vnn.vn*

Trần Thế Lâm, *Head of Binh Thuan Province Office, Vietnam Cotton Company (VCC), Binh Thuan, Vietnam. lam_ricfc@yahoo.com*

Trần Thị Cúc Hoà, *Head of Department of Biotechnology, Cuu Long Delta Rice Research Institute (CLRRI), MARD, Thoi Thanh, O Mon, Can Tho, Vietnam. cuchoa@hcm.vnn.vn*

Trương Nam Hải, *Deputy Director, Vietnam Institute of Biotechnology (IBT), Vietnamese Academy of Science and Technology (VAST), 18 Hoang Quoc Viet Road, Cau Giay, Hanoi, Vietnam. tnhai@hn.vnn.vn*

Underwood, Evelyn, *Institute of Integrative Biology, Swiss Federal Institute of Science and Technology ETHZ, Universitätsstrasse 16, Zurich CH-8092, Switzerland. evelyn.underwood@env.ethz.ch*

Vũ Đức Quang, *Head, Department of Molecular Biology, Institute of Agricultural Genetics (IAG), MARD, Co Nhue, Tu Liem, Hanoi, Vietnam. vdquanghn@gmail.com*

Vương Thị Phấn, *Biotechnology, Western Highlands Agroforestry Science and Technical Research Institute (WASI), MARD, Road 27, Hoa Thang, Buon Ma Thuot, Dac Lac, Vietnam. hongphan@dng.vnn.vn*

Wals, Arjen, *Associate Professor, Education & Competence Studies Group, Social Science Department, Wageningen University, PO Box 8130, Wageningen NL-6700 EW, Netherlands. Arjen.Wals@wur.nl*

Wan, Fang-Hao, *Senior Principal Research Scientist, State Key Laboratory for Biology of Plant Diseases and Insect Pests, Institute of Plant Protection, Chinese Academy of Agricultural Sciences (CAAS), 12, Zhong-Guan-Cun, South Street, Nan-Da-Jie, Beijing CN-100081, China (P.R.). wanfangh@public3.bta.net.cn*

Wheatley, Ron E., *Research Leader, Environment-Plant Interaction Programme, SCRI, Invergowrie, Dundee DD2 5DA, Scotland, UK. R.Wheatley@scri.ac.uk*

Wilson, Lewis J., *Senior Principal Research Scientist, CSIRO Plant Industry and Programme Leader, 'The Farm', Cotton Catchment Communities CRC, Locked Bag 59, Narrabri, New South Wales, NSW 2390, Australia. lewis.wilson@csiro.au*

Zhai, Baoping, *Department of Entomology, Nanjing Agricultural University, Weigang 1, Nanjing 210095, China (P.R.). bpzhai@njau.edu.cn*

Zwahlen, Claudia, *Research Associate, Institute of Biology, University of Neuchâtel, Case postale 2, Rue Emile Argand 11, Neuchâtel CH-2009, Switzerland. claudia.zwahlen@unine.ch*

Preface

The Cartagena Protocol on Biosafety (Biosafety Protocol) under the Convention on Biodiversity (CBD) identifies a need in both developing and developed countries for comprehensive, transparent, scientific methods for pre-release testing and post-release monitoring of transgenic plants to ensure their environmental safety and sustainable use. Most importantly, Article 22 of the Biosafety Protocol requires that Parties shall cooperate in the development and/or strengthening of human resources and institutional capacities in biosafety, especially in developing countries. Vietnam is a signatory to the Cartagena Protocol and committed to implementing a clear system for biosafety risk assessment, management and monitoring. Vietnam has made considerable progress towards this goal in the past few years (see Chapter 1, this volume). At the same time, the country is investing strongly in the development of agricultural biotechnology.

This Vietnam case study is a product of the GMO ERA Project, 'International Project on GMO Environmental Risk Assessment Methodologies'. This project is a continuation of the GMO Guidelines Project, which was launched by scientists of the International Organization for Biological Control (IOBC) Global Working Group on 'Transgenic Organisms in Integrated Pest Management and Biological Control'. It is funded by the Swiss Agency for Development and Cooperation (SDC) as a part of the Swiss government's commitment to the Biosafety Protocol. The project is advised by an advisory board representing a wide array of organizations from around the world. The board members function both as scientific advisors and as international mediators to the policy environment and relevant decision making processes. The project is governed by a steering committee, which is responsible for all significant decisions taken by the project.

The project addresses the environmental and agricultural effects of transgenic crops, but does not evaluate human health impacts or ethical implications. It has focused on the transgenic crop plants already available because there is more information on this class of transgenic crop than any other, plus it is possible to mobilize considerable expertise in this area. One of the aims of the project is to improve the capacity of scientists in developing countries to support environmental risk assessment of transgenic crop plants. To accomplish this, the project concentrates on scientist-to-scientist exchange, because these personal connections are likely to persist over time. To leverage these efforts, the project has focused on a few developing countries with sufficient scientific infrastructures, a desire to develop the scientific basis to support environmental risk assessment and a need to do so. By strengthening the scientific capacities for risk assessment in these countries, expertise should be able to diffuse more readily to neighbouring countries. Kenya and Brazil were the first two focal countries of the project (Hilbeck and Andow, 2004; Hilbeck *et al.*, 2006), with Vietnam being the third. The work conducted in Vietnam forms the basis of this book. Several of the South-east Asian countries have well-developed scientific infrastructures, but the project selected Vietnam as a partner based on its need and desire to develop the scientific basis to support environmental risk assessment.

This book is the result of a series of workshops held in Vietnam in 2004, 2006 and 2007 that involved the participation of more than 65 Vietnamese scientists and many public sector scientists from other countries, including Brazil, China, the Philippines, Kenya, Australia, the USA and a number of European countries. During 1–5 April, we convened a workshop in Ho Chi Minh City focusing on the environmental risk assessment of Bt cotton, which framed the analysis and provided the major results in this book. After completing drafts of nearly all the chapters, we convened a workshop during 22–25 May 2006 in Hanoi, which focused on insect resistance management (IRM) of Bt crops. This workshop finalized the content for Chapter 12 (this volume) and provided IRM plans for Bt maize and Bt rice. Finally, during 21–26 May 2007, we convened a workshop in Nha Trang on non-target risk assessment methodologies. This workshop finalized the methodologies and results in Chapters 5–10 (this volume) and also structured approaches to non-target risk assessment for Bt maize and Bt rice. This book is the final product from the case study of Bt cotton in Vietnam. An earlier draft of these results was printed in Vietnamese in 2004 (Nguyen, 2004).

We would first like to thank Dr Bùi Bá Bổng, the Vice Minister of Research in the Vietnam Ministry of Agriculture and Rural Development (MARD) and Dr K.L. Heong, International Rice Research Institute for providing the spark and guidance for the work we have accomplished in Vietnam.

We are indebted to Dr Nguyễn Văn Tuất, Dr Nguyễn Hồng Sơn, Mrs Phạm Thị Hồng Hạnh and their team from the Plant Protection Research Institute of MARD, who organized all of the project activities in Vietnam, supported by the other members of the Vietnam Steering Committee of the GMO ERA Project (most of whom are co-authors on this book). We thank them for their foresight and leadership: Professor Nguyễn Văn Uyển (Yersin University), Dr Lê Minh Sắt (Ministry of Science and Technology), Dr Lê Thanh Bình (Ministry of Natural Resources and Environment), Drs Lê Quang Quyến (Nhaho Research Institute for Cotton and Agricultural Development, MARD), Vũ Đức Quang

(Vietnam Institute of Agricultural Genetics, MARD) and Dr Trần Thị Cúc Hoà (Cuu Long Delta Rice Research Institute, MARD). We also thank Dr Nguyễn Văn Bộ (Vietnamese Academy of Agricultural Sciences), Dr Phan Văn Chi (Institute of Biotechnology), Dr Nguyễn Thơ (Vietnam Cotton Company) and Dr Bùi Cách Tuyển (Ho Chi Minh Agriculture and Forestry University) for their support. The project activities in Vietnam were made possible by the support and endorsement of Dr Lê Trần Binh (Director, Vietnam Institute of Biotechnology), member of the Advisory Board to the GMO ERA Project. We would also like to thank all the members of the Advisory Board for their continued support: Dr Katharina Jenny (Swiss Agency for Development Cooperation), Dr José Geraldo Eugênio de França (Embrapa, Brazil), Dr Marcus Vinicius Segurado Coelho (Brazil Ministry of Agriculture), Dr Paulo Jose Peret de Sant'Ana (Brazil Ministry of Science and Technology), Dr Stefano Colazza (Secretary General, International Organization of Biological Control), Dr Chris Briggs (Director, UNEP GEF Biosafety Programme), Dr Helmut Gaugitsch (Austrian Federal Environment Agency), Dr Julian Smith (Central Science Laboratories, UK), Dr Ahmed Djoghlaf (Executive Secretary, Secretariat of the CBD) and Dr Christian Borgemeister (Director General, ICIPE, Kenya). We also acknowledge the considerable efforts of the Steering Committee of the Project, without whom the project could not exist. In addition to the editors, these members are Drs Trịnh Khắc Quang (MARD), Pham Van Toan (Biotechnology Office, MARD), Gary Fitt (CSIRO Entomology), Kristen C. Nelson (Forest Resources, University of Minnesota), Deise Capalbo (Embrapa Meio Ambiente), Eliana Fontes (Embrapa Recursos Genéticos Biotecnologia), Celso Omoto (Entomology, University of São Paolo ESALQ), K.L. Heong (International Rice Research Institute), Nick Birch (Scottish Crop Research Institute), Gabor Lövei (Integrated Pest Management, Aarhus University), Salvatore Arpaia (ENEA – Italian National Agency for New Technologies, Energy and Environment, Research Centre Trisaia), Richard Edema (Faculty of Agriculture, Makerere University, Uganda), Josephine Songa (Kenya Agricultural Research Institute) and Fang-Hao Wan (Institute for Biological Control, Chinese Academy of Agricultural Science). Most importantly, we acknowledge Evelyn Underwood and Areca Treon, without whose help and enthusiasm the Vietnam activities and this book would not have been possible.

We thank the chapter reviewers for their careful anonymous reviews. These were: Salvatore Arpaia (ENEA Research Centre Trisaia), Julio Bernal (Texas A&M University), Jim Bever (Indiana University), Michael Cohen (Canadian Food Inspection Agency), Stefano Colazza (University of Palermo), Les Ehler (University of California, Davis), Cesar V. Galvan (Food and Agriculture Organization), Marc Giband (CIRAD), Bryan Griffiths (Scottish Crop Research Institute), Julia S. Guivant (University of Santa Catarina), Jack Heinemann (University of Canterbury), Jean-Luc Hofs (CIRAD), Claudia Jacobi (University of Minas Gerias), Lim Li Ching (Third Word Network), Danny Llewellyn (CSIRO), Jon Lundgren (USDA-ARS), Carol Mallory-Smith (Oregon State University), Louise A. Malone (Horticulture and Food Research Institute of New Zealand), Melodie A. McGeoch (University of Stellenbosch), Kristin Mercer (Ohio State University), Panos Milonas (Benaki Phytopathological Institute),

John Obrycki (University of Kentucky), Angelo Pallini (University of Viçosa), Débora Pires Paula (Embrapa), Alan Raybould (Syngenta), John Ruberson (University of Georgia), Nancy Schellhorn (CSIRO), Herbert Siqueira (University of Pernambuco), David Somers (Monsanto), Doreen Stabinsky (Greenpeace), Doug Sumerford (USDA-ARS), Maurice Vaissayre (CIRAD) and Mary Whitehouse (CSIRO). We also thank the Series Editors, Dr Anne Kapuscinski (University of Minnesota) and Dr Peter Schei (Fridjof Nansen Institute) and the three additional anonymous reviewers who reviewed the book on behalf of the Series Editors. Their comments were all extremely useful and greatly improved the contents of the book.

We thank the IOBC Global Working Group on Transgenic Crops for their continued intellectual support of our work. We also thank the MARD, the UNEP GEF Project of MONRE and the SDC for their financial and in-kind support of the work on which this book is based. Finally, all of the authors thank partners and families for their support and understanding when meeting the many tight deadlines.

David A. Andow
St Paul, Minnesota
Angelika Hilbeck
Zurich, Switzerland
Nguyễn Văn Tuất
Hanoi, Vietnam
26 November 2007

References

Hilbeck, A. and Andow, D.A. (eds) (2004) *Environmental Risk Assessment of Transgenic Organisms, Volume 1: A Case Study of Bt Maize in Kenya.* CAB International, Wallingford, UK.

Hilbeck, A., Andow, D.A. and Fontes, E.M.G. (eds) (2006) *Environmental Risk Assessment of Genetically Modified Organisms, Volume 2: Methodologies for Assessing Bt Cotton in Brazil.* CAB International, Wallingford, UK.

Nguyen Van Tuat (2004) Report on the results of the GMO Guidelines Project Workshop in Ho Chi Minh City, April 2004. Unpublished.

Series Foreword

The advent of genetically modified organisms (GMOs) offers new options for meeting food, agriculture and aquaculture needs in developing countries, but some of these uses of GMOs can also affect biodiversity and natural ecosystems. These potential environmental risks and benefits need to be taken into account when making decisions about the use of GMOs. International trade and the unintentional trans-boundary spread of GMOs can also pose environmental risks depending on the national and regional contexts.

The complex interactions that can occur between GMOs and the environment heighten the need to strengthen worldwide scientific and technical capacity[1] for assessing and managing environmental risks of GMOs.

Global environmental management of GMOs and the strengthening of scientific and technical capacity for biosafety will require building policy and legislative biosafety frameworks. The latter is especially urgent for developing countries, as the Cartagena Protocol on Biosafety of the Convention on Biological Diversity makes clear. And the World Summit on Sustainable Development also identified the importance of improved knowledge transfer to developing countries on biotechnology. This point was also stressed in recent international fora such as the Norway/UN Conference on Technology Transfer and Capacity Building, and the capacity building decisions of the first meeting of the parties to the Cartagena Protocol on Biosafety.

The Scientific and Technical Advisory Panel (STAP) of the Global Environment Facility (GEF), in collaboration with a number of international scientific networks, initiated this book series and supported production of the first three volumes. The mandate of the STAP covers *inter alia* providing a forum

---

[1] By 'scientific and technical capacity' we mean "the ability to generate, procure and apply science and technology to identify and solve a problem or problems" including "the generation and use of new knowledge and information as well as techniques to solve problems." (Mugabe, J. 2000. Capacity Development Initiative, Scientific and Technical Capacity Development, Needs and Priorities. GEF-UNDP Strategic Partnership, October 2000.)

for integrating expertise on science and technology, and synthesizing, promoting and galvanizing state of the art contributions from the scientific community in a number of focal areas, including biodiversity. The book series complements the projects undertaken by the United Nations Environment Programme and the GEF to help developing countries design and implement national biosafety frameworks.

The purpose of this series is to provide scientifically peer-reviewed tools that can help developing countries strengthen their own scientific and technical capacity in biosafety of GMOs. Each book in the series examines a different kind of GMO. The workshops and writing teams used to produce each book are also capacity building activities in themselves because they bring together scientists from developing countries and developed countries to analyse and integrate the relevant science and technology into the book. This fourth book, on assessing *Bt* cotton in Vietnam was written by more than 70 contributors from 13 countries. The first book, a case study of *Bt* maize in Kenya, was published in 2004; a second book, on methodologies for assessing *Bt* cotton in Brazil, was published in 2006; and a third book, on methodologies for risk assessment and management of transgenic fish, was published in 2007. Each book underwent independent, international and anonymous peer review led by the series editors. Each book provides methods and relevant scientific information for environmental risk assessment, rather than drawing conclusions. Relevant organisations in each country will therefore need to conduct their own scientific risk assessments in order to inform their own biosafety decisions.

We hope that this book will help governments, scientists, potential users of GMOs and civil society organizations in Vietnam, other countries in the region and in the rest of the world to strengthen their understanding of the scientific knowledge and methods that are available for conducting environmental risk assessments of GMOs. We encourage readers to draw their own insights in order to help them devise and conduct robust environmental risk assessments for their own countries.

Anne R. Kapuscinski
University of Minnesota
St. Paul, Minnesota, U.S.A.

Peter J. Schei
Fritdjof Nansen Institute
Lysajer, Norway

June 5, 2008

Foreword

There is no doubt that biotechnology provides immense opportunities for increasing agricultural production, particularly in developing countries where agriculture is still a major economic sector and supports a major part of society. However, to harvest the benefits of biotechnological advances in agriculture, developing countries must build capacity in science and, at the same time, develop regimes for biosafety regulations and controls and the protection of intellectual property rights. The delayed or incomplete establishment of such regulations and practices will hamper the adoption of biotechnology advances in agriculture and will, as a consequence, adversely affect the desired pace of agricultural modernization and transformation of the rural economy.

From this viewpoint, the GMO ERA Project funded by the Swiss Agency for Development Cooperation has been highly successful. The project has been implemented for the past 5 years in Kenya, Brazil and Vietnam to help develop their national capacity for sound environmental risk assessment (ERA) and management of genetically modified (GM) crops. It has succeeded in developing scientific methods, in raising the awareness of scientists and policy makers and in building expert training teams who can train their countries' regulators.

This peer-reviewed volume 'Environmental Risk Assessment of Genetically Modified Organisms: Challenges and Opportunities with Bt Cotton in Vietnam' has been compiled by international and Vietnamese scientists as a result of their cooperation and partnership under the direction of the GMO ERA Project. It will be a very good reference for Vietnam and for other developing countries also. I am honoured to introduce this book to its readers and congratulate the authors wholeheartedly on their excellent achievement.

Dr Bùi Bá Bổng
Vice Minister, Vietnam Ministry of Agriculture
and Rural Development

1 Challenges and Opportunities with GM Crops in Vietnam: the Case of Bt Cotton

Nguyễn Văn Tuất, David A. Andow, Gary P. Fitt, Edison R. Sujii, Eliana Fontes, David R. Just and Lê Quang Quyến

The Vietnamese economy still relies heavily on agriculture and is being integrated into the global economy. The study of GM organisms in general, and GM plants in particular, together with their management and oversight, is very important to the country (Resolution 18/CP, 1994;[1] Vietnam Agenda 21,[2] 2004). According to Nguyen Van Uyen (Biotechnology Institute, 2003a), the organization of government agencies to manage GM organisms is urgent because: (i) the government has committed to make biotechnology a leading industry, with GM technology and GM products at the centre of biotechnology activities; (ii) when implementing global economic policies, GM organisms will be introduced into agricultural production and health and environment protection is essential – this requires the government to develop effective risk assessment and management of GM organisms in advance; (iii) it is necessary to provide the public with sufficient information about GM organisms because, in Vietnam, biotechnology and GM technologies are in the initial period of development and the government should keep the public informed.

Despite persistent controversies, countries have agreed on the Cartagena Protocol on Biosafety as a framework for the management, implementation and trade of GM plants and products. Vietnam is a signatory to the Cartagena Protocol and thus must participate in developing policies and regulations on biological safety and must control the import and export of GM plants and products, which currently are uncontrolled. An active approach and appropriate management of biotechnology will be essential to promote the safe use of

---

[1] Resolution 18/CP on Development of Biotechnology in Vietnam to 2010, signed 11 March 1994.
[2] Decision No. 153/2004/QD-TTg issuing the Strategic Orientation for Sustainable Development in Vietnam (Vietnam Agenda 21), signed by the Prime Minister of Vietnam on 17 August 2004 (http://www.va21.org/eng/va21/va21_de.htm).

GM plants and products for both human health and the environment (Biotechnology Institute, 2003a).

1.1. Biosafety Regulations for Agricultural Biotechnology in Vietnam

In 2005, the Vietnamese Prime Minister signed Decision No. 212/2005/QD-TTg[3] on the management of the biological safety of genetically modified organisms, and products and goods originating from genetically modified organisms, which implemented the requirements of the Cartagena Protocol on Biosafety. This Decision made the Ministry of Natural Resources and Environment (MONRE) the focal point for the management of the biosafety of GMOs in Vietnam and gave it the task of assisting the Vietnamese government in performing unified state management of GMOs, in collaboration with all the relevant Ministries. MONRE, with the assistance of experts from relevant agencies, has produced a National Action Plan for Biosafety to 2010[4] and has commenced a project for the implementation of the national biosafety framework in Vietnam, funded by UNEP GEF.

Vietnam is developing biosafety regulations to implement the Biosafety Decision, so that production and trading of GM products are carried out in a regular and orderly way, as required by the Decision of July 2007.[5] All the industrialized and developed countries have biosafety regulations and oversight of commercial GM products, as well as laboratories to assess the impacts on human health and the environment. Some newly developed and less developed countries have also issued biosafety regulations or guidelines (BIOTEC, 1992a,b; DOAP, 2002; SADA, 2004). The Vietnamese Ministry of Agriculture and Rural Development (MARD) is responsible for managing and regulating agricultural biotechnology (crops and trees), including the management of field trials.[6] MARD is currently drafting regulations that require environmental risk assessments of GMOs covering: transgene flow, stability of transgene locus and trait, the effects on target organisms and on biodiversity, including non-target organisms, soil organisms and effects through food chains. Other regulations that are relevant to the management of GM crops are: the Ordinance on Plant

---

[3] Decision No. 212/2005/QD-TTg promulgating the Regulation on Management of Biological Safety of Genetically Modified Organisms, Products and Goods Originating from Genetically Modified Organisms, signed by Prime Minister Phan Van Khai, effective date 26 August 2005.
[4] Decision No. 79/2007/QD-TTg approving the National Action Plan on Biodiversity up to 2010 and Orientations towards 2020 for Implementation of the Convention on Biological Diversity and the Cartagena Protocol on Biosafety, signed by Prime Minister Nguyen Tan Dzung, effective date 31 May 2007.
[5] Decision No. 102/2007/QD-TTg promulgating the Comprehensive Plan to Strengthen Management Capacity in Biosafety of GMOs, Goods and Products Originating from GMOs until 2010, as well as Implementation of the Cartagena Protocol on Biosafety, signed by Prime Minister Nguyen Tan Dzung, effective date 11 July 2007.
[6] With the recent reorganization of the Ministries, MARD also has responsibility for fisheries.

Protection and Quarantine (2001),[7] which implements the WTO SPS agreement in Vietnam; the Ordinance on Plant Varieties (2004),[8] which implements the UPOV convention in Vietnam; and the Law on Environmental Protection (2005).

According to the 2005 and 2007 Biosafety Regulations and the Decree on Labelling (2006), GM products that are being circulated, or soon will be circulated, in Vietnam should be labelled. Initially, this may cause import and export losses to some businesses, but it will heighten the responsibility and competitiveness of Vietnamese scientists and businesses thereafter. Initial development strategies are expected to minimize this potential adverse effect (Biotechnology Institute, 2003b).

1.2. Status of GM Plants and Products in Vietnam

The use of biotechnology is considered key for agricultural development in Vietnam. Some crops in Vietnam that might benefit from transgenesis include: industrial crops, forest trees, flowers and agricultural crops. Useful traits include: pest resistance, fungal resistance, slow-ripening and yield quality enhancement (Biotechnology Institute, 2003a). Research on GM plants in Vietnam has been carried out mainly at the Biotechnology Institute, Agricultural Genetics Institute, Cuu Long Delta Rice Research Institute, Institute for Tropical Biology and several other organizations. Recently, substantial investments have been made in high-tech biotechnology centres in Ho Chi Minh City[9] and Hanoi (the Hanoi Biotech Park, HaBiotech). Herbicide tolerance, pest resistance and pro-vitamin A genes have been transferred into rice, cabbage, maize, papaya and flowers on an experimental basis, but they have not yet been applied to production.

During 2006, Vietnam developed a long-term plan extending to the year 2020 for implementing GM crops.[10] The programme is being implemented by the Ministry of Agriculture and Rural Development. Up to the year 2013, MARD will focus its efforts to commercialize GM crops of cotton, soybean and maize. The main traits desired are transgenic insect resistance traits, targeted against lepidopteran leaf- and boll-feeders in cotton, pod-borers in soybean and stem-borers and ear-feeders in maize. These three crops are emphasized because they are not exported from Vietnam and efforts to improve production and productivity could offset the need to spend hard currency on imports or

[7] Socialist Republic of Vietnam (2001) Ordinance on Plant Protection and Quarantine No. 36/2001/PL-UBTVQH10. 25 July 2001. Standing Committee of the National Assembly.
[8] Socialist Republic of Vietnam (2004) Ordinance on Plant Varieties. Order No. 03/2004/ L-CTN of 5 April 2004. Standing Committee of the National Assembly. Official Gazette No. 16 (24-4-2004), pp. 3–20.
[9] Biotechnology Centre of Ho Chi Minh City, run by the People's Committee of Ho Chi Minh City (http://www.hcmbiotech.com.vn/ – accessed 10 November 2007).
[10] Decision No. 11/2006/QD-TTg approving the key programme on development and application of biotechnology in the domain of agriculture and rural development up to 2020, signed by Prime Minister Phan Van Khai, effective date 12 January 2006.

insecticides. The policy also indicates that genetically engineered export crops (e.g. rice, coffee and pepper) will not be considered for commercialization. After 2013, Vietnam will re-evaluate its strategy and possibly include export crops in its GMO development strategy. Partly as a consequence of this policy, and because the need is lacking, Vietnam is unlikely to commercialize pro-vitamin A rice anytime in the near future.

The plan for implementing GM crops in Vietnam focuses on Bt cotton as the first crop for commercialization. Because cottonseed oil is the only cotton product that is used for human food in Vietnam, the food safety assessment of cotton will be relatively restricted in scope, allowing more thorough and measured consideration of the environmental risks. Vietnam will use the experience gained from the environmental risk assessment and commercialization of this crop for the development of other GM crops.

In this book, we identify the issues that should be addressed in an environmental risk assessment of Bt cotton in Vietnam and propose methods and protocols for how some of the challenges can be met. Specifically, we provide a scope and framework for environmental risk assessment of Bt cotton in Vietnam, identifying potentially significant adverse effects of Bt cotton that may need to be characterized and, in some cases, characterizing these effects and associated management alternatives.

1.3. Cotton Production in Vietnam

Cotton is an important fibre crop and has high economic value in Vietnam, primarily because of the large number of textile products produced in the country (see Chapter 2, this volume, for a more thorough discussion of cotton in Vietnam). Currently, locally produced cotton fibre meets only about 10% of the Vietnamese textile industry's real demand, even under favourable weather, land and labour conditions. In principle, it should be possible to increase local production in Vietnam to supply the demand of the Vietnamese textile industry fully and defray the costs of importing raw cotton fibre. Based on this understanding, the Vietnamese government pursued a strategy to increase cotton production rapidly during 2001–2010,[11] aimed to provide sufficient fibre inputs for the textile industry, not only to reduce State use of foreign currency to purchase imported cotton, but also to facilitate changes in the structure of agricultural cultivation and to improve the production of goods and increase people's incomes, thereby ensuring socio-economic stability. However, since 2004, the national economy has performed increasingly better and the worldwide cotton price has declined, such that the profitability of domestic cotton production is threatened. Bt cotton is being evaluated in the hope that it will improve the profitability of cotton in Vietnam.

---

[11] Decision No. 17/2002/QD-TTg on the orientations and solutions for the development of industrial cotton plant in the 2001–2010 period, signed by Prime Minister Phan Van Khai, effective date 31 January 2002.

Cotton cultivation in Vietnam originated at least 700 years ago, when *Gossypium arboreum*, which is called *Co* cotton, was introduced from India (Vũ, 1962). During the late 19th and early 20th centuries, *G. hirsutum*, which is called *Luoi* cotton, was introduced from Cambodia and China (Vũ, 1962, 1971) and gradually replaced *Co* cotton, except for subsistence production in some locations, such as the northern mountainous regions of Vietnam (Tran and Nguyen, 1995). *G. barbadense* (*Hai Dao* cotton) was also introduced around the same time, but is no longer grown commercially, instead being used primarily in breeding programmes (Vũ, 1971).

Vietnam grows about 15,000–20,000 ha of cotton, most of which (~80%) is improved *G. hirsutum* cotton hybrids with high yield potential, high quality fibre and good resistance to leafhopper *Amrasca devastans* Distant (Homoptera: Cicadellidae) and bollworm *Helicoverpa armigera* Hübner (Lepidoptera: Noctuidae). In addition, a series of advanced technologies, such as integrated pest management (IPM) measures to reduce pesticide costs, cultivation methods like crop rotation, intercropping and cover cropping, and using crop growth regulators, have raised cotton productivity and quality and created economic value (Nguyen, 1994a,b), doubling yields from the mid-1990s to over 1 t/ha and reducing pesticide use during the rainy season to 3–4 sprays plus seed treatment (Chapter 2, this volume). Nearly all cotton (~90%) is grown during the rainy season, but hybrid cotton can be grown during the dry season when irrigated. During the dry season, cotton can yield higher quantity and quality of fibre than during the rainy season, and receives a higher price. Average yields from dry season production have been increasing steadily since 2000 to over 2 t/ha.

Most of the cotton in Vietnam is intercropped with other annual or perennial crops. Intercropping is preferred over cotton monoculture because of the higher economic value per land area and lower risk to producers than the cotton crop by itself. Cotton can add net income for a farmer as a supplementary crop (Nhaho RICOTAD, 2004). Rainfed cotton production costs may range between VND 1–3 million/ha,[12] depending on fertilizer use. With average yields of 1.1 t seed cotton/ha and an average cotton area of 0.6 ha, a grower could obtain total revenue of VND 4.6 million, resulting in a net revenue of VND 1.6–3.6 million.

Despite these potentially favourable conditions for cotton production, the cotton area in Vietnam has decreased by half since 2002. Some factors that may have contributed to this decrease, despite the government policy to increase cotton area, include insufficient capital investment, high production risk and low returns per land area. For example, material investments in irrigation systems may fail to meet farmers' needs. Rice irrigation systems are not adapted readily for furrow irrigation, which is generally preferred for cotton

---

[12] The seed is offered fully on credit at 650,000 VND/ha (at a seeding rate of 5 kg/ha); pesticides are offered fully on credit at 200,000–500,000 VND/ha, depending on pest control recommendations for the region. The full recommended amount of fertilizer costs VND 2 million/ha but farmers normally use much less and are not offered credit for the full amount.

during the dry season. In addition, even though cotton can attain higher net income than rice on a per crop basis, overall annual net income is usually higher in a rice triple-crop system than a cotton double-crop system. Cotton generally requires 5–6 months per crop cycle, while rice generally requires only 4–5 months per crop cycle, allowing two rice crops and one other short-season crop (Minot and Goletti, 2000).

In addition, there are significant production risks that farmers need to learn to manage (Biotechnology Institute, 2003b). Most cotton producers do not know how to use appropriate crop management and the cotton plant may fail to attain acceptable quantity and quality of yield. Another important risk is losses to pest arthropods, diseases and weeds. Cotton blue disease (CBD) is a problem in all production regions, causing crop losses in both rainy and dry season cotton. To control CBD, farmers spray insecticides against aphids once or twice during the first 30 days of crop growth, despite a low density of aphids. These sprays can lead to outbreaks of *H. armigera* in the midseason period, because natural enemies are killed. Varieties resistant to CBD and *H. armigera* are needed (Nguyen, 1999). Without effective resistance to the disease, more pesticides will need to be used and pesticide-resistant aphids may occur. Varieties resistant to *H. armigera* could help to reduce the risk of losses to pesticide-induced outbreaks of this lepidopteran pest.

1.4. Genetically Modified Cotton Worldwide

Genetically modified (GM) cotton was grown commercially for the first time in the USA and Australia in 1996 and, within 5 years, more than 5 million ha were grown worldwide. There are three kinds of genetically modified cotton that have been commercialized: cotton with resistance against herbicides, resistant against lepidopteran pests and resistant against both herbicides and lepidopteran pests. In 2006, GM cotton was grown in at least 11 countries, with the largest amounts grown in the USA, India, China, Australia, Brazil, Argentina, Mexico, Colombia and South Africa (Falck-Zepeda *et al.*, 2007). In this book, we focus on GM cotton that is resistant to lepidopteran pests based on protein toxins from *Bacillus thuringiensis*. These include both crystal protein (Cry) toxins and vegetative insecticidal protein (VIP) toxins. In the following section, we review the experiences of some of the countries that already use Bt cotton and draw conclusions to identify possible opportunities and challenges with Bt cotton for Vietnam.

USA

The first Bt cotton commercialized in the USA was Cry1Ac (Bollgard®) cotton, which provided excellent control of *Heliothis virescens* (cotton budworm) and *Pectinophora gossypiella* (pink bollworm) and some control of other important lepidopteran pests (US EPA, 2001). Currently, several additional Bt transgenes have been commercialized and the total area planted to Bt cotton in

2006 was 3.45 million ha (NASS, 2006). Most of the published information on the benefits and risks of growing Bt cotton is on Cry1Ac cotton, although several others, such as Cry1Ac/Cry2Ab and Cry1Fa/Cry1Ac cottons, are used widely.

Yield studies have demonstrated that Bt cotton has higher average lint yields than non-Bt cotton in the large-scale commercial production systems in the USA (Marra *et al.*, 2002; Cattaneo *et al.*, 2006). This yield benefit, however, is variable and depends on the level of budworm pressure, other pests and environmental conditions. When the target pest pressure is low, the yield benefit is lower, and when other pests are common or environmental conditions are poor, the yield benefit again is lower. Average yield benefits (Bt yield minus non-Bt yield) ranged from 83.2 to 130 kg lint/ha when examined in 47 studies throughout the USA; however, particular yield trials varied from a 330 kg lint/ha loss to a 1030 kg lint/ha gain. In other words, Bt cotton did not always result in higher yields. Bt cotton received fewer applications of insecticides than non-Bt cotton, ranging from 1.1 to 2.5 fewer applications (Marra *et al.*, 2002; Head *et al.*, 2005; Cattaneo *et al.*, 2006). In the south-east and mid-south, there were 2.3 applications on non-Bt cotton and only 1.2 applications on Bt cotton. These low rates of insecticide application on both Bt and non-Bt cotton have been possible because of the (unrelated) successful eradication of boll weevil. In the south-west (Arizona), there were 6.7 applications on non-Bt cotton and 4.3 applications on Bt cotton.

Many potential adverse environmental effects of Bt cotton have been investigated in the USA, including reductions in pollinators, potential gene flow to crop relatives, resistance evolution in the target pests and disruption of natural pest population regulation by the reduction in parasitoids and predators. For Bt cotton in the USA, resistance risks are assessed and considered serious enough to require management using non-Bt refuges (Chapter 12, this volume). Compliance with this requirement has been high (Carrière *et al.*, 2005). To minimize risks arising from gene flow, planting is not allowed in locations where there are wild relatives with which cotton can interbreed. These restrictions have had no commercial consequence because the restricted areas are not suitable for cotton production. Some non-target herbivores have become significant pests in Bt cotton. These include stink bugs in the south-east (Greene *et al.*, 2001, 2006), *Lygus* bugs in the mid-south (Hardee and Bryan, 1997) and *Lygus* bugs and tobacco whitefly in the south-west (Wilson *et al.*, 1992, 1994) of the USA.

More recently, risks associated with seed industry practices have surfaced (NRC, 2002). Commercial companies do not want to maintain inventories of large numbers of varieties. In the USA, Bt cotton is most often available with a herbicide tolerance (HT) gene. This has resulted in overuse of the Bt trait because many growers require only the HT trait, but cannot obtain sufficient high quality seed with this trait only. In addition, while the base genetics of the transgenic varieties continue to improve, the base genetics of the non-transgenic varieties are lagging behind. This is expected to affect the economics of resistance management by increasing the opportunity costs of planting a refuge.

Australia

INGARD® (= Bollgard®) was the first commercialized Bt cotton in Australia. It provided economic control against *H. armigera* and significant reductions in pesticide requirement (~50%) but, because Cry1Ac concentrations declined during the season, control efficacy also declined (Fitt *et al.*, 1994, 1998). There were also significant concerns about resistance risk. Consequently, in 2004/05, INGARD® varieties were removed from the market once two-gene Bollgard II® varieties were available. Bollgard II® contains both Cry1Ac and Cry2Ab, which provide season-long control of *H. armigera*. In 2006, approximately 80% of cotton grown in Australia was Bt varieties, although the total area was suppressed dramatically by drought. Most of the published literature from Australia is on INGARD® cotton, but newer work on Bollgard II® and VIP3A cotton is emerging (Llewellyn *et al.*, 2007).

The major benefit of Bt cotton in Australia has been a reduction in insecticide use. The average number of insecticide applications for the control of *H. armigera* was 9.7 sprays/season for conventional cotton and only 4.2 sprays/season for Bt cotton (INGARD®), a 57% reduction (Fitt, 2004). During the first several years of use, Bt cotton required 40–60% fewer insecticide applications than conventional, non-Bt cotton (Fitt, 2004). With Bollgard II varieties, pesticide reductions of 85% have been achieved and, in situations of high pest pressure, economically significant yield increases have been realized (Gary Fitt, Brisbane, 2007, personal communication).

The potential environmental risks of Bt cotton have been examined carefully in Australia (Fitt and Wilson, 2002; Whitehouse *et al.*, 2005). Resistance evolution in *H. armigera* is considered the most serious risk and Australia has implemented a comprehensive resistance management strategy to delay it (Fitt, 2004). This includes a requirement for refuges, which include non-Bt cotton and alternative host plants such as pigeon pea, sorghum and maize. Growers pay a significant fee to use the technology and are audited several times during the growing season to document compliance with the management strategy. Compliance is very high. Initially, growers received monetary rebates in recognition of compliance, although now, with Bollgard II varieties, the pricing structure no longer includes rewards per se. Bt cotton production is also now approved in Northern Australia after initial concerns about the risk of gene flow to native Australian cottons and about potential weediness were demonstrated to be negligible risks (OGTR, 2006). Although recognized as a potential issue (Ward, 2005), no sucking pests have become secondary pests of Bt cotton in Australia, although higher populations of two-spotted spider mites have been observed associated with Bt cotton production systems (L. Wilson, Narrabri, 2007, personal communication). No effects on natural enemies have been observed (Whitehouse *et al.*, 2005). On balance, Bt cotton has been accepted widely by Australian growers, who have benefited from reduced insecticide use and from the positive community recognition of reduced environmental impact.

China

China has grown Bt cotton since 1997, based on three different transformation events – *cry1Ac*, *cry1A* (a fusion of *cry1Ab* and *cry1Ac*) and *CpTI* (cowpea trypsin inhibitor), an inhibitor of protein digestion. *CpTI* often occurs in commercial varieties with one of the other genes. These Bt cottons provide control against *H. armigera*, *P. gossypiella* and *Anomis flava* (cotton looper) (Cui and Xia, 2000; Wan *et al.*, 2004, 2005). The Chinese Agricultural Yearbook reports 3.7 million ha of Bt cotton were grown in 2004. A third of Chinese cotton is produced from large, subsidized state-run farms of irrigated cotton in north-western China, but millions of smallholder farmers in the Yellow River and Yangtze River cotton growing regions also produce a large amount of cotton (Wu and Guo, 2005).

Early work demonstrated that Chinese smallholder farmers benefited from Bt cotton. Farm budget analysis based on surveys of 300–400 farmers in five provinces during 1999–2001 showed a complex pattern of adjustments of pesticide use and yield associated with Bt cotton (Huang *et al.*, 2002; Pray *et al.*, 2002). In the Shandong Province during 1999, yields of Bt cotton were 9% higher and the amount of insecticide applied was 71% lower. In the Henan Province, yields were 17.7% higher during 2000 and 6.7% higher during 2001, and insecticide use was 63% lower during 2000 and 55% lower during 2001. In the other provinces in 2001, yields were 6.0–7.2% higher and insecticide was 14–47% lower. Surprisingly, labour expenditures did not parallel insecticide use patterns, even though insecticide application was one of the main labour costs of cotton production. In all 3 years, use of Bt cotton resulted in increased yield and reduced amounts of insecticide, although this reduction in insecticide use appeared to decline over time. The studies conclude that farmers had a positive net income from Bt cotton, mainly through reduced input and labour costs (Huang *et al.*, 2003).

A follow-up study of > 400 farmers in five provinces during 2004 showed a striking change to the economics of Bt cotton (Wang *et al.*, 2006). Using an econometric model, the authors confirmed that farmers benefited substantially from Bt cotton during 1999–2001, having a higher net revenue of US$121/ ha. In 2004, this was reversed and farmers using Bt cotton had a net revenue 8% lower than non-Bt cotton growers (approximately –US$250/ha). They showed that farmers had slightly higher insecticide expenditures than non-Bt farmers, probably for the control of secondary pests, such as mirid bugs (Wu *et al.*, 2002a). It is possible that 2004 was an atypical year, but it illustrates that Bt cotton may not always be beneficial to the farmers who use it.

It is not certain if Chinese Bt cotton farmers will be able to maintain its benefits without institutional support to guarantee the quality of Bt cottonseed, discourage pesticide use and provide IPM training (Huang *et al.*, 2003; Yang *et al.*, 2005a,b; Pemsl and Waibel, 2007). Efficacy of Bt cotton has been variable. In 2002, < 15% of farmers in Shandong purchased official certified Bt seed and most of the rest used saved seed or unofficial, inexpensive Bt seed from informal sources (Pemsl, 2006; Pemsl and Waibel, 2007). Efficacy of the official Bt cotton had high variability and the variability of the saved and unofficial seed was even higher. Pesticide use on both Bt and non-Bt cotton exceeded the economically optimal level greatly (by 10–40 kg active ingredient/ha) and farmers probably could reduce pesticide use without affecting yields of either Bt or non-Bt cotton (Huang *et al.*, 2002; Pemsl *et al.*, 2005; Yang *et al.*, 2005a). Farmers in the Shandong Province sprayed Bt cotton 10–12 times (Pemsl *et al.*, 2005; Yang *et al.*, 2005a) and farmers in five other provinces sprayed Bt cotton 18 times (Wang *et al.*, 2006). IPM training may be essential to reduce pesticide use in both Bt and non-Bt cotton (Yang *et al.*, 2005b). Farmers without IPM training used three times more pesticide and sprayed twice as frequently on their Bt cotton plots as the farmers with IPM training. Thus, institutional support is needed to ensure that smallholder farmers in China obtain the benefits Bt cotton offers.

Resistance risk is the main risk of Bt cotton that is managed in China. Resistance risk management relies on the combined use of the fused Cry1A toxin and CpTI in the same varieties to delay resistance evolution in *H. armigera* and *P. gossypiella*. Refuges are unstructured and rely on maize and other crops for *H. armigera* (Wu *et al.*, 2002b, 2004; Wu and Guo, 2005). No refuge is available for *P. gossypiella*. If Bt maize is commercialized, there will be a need for structured refuges for Bt cotton (Wan *et al.*, 2005, Wu and Guo, 2005; Wu, 2007). Several non-target pests have emerged, including mirids (Wu *et al.*, 2002a; Wang *et al.*, 2006) and possibly aphids (Wu and Guo, 2003) and leafhoppers (Men *et al.*, 2005). Gene flow risks are considered unlikely in China (Zhang *et al.*, 2002).

Argentina

Bt cotton was commercialized in Argentina by Monsanto in 1998 and, in 2005/06, it was planted on 61,000 ha, about 20% of the cotton area in Argentina (Falck-Zepeda *et al.*, 2007). Two varieties containing a *cry1Ac* gene are planted and have good resistance against the two major pests, tobacco budworm (*H. virescens*) and the cotton bollworm (*H. gelotopoeon* (Dyar)), and also provide protection against the cotton leafworm (*Alabama argillacea*), the pink bollworm (*P. gossypiella*) and, to a lesser extent, the armyworms (*Spodoptera* spp.) (Qaim and de Janvry, 2003).

Field surveys of 89 large-scale farmers (Qaim and de Janvry, 2003; Qaim *et al.*, 2003) showed that, over two cropping seasons, farmers applied 55% and 43% less insecticide (kg a.i./ha) to their Bt cotton compared to their non-Bt cotton and had significantly higher yields (32%) from the Bt cotton (Qaim *et al.*, 2003). Yield of Bt cotton was predicted to be higher for smallholder farmers (~42%), and total gross benefit/ha was also expected to be higher (17%) than for large-scale producers. The authors attribute this prediction to insufficient control of pests by smallholders, who apply little to no insecticide (average of 2.9 sprays). However, relatively few large-scale farmers and no smallholder farmers have adopted Bt cotton in Argentina, due to a substantial technology fee charged for Bt cottonseeds, which doubles input expenses for large-scale farmers and quadruples input expenses for smallholder farmers (Qaim and de Janvry, 2003, 2005). Around 40% of the large-scale producers that used Bt cotton between 1999 and 2001 had negative net returns and stopped buying Bt cottonseed (Qaim and de Janvry, 2003). Surveys should be repeated to provide multi-year data on market acceptance, yield performance and pest response to the technology.

Brazil

In Brazil, cotton is one of the ten most important cash crops. It was cropped on almost 900,000 ha in 2006, producing 1.7 million t lint (Conab, 2007). Pest control is one of the most significant items in total crop production costs, without which farmers can lose all their profits. Current pest control strategies rely heavily on chemical pesticides, sometimes associated with the use of resistant cultivars and cultural control methods. There is an interest among farmers to use Bt cotton, as it may contribute to reducing the large amount of insecticide applied each year. In March 2005, the National Technical Biosafety Commission issued an authorization for the commercial use of Bollgard® cotton expressing Cry1Ac protein and targeting lepidopteran insects (CTNBio, 2005). The most important lepidopteran pests on cotton in Brazil are cotton leafworm (*A. argillacea* (Hübner)), pink bollworm (*P. gossypiella* (Saunders)), tobacco budworm (*H. virescens* (F.)) and fall armyworm (*S. frugiperda* (J.E. Smith)). Bollgard® cotton is effective against leafworm, pink bollworm and tobacco budworm, but has little effect on fall armyworm (Chitkowski *et al.*, 2003) and will not control two of the most serious primary cotton pests, the boll weevil and the cotton aphid. Since Bollgard® seeds have been available in the market in Brazil for only one year, no information is available yet on its performance and impact on cotton cropping.

Other Bt cotton varieties commercialized in other countries, such as Bollgard II™ (Cry1Ac + Cry2Ab) and WideStrike™ (Cry1Ac + Cry1Fa), are targeted to the same lepidopteran pests and also control the fall armyworm (Chapter 4, this volume) and therefore should be of greater benefit to Brazilian farmers. These Bt varieties, and one herbicide-tolerant variety, are under evaluation for commercial release in Brazil. The use of these Bt cottons might reduce the amount of insecticides used on cotton, but it is hard to predict how much reduction will be possible as farmers will still have to apply a considerable amount of insecticides to control virus disease vectors, such as aphids and whiteflies, and the worst pest of cotton in Brazil, the boll weevil (Fontes *et al.*, 2006). Smallholder farmers are more likely to benefit from this technology through crop yield increase, as presently they use less insecticide than large-scale farmers.

The environmental risk assessments of Bt cotton in Brazil identified potential gene flow to wild relatives as the most serious risk, followed by the risk of resistance evolution in lepidopteran pests (CTNBio, 2005). One native, one naturalized *Gossypium* species and one landrace of cultivated cotton occur in Brazil which are sexually compatible and can form fertile hybrids with the predominant cultivated cotton varieties (Johnston *et al.*, 2006). Based on the history of resistance to insecticides, *H. virescens*, *S. frugiperda* and *A. argillacea* were identified as having the highest risk of resistance to Bt cotton; *P. gossypiella* also poses a significant resistance risk (Fitt *et al.*, 2006). Non-Bt refuges to delay the risk of resistance evolution are required in Brazil. Furthermore, to protect cotton genetic diversity and prevent gene flow to cotton wild relatives and cotton landraces, a plan of action was established to map the wild cotton populations across the country, collect and preserve the germplasm in *ex situ* seed banks and study the reproductive biology and phenology of the cotton species. Based on this information, mandatory isolation zones were established where GM cotton could not be cultivated (Barroso *et al.*, 2005; CTNBio, 2005). Programmes to monitor compliance with the gene flow and resistance risk management measures still need to be implemented (Celso Omoto, Sao Paulo, 2007, personal communication; Fontes, 2007).

South Africa

Bt cotton was first grown commercially during 1998 and *cry1Ac* was the main Bt cotton used, targeted against the lepidopteran pests *H. armigera*, *Diparopsis castanea* Hampson and *Earias* spp. About 300 large-scale farmers (Hofs *et al.*, 2006a) and many smallholder farmers in the Makhathini Flats (Shankar and Thirtle, 2005) adopted Bt varieties.

Studies on large-scale irrigated farms suggested that Bt cotton might increase yield by 3.5–24% (Hofs *et al.*, 2006a). Results for smallholder farmers are markedly different. Initial results from 100 smallholder farmers in the Makhathini Flats suggested that there were significant financial benefits to growing Bt cotton, above the higher cost of the Bt cottonseed and licence fee (Ismael *et al.*, 2002a,b; Thirtle *et al.*, 2003; Shankar and Thirtle, 2005), and significantly less use of insecticides (Bennett *et al.*, 2004; Morse *et al.*, 2006). Subsequent analysis has shown that these financial benefits were probably a result of generous terms of credit and debt relief given preferentially to growers who switched to a new ginning operation in 2002 which required the use of Bt varieties (Witt *et al.*, 2006). Careful analysis of the available data indicated that there were no yield increases associated with the use of Bt cotton in this region (Hofs *et al.*, 2006b; Witt *et al.*, 2006). Bt cotton farmers reduced pesticide use by an average of two pyrethroid sprays, but this was not sufficient to offset the Bt cottonseed price plus licence fee (Hofs *et al.*, 2006c). As previous credit and debt policies have returned, the financial benefits of Bt cotton have also disappeared. Many farmers have been unable to obtain credit and have stopped growing cotton (Hofs *et al.*, 2006c; Morse *et al.*, 2006; Witt *et al.*, 2006). Hofs *et al.* (2006c) conclude that unless smallholder farmers can improve other management practices to raise yields, such as the use of fertilizer or early season weeding, they will not be able to benefit from Bt cotton.

Little research has been conducted on the possible environmental effects of Bt cotton in South Africa (Hofs *et al.*, 2004; McGeoch and Pringle, 2005). Southern Africa has three wild and cultivated diploid cottons: *G. herbaceum*,

G. *anomalum* Wawra and G. *triphyllum* (Harv.) Hochr., which are unlikely to cross with tetraploid G. *hirsutum*, although this has not been tested experimentally. Some non-target herbivore pests, such as leafhoppers and stink bugs, may be more abundant on Bt cotton (Hofs and Kirsten, 2001; Witt *et al.*, 2006). Smallholder farmers are supposed to plant a refuge of non-Bt cotton and alternative hosts of the target pests also provide unstructured refuges (Green *et al.*, 2003; Morse *et al.*, 2006).

1.5. Non-target Effects of Bt Cotton

Several studies have been published examining the effects of Bt cotton on non-target species in the field but, until now, there has been no systematic meta-analysis of these studies (Marvier *et al.*, 2007). Using the data set compiled by Marvier *et al.* (2007), we selected all cotton field studies measuring arthropod abundance (Table 1.1), and identified 1521 observations of non-target species response to Bt cotton. Eliminating those without true replication or without a reported standard error or standard deviation, we found 1129 observations. Most of the records have been collected from the USA and Australia (Table 1.1), originating from Head *et al.* (2005), Torres and Ruberson (2005), Naranjo (2005) and Whitehouse *et al.* (2005), so care should be taken in extrapolating the results to Vietnam. The results on Cry1Ac/Cry2Ab cotton are based almost entirely on a single study from Australia (Table 1.1) and additional studies are needed on this Bt cotton before we can draw conclusions for Vietnam with confidence. There are too little data on the Chinese Bt cotton events and none on the Cry1Ac/Cry1F cotton to allow for meta-analysis. We focused a formal meta-analysis on the Cry1Ac and Cry1Ac/Cry2Ab data, which included 1116 observations.

The meta-analysis classified the data by type of Bt cotton, type of control and by ecological functional groups related to our non-target risk assessment

Table 1.1. Number of field studies and observations of different non-target species' response to Bt cotton.[a]

Location	Number of studies[b]	Cry1Ac	Cry1Ac/ Cry2Ab	Cry1Ab	Fused Cry1A	Fused Cry1A & CpTI	Total number of observations
China	3	24	0	0	3	1	28
India	1	6	0	0	0	0	6
Australia	1	368	144	0	0	0	512
USA	8	553	21	9	0	0	583
Total	13	951	165	9	3	1	1129

Notes: [a]Data are from http://delphi.nceas.ucsb.edu/btcrops/ (Marvier *et al.*, 2007).[b]Wilson *et al.*, 1992; Armstrong *et al.*, 2000; Sumerford and Solomon, 2000; Wu *et al.*, 2002a; Men *et al.*, 2003, 2005; Bambawale *et al.*, 2004; Sisterson *et al.*, 2004; Hagerty *et al.*, 2005; Head *et al.*, 2005; Naranjo, 2005; Torres and Ruberson, 2005; Whitehouse *et al.*, 2005.

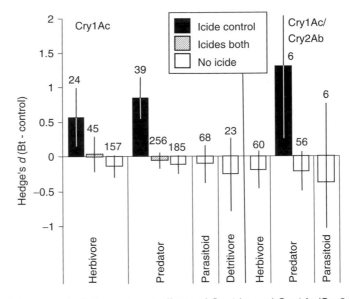

Fig. 1.1. Meta-analysis[13] of non-target effects of Cry1Ac and Cry1Ac/Cry2Ab cotton, compared to three kinds of controls for four functional groups of non-target organisms.[14] 'Icide control' is a non-Bt control sprayed with insecticides according to conventional recommendations. 'Icides both' means that both the non-Bt control and the Bt treatment were sprayed with insecticides. 'No icide' control is a non-Bt control with no insecticides applied. The numbers above each bar are the number of observations. The lines associated with each bar are 95% confidence intervals.

processes described in Chapters 6–10 (this volume). Because non-Bt cotton in Vietnam typically is sprayed with insecticides, the most relevant control comparisons are the insecticide treated control or insecticides applied to both the non-Bt control and the Bt treatment. There are not many relevant comparisons (Fig. 1.1): only 69 observations with the insecticide control ('Icide control') and 301 comparisons for insecticides on both ('Icides both'; nearly all of these are of predators). There are no relevant comparisons for parasitoids, detritivores or flower visitors, including pollinators.

Most observations are on non-target herbivores and predators (Fig. 1.1). The results suggest that non-target herbivores are likely to be more abundant in unsprayed Cry1Ac Bt cotton than in an insecticide control, but these non-target herbivores can be controlled if appropriate pest management is in place (Fig. 1.1, herbivore cross-hatched bar –'Icides both' – for herbivores on Cry1Ac

[13] The meta-analysis used Hedge's *d* as the summary statistic. This converts the published data into standard normal variates, weighted by sample size. Average Hedge's *d* is a weighted mean, weighted by the inverse of the variance of the individual estimates of Hedge's *d*.
[14] Several of the observations in Table 1.1 are not included in this figure because the number of observations for the cross-classification is < 5.

cotton is not significantly different from 0). Some non-target herbivores reported by the authors to be significantly more abundant in the Bt field than the insecticide control were *Empoasca biguttula* (Shiraki) (= *A. devastans*) and Miridae (Men *et al.*, 2005). We suggest that Vietnam should evaluate risks associated with non-target herbivores and be prepared to implement new pest management practices if Bt cotton is introduced. Predators are likely to be more abundant in unsprayed Cry1Ac Bt cotton than an insecticide control, but this potential benefit disappears when non-target herbivores are controlled (Fig. 1.1, predator cross-hatched bar – 'Icides both' – for predators in Cry1Ac cotton is not significantly different from 0). Some species of predators reported by the authors to be significantly less abundant in Bt fields than an insecticide control were *Hippodamia convergens* Guerin (Naranjo, 2005; Torres and Ruberson, 2005), *Geocoris uliginosus* (Say) (Torres and Ruberson, 2005) and *Nabis alternatus* Parshley (Naranjo, 2005). These results suggest that Vietnam should evaluate some worst-case scenarios for predators.

1.6. Prospects for Vietnam

A careful examination of the evidence on the benefits of Bt cotton illustrates that large scale growers in the USA and Australia have benefited economically from the use of Bt cotton. The benefits include higher lint yield in the USA, but not in Australia. Insecticide use was reduced by about 30–40% in the USA and by 85% in Australia on two-gene Bt cotton. These benefits have persisted over several years. In Argentina, large-scale producers of Bt cotton obtained higher yields and lower insecticide use, but the high cost of Bt seed often dwarfed these benefits.

The picture is not so clear for smallholder farmers in developing countries and economies in transition. Yield increases, insecticide reduction and economic benefits for smallholder farmers were inconsistent in China and undetectable in South Africa. In China, institutional support is needed to ensure that smallholder farmers obtain the benefits of Bt cotton, including guarantee of the quality of Bt cottonseed, incentives to reduce pesticide use and support for IPM training. In South Africa, Bt cotton is an increased financial risk due to the cost of the seed and licence fee and the fact that, under current crop management conditions, smallholder farmers are unable to obtain consistent benefits (Hofs *et al.*, 2006c).

It is likely that yield, insecticide use and economic benefits of Bt cotton in Vietnam will follow those of smallholders in China or South Africa rather than those of large-scale growers in the USA, Australia or Argentina (Biotechnology Institute, 2003a). In China and South Africa, smallholders obtained benefits only when appropriate institutional support was available. Thus, Vietnam may need to determine the necessary institutional support to ensure that smallholder farmers can experience yield increases, greater yield stability, reduced insecticide use and lower labour demands through the use of Bt cotton. Chapter 3 (this volume) provides a process to identify the needed institutional support so that Vietnamese smallholder cotton farmers can benefit.

Australia, Argentina, South Africa and the USA are not centres of diversity for cotton, so gene flow risks are minimal in these countries. Brazil is a centre of diversity for *Gossypium* and this has resulted in several exclusion zones to protect key centres of cotton biodiversity (Barroso *et al.*, 2005). Secondary pests of Bt cotton have emerged in the USA, China and possibly South Africa, but have not been observed in Australia. In China, these secondary pests may be severe enough to eliminate the economic benefits of Bt cotton. Resistance risks are recognized as significant worldwide and are managed actively in most countries. The experiences of other countries indicate that it will be crucial for Vietnam to determine if it harbours a centre of cotton biodiversity (see Chapter 11, this volume), to manage and monitor resistance evolution in the target pests (see Chapter 12, this volume) and to protect against potential secondary pests (see Chapter 6, this volume). In addition, Bt cotton should not prejudice ecosystem functions essential to smallholder farming or biodiversity (see Chapters 7, 8, 9 and 10, this volume).

1.7. Scope of This Case Study

This book focuses on the potential environmental effects of Bt cotton in Vietnam within a broader context of setting procedures and methods for regulating the introduction of GM crops. The environmental effects considered are those on non-target organisms, gene flow to cotton relatives and the evolution of resistance in the target pests. In addition, we examine processes for evaluating the potential benefits of GM plants in Vietnam and the necessary requirements for characterizing transgenes and transgenic traits. The analysis draws on previously published case studies on Bt maize in Kenya (Hilbeck and Andow, 2004) and Bt cotton in Brazil (Hilbeck *et al.*, 2006a). An early draft of the results presented in this book was submitted to MARD to seek regulatory approval for Bt cotton in Vietnam (Nguyen, 2004).

This book is the first scientific effort to synthesize information relevant to environmental risk assessment of a GM plant in Vietnam, taking Bt cotton as an example. It provides scientific tools that can be applied readily to assess Bt cotton and it forms a foundation that can be used to assess other GM plants. Through the process of assembling the book, many gaps of knowledge were identified and we have proposed a variety of methods to address these uncertainties. Some of these gaps will be addressed in scientific studies in Vietnam in the near future. Nearly 50 Vietnamese scientists have used and evaluated these methods and this book should be beneficial as a technical tool that can be used with biosafety regulations of GM plants to ensure that they are assessed, managed and monitored in an effective manner.

The methodologies described in this book focus on issues that must be considered in environmental risk assessment. There is much discussion about things that are 'nice to know' versus 'need to know' for risk assessment and this distinction is doubly important for developing countries, which are severely strapped for resources, to conduct and evaluate environmental risk assessments. We take a broad view of the kinds of environmental risks that should be considered and provide methodologies that are relatively simple and inexpensive to use to determine which of these may merit additional investigation. This allows a developing country, such as Vietnam, to focus its limited resources on the most significant issues in a cost-effective way.

This book begins by describing cotton production in Vietnam (Chapter 2), identifying the major production regions, describing the main production practices and the key biological factors affecting production. This is followed by problem formulation and options assessment (Chapter 3), which concentrates on how the formulation of the environmental risk assessment problem may be used as a component of risk communication. Chapter 4 addresses how a transgene can be designed to minimize the cost of risk assessment and identifies the features of a transgene that are essential to characterize for risk assessment. We emphasize that this chapter focuses on features that are necessary to support risk assessment. Chapters 5–10 address non-target and biodiversity risks, including non-target herbivores, predators, parasitoids, flower visitors and soil processes. In these chapters, we complete the prioritization process initiated in previous papers and volumes (Andow and Hilbeck, 2004; Birch *et al.*, 2004; Hilbeck *et al.*, 2006b). Chapter 11 completes a preliminary risk assessment on gene flow and its consequences, and Chapter 12 does the same for resistance risks and risk management. Chapter 13 summarizes the main technical conclusions of the previous chapters.

Cotton is grown on a limited area in Vietnam, despite efforts to increase the cropping area. Improved irrigation, better methods to control the aphid-transmitted cotton blue disease (CBD) and new methods to control pests disrupted by CBD control are some of the technical developments needed to enable cotton expansion, but low prices for cotton fibre on worldwide markets are a major limitation to expansion. Bt cotton might help stabilize dry season cotton production by controlling *H. armigera* disrupted by CBD controls. Because cotton is a small-scale crop and only cottonseed oil is used in human foods, Bt cotton is a useful case for Vietnam to determine how it will address the potential environmental risks of a GM crop.

This book will be used as a technical manual to enable Vietnamese scientists to evaluate the potential environmental impacts of Bt cotton prior to commercialization. This is part of a larger effort by Vietnam to increase investment in studying GM plants, managing GM products safely and building modern laboratories to assess their safety. Vietnam is committed to the development of biotechnology and GM crops and products to increase economic value for the country. Scientific assessment of the potential risks to the agricultural ecological system is a foundation for creating safe, sustainable and effective agriculture in the future.

References

Andow, D. and Hilbeck, A. (2004) Science-based risk assessment for non-target effects of transgenic crops. *BioScience* 54, 637–649.

Armstrong, J.S., Leser, J. and Kraemer, G. (2000) An inventory of the key predators of cotton pests on Bt and non-Bt cotton in west Texas. *Proceedings of the 2000 Beltwide Cotton Conferences*. National Cotton Council of America, Memphis, Tennessee, pp. 1030–1033.

Bambawale, O.M., Amerika Singh, O.P., Bhosle, B.B., Lavekar, R.C., Dhandapani, A., Kanwar, V., Tanwar, R.K., Rathod, K.S., Patange, N.R. and Pawar, V.M. (2004) Performance of Bt cotton (MECH-162) under integrated pest management in farmers' participatory field trial in Nandad district, Central India. *Current Science* 86(12), 1628–1633.

Barroso, P.A.V., Freire, E.C., Amaral, J.A.B. do and Silva, M.T. (2005) *Zonas de Exclusão de Algodoeiros Transgênicos para Preservação de Espécies e Gossypium Nativas ou Naturalizadas*. Série Comunicado Técnico, Embrapa Algodão. Campina Grande, Brazil.

Bennett, R., Ismael, Y., Morse, S. and Shankar, B. (2004) Reductions in insecticide use from adoption of Bt cotton in South Africa: impacts on economic performance and toxic load to the environment. *Journal of Agricultural Science* 142(6), 665–674.

BIOTEC (National Centre for Genetic Engineering and Biotechnology) (1992a) Biosafety Guidelines in Genetic Engineering and Biotechnology for Laboratory Work. Ad Hoc Biosafety Sub-Committee, National Centre for Genetic Engineering and Biotechnology, National Science and Technology Development Agency, Thailand. http://www.unescobkk.org/fileadmin/user_upload/shs/BiosafetyRegs/THAILAND_Biosafety_Guidelines_in_Genetic_Engineering_and_Biotechnology_for_Laboratory_Work.pdf (accessed 8 September 2007).

BIOTEC (National Centre for Genetic Engineering and Biotechnology) (1992b) Biosafety Guidelines in Genetic Engineering and Biotechnology for Field Works and Planned Release. Ad Hoc Biosafety Sub-Committee, National Centre for Genetic Engineering and Biotechnology, National Science and Technology Development Agency, Thailand. http://www.opbw.org/nat_imp/leg_reg/Thailand/biosafety.pdf (accessed 8 September 2007).

Biotechnology Institute (2003a) Biosafety: Unity mechanism on valuation process of risk, recognition and information processing. Short Report. UNEP-GEF. Biotechnology Institute, Hanoi.

Biotechnology Institute (2003b) Research on management, use of biology technology and safety. Short Report. UNEP GEF. Biotechnology Institute, Hanoi.

Birch, A.N.E., Wheatley, R.E., Anyango, B., Arpaia, S., Capalbo, D., Getu Degaga, E., Fontes, E., Kalama, P., Lelmen, E., Lövei, G., Melo, I.S., Muyekho, F., Ngi-Song, A., Ochieno, D., Ogwang, J., Pitelli, R., Schuler, T., Sétamou, M., Sithanantham, S., Smith, J., Van Son, N., Songa, J., Sujii, E., Tan, T.Q., Wan, F.-H. and Hilbeck, A. (2004) Biodiversity and non-target impacts: a case study of Bt maize in Kenya. In: Hilbeck, A. and Andow, D.A. (eds) *Environmental Risk Assessment of Genetically Modified Organisms, Volume 1: A Case Study of Bt Maize in Kenya*. CAB International, Wallingford, UK, pp. 117–186.

Carrière, Y., Ellers-Kirk, C., Kumar, K., Heuberger, S., Whitlow, M., Antilla, L., Dennehy, T.J. and Tabashnik, B.E. (2005) Long-term evaluation of compliance with refuge requirements for Bt cotton. *Pest Management Science* 61(4), 327–330.

Cattaneo, M.G., Yafuso, C., Schmidt, C., Huang, C.Y., Rahman, M., Olson, C., Ellers-Kirk, C., Orr, B.J., Marsh, S.E., Antilla, L., Dutilleu, P. and Carrière, Y. (2006) Farm-scale evaluation of the impacts of transgenic cotton on biodiversity, pesticide use, and yield. *Proceedings of the National Academy of Sciences of the USA* 103, 7571–7576.

Chitkowski, R.L., Turnipseed, S.G., Sullivan, M.J. and Bridges, W.C. Jr (2003) Field and laboratory evaluations of transgenic cottons expressing one or two *Bacillus thuringiensis* var. *kurstaki* Berliner proteins for management of noctuid (Lepidoptera) pests. *Journal of Economic Entomology* 96(3), 755–762.

Conab (2007) Série Histórica de Produção. http://www.conab.gov.br/conabweb/download/safra/BrasilProdutoSerieHist.xls (accessed 2 March 2007).

CTNBio (2005) Previous conclusive technical opinion No. 513/2005. http://www.ctnbio.gov.br/index.php/content/view/3663.html (accessed 2 March 2007).

Cui, J.-J. and Xia, J.-Y. (2000) Effects of Bt (*Bacillus thuringiensis*) transgenic cotton on the dynamics of pest population and their enemies. *Acta Phytophylacica Sinica* 27, 141–145 (in Chinese with English abstract).

DOAP (Department of Agriculture – Republic of the Philippines) (2002) Rules and regulations for the importation and release into the environment of plants and plant products derived from the use of modern biotechnology. Administrative Order No. 08, 3 April 2002. Department of Agriculture – Republic of the Philippines. http://www.da.gov.ph/agrilaws/AO2002/AO_08.html (accessed 8 September 2007).

Falck-Zepeda, J., Horna, D. and Smale, M. (2007) The economic impact and the distribution of benefits and risk from the adoption of insect resistant (Bt) cotton in West Africa. IFPRI Discussion Paper 00718. International Food Policy Research Institute, Washington, DC. http://www.ifpri.org/pubs/dp/IFPRIDP00718.pdf (accessed 13 November 2007).

Fitt, G.P. (2004) Implementation and impact of transgenic Bt cottons in Australia. In: *Cotton Production for the New Millennium. Proceedings of the Third World Cotton Research Conference*, 9–13 March 2003, Cape Town, South Africa. Agricultural Research Council – Institute for Industrial Crops, Pretoria, South Africa, pp. 371–381.

Fitt, G.P. and Wilson, L.J. (2002) Non-target effects of Bt-cotton: a case study from Australia. In: Akhurst, R.J., Beard, C.E. and Hughes, P.A. (eds) *Biotechnology of Bacillus thuringiensis and Its Environmental Impact: Proceedings of the 4th Pacific Rim Conference*. CSIRO, Canberra, Australia, pp. 175–182.

Fitt, G.P., Mares, C.L. and Llewellyn, D.J. (1994) Field evaluation and potential impact of transgenic cottons (*Gossypium hirsutum*) in Australia. *Biocontrol Science and Technology* 4, 535–548.

Fitt, G.P., Daly, J.C., Mares, C.L. and Olsen, K. (1998) Changing efficacy of transgenic Bt cotton – patterns and consequences. In: Zalucki, M.P., Drew, R.A.I. and White, G.G. (eds) *Pest Management – Future Challenges*. University of Queensland Press, Brisbane, Australia, pp. 189–196.

Fitt, G.P., Omoto, C., Maia, A.H., Waquil, J.M., Caprio, M., Okech, M.A., Cia, E., Nguyen Huu Huan and Andow, D.A. (2006) Resistance risks of Bt cotton and their management in Brazil. In: Hilbeck, A., Andow, D.A. and Fontes, E.M.G. (eds) *Environmental Risk Assessment of Genetically Modified Organisms, Volume 2: Methodologies for Assessing Bt Cotton in Brazil*. CAB International, Wallingford, UK, pp. 300–345.

Fontes, E. (2007) Policy briefs: a healthy mix: strategies for GM and non-GM crop coexistence. *SciDevNet* April 2007. http://www.scidev.net/dossiers/index.cfm?fuseaction=policybrief&dossier=6&policy=137 (accessed 10 November 2007).

Fontes, E.M.G., Ramalho, F. de S., Underwood, E., Barroso, P.A.V., Simon, M.F., Sujii, E.R., Pires, C.S.S., Beltrão, N., Lucena, W.A. and Freire, E.C. (2006) The cotton agricultural context in Brazil. In: Hilbeck, A., Andow, D.A. and Fontes, E.M.G. (eds) *Environmental Risk Assessment of Genetically Modified Organisms, Volume 2: Case Study of Bt Cotton in Brazil*. CAB International, Wallingford, UK, pp. 21–66.

Green, W.M., de Billot, M.C., Joffe, T., van Staden, L., Bennet-Nel, A., Toit, C.N.L. du and van der Westhuizen, L. (2003) Indigenous plants and weeds on the Makhathini Flats as refuge hosts to maintain bollworm population susceptibility to transgenic cotton (Bollgard®). *African Entomology* 11, 21–29.

Greene, J.K., Turnipseed, S.G., Sullivan, M.J. and May, O.L. (2001) Treatment thresholds for stink bugs in cotton. *Journal of Economic Entomology* 94, 403–409.

Greene, J.K., Bundy, C.S., Roberts, P.M. and Leonard, B.R. (2006) Identification and management of common boll feeding bugs in cotton. EB158 Clemson Extension. http://www.clemson.edu/psapublishing/Pages/Entom/EB158.pdf (accessed 1 November 2007).

Hagerty, A.M., Kilpatrick, A.L., Turnipseed, S.G., Sullivan, M.J. and Bridges, W.C. Jr (2005) Predaceous arthropods and lepidopteran pests on conventional, Bollgard, and Bollgard II cotton under untreated and disrupted conditions. *Environmental Entomology* 34(1), 105–114.

Hardee, D.D. and Bryan, W.W. (1997) Influence of *Bacillus thuringiensis*-transgenic and nec-
    tariless cotton on insect populations with emphasis on the tarnished plant bug (Heteroptera:
    Miridae). *Journal of Economic Entomology* 90, 663–668.
Head, G., Moar, W., Eubanks, M., Freeman, B., Ruberson, J., Hagerty, A. and Turnipseed, S.
    (2005) A multiyear, large-scale comparison of arthropod populations on commercially man-
    aged Bt and non-Bt cotton fields. *Environmental Entomology* 34, 1257–1266.
Hilbeck, A. and Andow, D.A. (eds) (2004) *Environmental Risk Assessment of Transgenic
    Organisms, Volume 1: A Case Study of Bt Maize in Kenya*. CAB International, Wallingford,
    UK.
Hilbeck, A., Andow, D.A. and Fontes, E.M.G. (eds) (2006a) *Environmental Risk Assessment of
    Genetically Modified Organisms, Volume 2: Methodologies for Assessing Bt Cotton in
    Brazil*. CAB International, Wallingford, UK.
Hilbeck, A., Andow, D.A., Arpaia, S., Birch, A.N.E., Fontes, E.M.G., Lövei, G.L., Sujii, E.,
    Wheatley, R.E. and Underwood, E. (2006b) Methodology to support non-target and bio-
    diversity risk assessment. In: Hilbeck, A., Andow, D.A. and Fontes, E.M.G. (eds)
    *Environmental Risk Assessment of Genetically Modified Organisms, Volume 2:
    Methodologies for Assessing Bt Cotton in Brazil*. CAB International, Wallingford, UK,
    pp. 108–132.
Hofs, J.L. and Kirsten, J. (2001) Genetically modified cotton in South Africa: the solution for
    rural development? CIRAD/University of Pretoria Working Paper 2001–17. Department of
    Agricultural Economics, Extension and Rural Development, University of Pretoria, Pretoria,
    South Africa.
Hofs, J.L., Schoeman, A. and Vaissayre, M. (2004) Effect of Bt cotton on arthropod biodiversity
    in South African cotton fields. *Communications in Applied Biological Sciences* (Ghent
    University, Belgium) 69(3), 191–194.
Hofs, J.L., Hau, B. and Marais, D. (2006a) Boll distribution patterns in Bt and non-Bt cultivars
    1. Study on commercial irrigated farming systems in South Africa. *Field Crops Research*
    98(2–3), 203–209.
Hofs, J.L., Hau, B., Marais, D. and Fok, M. (2006b) Boll distribution patterns in Bt and non-Bt
    cotton cultivars II. Study on small-scale farming systems in South Africa. *Field Crops
    Research* 98(2–3), 210–215.
Hofs, J.L., Fok, M. and Vaissayre, M. (2006c) Impact of Bt cotton adoption on pesticide use by
    smallholders: a 2-year survey in Makhathini Flats (South Africa). *Crop Protection* 25(9),
    984–988.
Huang, J., Hu, R., Rozelle, S., Qiao, F. and Pray, C.E. (2002) Transgenic varieties and productiv-
    ity of smallholder cotton farmers in China. *Australian Journal of Agricultural and Resource
    Economics* 46(3), 367–387.
Huang, J., Hu, R., Pray, C., Qiao, F. and Rozelle, S. (2003) Biotechnology as an alternative to
    chemical pesticides: a case study of Bt cotton in China. *Agricultural Economics* 29(1),
    55–67.
Ismael, Y., Thirtle, C., Piesse, J. and Beyer, L. (2002a) Can GM-technologies help the poor? The
    efficiency of Bt cotton adopters in the Makhathini Flats of KwaZulu-Natal. *Agrekon* 41,
    62–74.
Ismael, Y., Bennett, R. and Morse, S. (2002b) Benefits from Bt cotton use by smallholder farmers
    in South Africa. *AgBioForum* 5(1), 1–5. http://www.agbioforum.org/v5n1/v5n1a01-
    morse.htm (accessed 10 November 2007).
Johnston, J.A., Mallory-Smith, C., Brubaker, C.L., Gandara, F., Aragão, F.J.L., Barroso, P.A.V.,
    Quang Vu Duc, Carvalho, L.P. de, Kageyama, P., Ciampi, A.Y., Fuzatto, M., Cirino, V. and
    Freire, E. (2006) Assessing gene flow from Bt cotton in Brazil and its possible consequences.
    In: Hilbeck, A., Andow, D.A. and Fontes, E.M.G. (eds) *Environmental Risk Assessment of*

Genetically Modified Organisms, Volume 2: Methodologies for Assessing Bt Cotton in Brazil. CAB International, Wallingford, UK, pp. 261–299.

Llewellyn, D.J., Mares, C.L. and Fitt, G.P. (2007) Field performance and seasonal changes in the efficacy against *Helicoverpa armigera* (Hübner) of transgenic cotton (VipCot) expressing the insecticidal protein VIP3A. *Agricultural and Forest Entomology* 9(2), 93–101.

McGeoch, M.A. and Pringle, K.L. (2005) Science and advocacy: the GM debate in South Africa. *South African Journal of Science* 101, 7–9.

Marra, M.C., Pardey, P.G. and Alston, J.M. (2002) The payoffs to transgenic field crops: an assessment of the evidence. *AgBioForum* 5(2), 43–50.

Marvier, M., McCreedy, C., Regetz, J. and Kareiva, P. (2007) A meta-analysis of effects of Bt cotton and maize on non-target invertebrates. *Science* 316, 1475–1477.

Men, X., Ge, F., Liu, X. and Yardim, E.N. (2003) Diversity of arthropod communities in transgenic Bt cotton and non-transgenic cotton agroecosystems. *Environmental Entomology* 32(2), 270–275.

Men, X., Ge, F., Edwards, C.A. and Yardim, E.N. (2005) The influence of pesticide applications on *Helicoverpa armigera* Hubner and sucking pests in transgenic Bt cotton and non-transgenic cotton in China. *Crop Protection* 24(4), 319–324.

Minot, N. and Goletti, F. (2000) Rice market liberalization and poverty in Vietnam. Research Report 114. International Food Policy Research Institute, Washington, DC. http://pdf.dec. org/pdf docs/Pnack940.pdf (accessed 4 November, 2007).

Morse, S., Bennett, R. and Ismael, Y. (2006) Environmental impact of genetically modified cotton in South Africa. *Agriculture Ecosystems and Environment* 117(4), 277–289.

Naranjo, S.E. (2005) Long-term assessment of the effects of transgenic Bt cotton on the abundance of non-target arthropod natural enemies. *Environmental Entomology* 34, 1193–1210.

NASS (National Agricultural Statistics Service) (2006) Acreage. Agricultural Statistics Board, USDA, Washington, DC. http://usda.mannlib.cornell.edu/usda/nass/Acre//2000s/2006/ Acre-06–30–2006.pdf (accessed 18 July 2007).

Nguyen Huu Binh (1994a) American bollworm (*Helicoverpa armigera* Hübner) and necessary attention in its control. *Journal of Plant Protection* 3, 35–37 (in Vietnamese).

Nguyen Huu Binh (1994b) Epidemic development of cotton insect pests and management measures. *Journal of Plant Protection* 4, 35–37 (in Vietnamese).

Nguyen Thi Thanh Binh (1999) Study on cotton blue disease and its management. PhD thesis, Vietnam Agricultural Science Institute, Hanoi (in Vietnamese).

Nguyen Van Tuat (2004) Report on the results of the GMO Guidelines Project workshop in Ho Chi Minh City, April 2004. Unpublished.

Nhaho RICOTAD (Nhaho Research Institute for Cotton and Agricultural Development) (2004) Developing Cotton Trees and Cotton Farming in Vietnam. Background Report for the GMO Guidelines Project Vietnam Workshop, 1–5 April 2004. Nhaho Research Institute for Cotton and Agricultural Development, Nha Trang, Vietnam.

NRC (National Research Council) (2002) *Environmental Effects of Transgenic Plants: The Scope and Adequacy of Regulation.* National Academy Press, Washington, DC.

OGTR (2006) Risk Assessment and Risk Management Plan DIR 066/2006. Commercial release of herbicide tolerant and/or insect resistant cotton lines north of latitude 22° south. Office of the Gene Technology Regulator, Australian Government, Canberra. http://www.ogtr. gov.au/pdf/ir/dir066rarmp2.pdf (accessed 2 December 2007).

Pemsl, D. (2006) *Economics of Agricultural Biotechnology in Crop Protection in Developing Countries – The Case of Bt-cotton in Shandong Province, China.* Pesticide Policy Project Publication Series Special Issue No. 11. University of Hannover, Hannover, Germany. http://www.ifgb.uni-hannover.de/401.html (accessed 10 November 2007).

Pemsl, D. and Waibel, H. (2007) Assessing the profitability of different crop protection strategies in cotton: case study results from Shandong Province, China. *Agricultural Systems* 95, 28–36.

Pemsl, D., Waibel, H. and Gutierrez, A.P. (2005) Why do some Bt-cotton farmers in China continue to use high levels of pesticides? *International Journal of Agricultural Sustainability* 3(1), 44–56.

Pray, C.E., Huang, J., Hu, R. and Rozelle, S. (2002) Five years of Bt cotton in China – the benefits continue. *The Plant Journal* 31(4), 423–430.

Qaim, M. and de Janvry, A. (2003) Genetically modified crops, corporate pricing strategies, and farmers' adoption: the case of Bt cotton in Argentina. *American Journal of Agricultural Economics* 85, 814–828.

Qaim, M. and de Janvry, A. (2005) Bt cotton and pesticide use in Argentina: economic and environmental effects. *Environment and Development Economics* 10, 179–200.

Qaim, M., Cap, E.J. and de Janvry, A. (2003) Agronomics and sustainability of transgenic cotton in Argentina. *AgBioForum* 6(1&2), 41–47.

SADA (South African Department of Agriculture) (2004) Guideline document for use by the advisory committee when considering proposals/applications for activities with genetically modified organisms (Genetically Modified Organisms Act, 1997). Notice 1047 of 2004. *State Gazette* No. 26422 (113), 11 June 2004. http://www.info.gov.za/gazette/notices/2004/26422d.pdf (accessed 1 November 2007).

Shankar, B. and Thirtle, C. (2005) Pesticide productivity and transgenic cotton technology: the South African smallholder case. *Journal of Agricultural Economics* 56(1), 97–116.

Sisterson, M.S., Biggs, R.W., Olson, C., Carrière, Y., Dennehy, T.J. and Tabashnik, B.E. (2004) Arthropod abundance and diversity in Bt and non-Bt cotton fields. *Environmental Entomology* 33, 921–929.

Sumerford, D.V. and Solomon, W.L. (2000) Growth of wild *Pseudoplusia includens* (Lepidoptera: Noctuidae) larvae collected from Bt and non-Bt cotton. *Florida Entomologist* 83(3), 354–357.

Thirtle, C., Beyers, L., Ismael, Y. and Piesse, J. (2003) Can GM-technologies help the poor? The impact of Bt cotton in Makhathini Flats, KwaZulu-Natal. *World Development* 31(4), 717–732.

Torres, J.B. and Ruberson, J.R. (2005) Canopy- and ground-dwelling predatory arthropods in commercial Bt and non-Bt cotton fields: patterns and mechanisms. *Environmental Entomology* 34, 1242–1256.

Tran Van Phong and Nguyen Tho (1995) *An Overview on Cotton in Vietnam*. Vietnam Cotton Company. Agriculture Publishing House, Ho Chi Minh City, Vietnam (in Vietnamese).

US EPA (2001) Biopesticides Registration Action Document – *Bacillus thuringiensis* Plant-Incorporated Protectants. USA Environmental Protection Agency. http://www.epa.gov/oppbppd1/biopesticides/pips/bt_brad.htm (accessed 10 November 2007).

Vũ Công Hậu (1962) *Cotton in Vietnam*. Agriculture Publishing House, Hanoi (in Vietnamese).

Vũ Công Hậu (1971) *Cotton production in Vietnam and cotton varieties*. Scientific and Technical Publishing House, Hanoi (in Vietnamese).

Wan, P., Wu, K.-M., Huang, M. and Wu, J. (2004) Seasonal pattern of infestation by pink bollworm *Pectinophora gossypiella* (Saunders) in field plots of Bt transgenic cotton in the Yangtze River valley of China. *Crop Protection* 23(5), 463–467.

Wan, P., Zhang, Y.-J., Wu, K.-M. and Huang, M. (2005) Seasonal expression profiles of insecticidal protein and control efficacy against *Helicoverpa armigera* for Bt cotton in the Yangtze River valley of China. *Journal of Economic Entomology* 98, 195–201.

Wang, S., Just, D.R. and Pinstrup-Andersen, P. (2006) Tarnishing silver bullets: Bt technology adoption, bounded rationality and the outbreak of secondary pest infestations in China. Selected Paper prepared for presentation at the American Agricultural Economics

Association Annual Meeting, Long Beach, California, 22–26 July 2006. www.grain.org/research/btcotton.cfm?id=374 (accessed 17 July 2007).

Ward, A.L. (2005) Development of a treatment threshold for sucking insects in determinate Bollgard II transgenic cotton grown in winter production areas. *Australian Journal of Entomology* 44, 310–315.

Whitehouse, M.E.A., Wilson, L.J. and Fitt, G.P. (2005) A comparison of arthropod communities in transgenic Bt and conventional cotton in Australia. *Environmental Entomology* 34, 1224–1241.

Wilson, F.D., Flint, H.M., Deaton, W.R., Fuschhoff, D.A., Perlak, E.J., Armstrong, T.A., Fuchs, R.L., Berberich, S.A., Parks, N.J. and Stapp, B.R. (1992) Resistance of cotton lines containing a *Bacillus thuringiensis* toxin to pink bollworm (Lepidoptera: Noctuidae) and other insects. *Journal of Entomological Science* 34, 415–425.

Wilson, F.D., Flint, H.M., Deaton, W.R. and Buehler, R.E. (1994) Yield, yield components, and fiber properties of insect-resistant cotton lines containing a *Bacillus thuringiensis* toxin gene. *Crop Science* 34, 38–41.

Witt, H., Patel, R. and Schnurr, M. (2006) Can the poor help GM crops? Technology, representation and cotton in the Makhathini flats, South Africa. *Review of African Political Economy* 109, 409–513.

Wu, K.M. (2007) Environmental impact and risk management strategies of Bt cotton commercialization in China. *Chinese Journal of Agricultural Biotechnology* 4(2), 93–97.

Wu, K.M. and Guo, Y.Y. (2003) Influences of *Bacillus thuringiensis* Berliner cotton planting on population dynamics of the cotton aphid, *Aphis gossypii* Glover, in Northern China. *Environmental Entomology* 32(2), 312–318.

Wu, K.M. and Guo, Y.Y. (2005) The evolution of cotton pest management practices in China. *Annual Review of Entomology* 50, 31–52.

Wu, K.M., Lin, K., Feng, H. and Guo, Y. (2002a) Seasonal abundance of the mirids, *Lygus lucorum* and *Adelphocoris* spp. (Hemiptera: Miridae) on Bt cotton in northern China. *Crop Protection* 21, 997–1002.

Wu, K.M., Guo, Y.Y. and Gao, S.S. (2002b) Evaluation of the natural refuge function for *Helicoverpa armigera* (Lepidoptera: Noctuidae) within *Bacillus thuringiensis* transgenic cotton growing areas in North China. *Journal of Economic Entomology* 95(4), 832–837.

Wu, K., Feng, H. and Guo, Y. (2004) Evaluation of maize as a refuge for management of resistance to Bt cotton by *Helicoverpa armigera* (Hübner) in the Yellow River cotton-farming region of China. *Crop Protection* 23(6), 523–530.

Yang, P., Iles, M., Yan, S. and Jolliffe, F. (2005a) Farmers' knowledge, perceptions and practices in transgenic Bt cotton in small producer systems in Northern China. *Crop Protection* 24(3), 229–239.

Yang, P., Li, K., Shi, S., Xia, J.-Y., Guo, R., Li, S. and Wang, L. (2005b) Impacts of transgenic Bt cotton and integrated pest management education on smallholder cotton farmers. *International Journal of Pest Management* 51(4), 231–244.

Zhang, Y.-J., Wu, K.M., Peng, Y.-F. and Guo, Y.Y. (2002) Progress in ecological safety of insect-resistant transgenic plants. *Entomological Knowledge* 39(5), 321–327 (in Chinese with English abstract).

2 Cotton Production in Vietnam

LÊ QUANG QUYẾN, NGUYỄN THỊ HAI, TRẦN ANH HÀO, MAI VĂN HÀO, NGUYỄN THỊ THANH BÌNH, ĐÀNG NĂNG BỬU, DƯƠNG XUÂN DIÊU AND EVELYN UNDERWOOD

Cotton is an established part of Vietnamese agriculture and meets part of the raw material demand of the Vietnamese textile industry. It provides a cash income for small-scale farmers and harvesting and processing of the cotton fibre provides jobs for agricultural labourers and workers. However, despite the successful recovery and development of cotton cultivation in Vietnam, only about 10% of the raw material demand of the expanding Vietnamese textile industry can be met. Vietnam is looking for ways to increase the yield and quality of cotton production, without losing the achievements gained through the implementation of integrated pest management (IPM).

With 85 million people, Vietnam is one of the most densely populated countries in the world and the majority of the population is rural, each household farming on small plots of land less than 1 ha in size. Agriculture in Vietnam accounts for half of the country's employment and economic production. Vietnam's role in rice farming is significant. With 6.3 million ha devoted to rice, it is the most important crop in Vietnam and a major export commodity. In comparison, cotton is currently a minor crop: it takes up 0.3% of the area that rice occupies and 1.7% of the hybrid maize area. Increasing Vietnamese cotton production will require replacement by cotton on some of the areas where rice or other crops are grown currently.

This chapter describes the history and current situation of cotton production, the environmental and economic context for cotton growing in the different regions in Vietnam, the main diseases and pests and IPM for cotton under both rainfed and irrigated conditions. It also describes plans to increase cotton production in Vietnam and challenges to expansion. Bt cotton is one possible option that is being examined for sustainable development of cotton cultivation (see Chapter 3, this volume) and this chapter provides the context for this introduction. This chapter also provides some of the necessary context for the environmental risk assessment of Bt cotton in Vietnam, which is addressed in Chapters 6–12 (this volume).

2.1. History and Achievements of Cotton Production in Vietnam

Cotton has been grown in Vietnam probably for thousands of years, with artisanal or subsistence production using local varieties of Asian tree cotton (*Co* cotton) *Gossypium arboreum* L., which gave low yields but had good tolerance to high humidity and rainfall. Commercial cotton production using *G. hirsutum* varieties (*Luoi* cotton), and also some *G. barbadense* varieties (*Hai Dao* cotton), was developed during the French colonial period in the early 20th century (Vũ, 1971). During this period, the cotton area rose to around 23,000 ha and Vietnam exported cotton to Japan and Hong Kong. During the American war, cotton production collapsed in southern Vietnam, but was continued in the north on a small area and, after the war, cotton growing in central coastal and southern Vietnam was resumed (Napompeth, 1987).

However, cotton production was not stable in the face of changing environmental or climatic conditions, or under increasing pressure from insect pests. Dry season cotton production under irrigation was tried but the introduced varieties, such as the US *G. hirsutum latifolium* variety Deltapine 16, had smooth leaves and were very susceptible to damage by the leafhopper *Amrasca devastans* Distant (Homoptera: Cicadellidae). By the 1980s, cotton production had reached crisis point (Napompeth, 1987). Pesticide use had increased to 15–20 applications/season, mainly to try to control leafhoppers and *Helicoverpa armigera* Hübner (Lepidoptera: Noctuidae), which probably had developed resistance to insecticides. The pesticide applications also caused outbreaks of the armyworm *Spodoptera exigua* Hübner (Lepidoptera: Noctuidae). The raw (seed)[1] cotton yield was low at 0.3 to 0.4 t/ha. The area of cotton production in Vietnam decreased, insects and diseases damaged crops and there were no varieties suitable for the different conditions in each cotton growing region. Cotton growers suffered losses because of the high insecticide costs and damage to their health from applying toxic pesticides, which polluted the environment. The Vietnam cotton industry faced considerable difficulties.

In 1977, the Nhaho Research Institute for Cotton and Agricultural Development (Nhaho – RICOTAD) was founded as an agricultural technology centre for central southern Vietnam. In 1982, it was assigned to the newly created Vietnam Cotton Company (VCC). The Nhaho – RICOTAD has bred a number of new, high yielding and good quality cotton varieties well adapted for Vietnam (Table 2.1) and now nearly all cotton growers in Vietnam use hybrid seeds. They also developed an innovative IPM strategy, based on the use of hairy cotton varieties and intercropping, which has been implemented successfully on cotton in Vietnam for more than 10 years.

Education and training have been effective tools for the implementation of IPM on cotton. The training of scientists and research on IPM strategies in Vietnam started in 1985 under the sponsorship of the United Nations Food

---

[1] Raw or seed cotton is used in this chapter to refer to the cotton fibre with the seed before ginning.

Table 2.1. Fibre quality of hybrid cotton cultivars bred and grown in Vietnam since 1990.

Cultivar	Release year	% of area in 2005	Grade (% GOT)	Staple length (mm)	Fibre strength g/tex	Micronaire
L18	1995	1	38.0	29.5	30	4.3
VN20	1997	10	37.5	30	31	4.1
VN35	1999	5	36.5	29	32	4.3
VN15	2002	60	36.5	30	31	3.8
VN01-2	2002	24	36.5	30	31	4.3

and Agriculture Organization (FAO) and the European Union (EU). The VCC and the Plant Protection Department (PPD) developed curricula for Farmer Field Schools for the IPM of cotton (FAO-EU, 2002). Initial, full-season Farmer Field Schools were held and the trained facilitators then organized and facilitated training every year. Since 2000, the PPD and VCC have been scaling up cotton IPM training and development activities, supported by FAO field staff and, at the end of the FAO-EU project, 424 Farmer Field Schools had been organized and 10,615 cotton farmers trained on IPM. Before the 1990s, the principal cotton cropping system in Vietnam was cotton monoculture, whereas now all of the cotton area in Vietnam is in mixed, small-scale cropping systems and 90% of the cotton area is rainfed cotton. Due to the use of systemic seed treatments, cotton pests are kept below their economic threshold and rarely have an outbreak until 60–70 days after sowing. However, external funding for IPM training activities in Vietnam ceased in 2005 and, at present, the provincial plant protection departments in Vietnam have not committed funds to continue IPM training for cotton growers.

In 2005, the VCC entered into collaboration with the Chinese Biocentury Transgene Company[2] to test new Chinese Bt varieties and to breed the Bt transgene into Vietnamese varieties. With these new varieties, the VCC and the Nhaho – RICOTAD initiated dry season cotton production in new regions in Vietnam and hope to increase cotton production significantly in the future.

2.2. Organization of Cotton Production in Vietnam

Cotton in Vietnam is grown by small-scale farmers on 0.1–1.0 ha of land. Every season, farmers sign contracts with the VCC, or other joint stock companies, under which the companies provide treated seeds and agricultural inputs on credit and guarantee a price in the early season. They also give the farmers technical support. Most farmers buy their seed from these companies. For the farmer, the seed price represents about one quarter, pesticide one quarter and fertilizer one half of costs. Most farmers do not buy or use the full recommended amount of fertilizer for cotton and often credit is not offered for

[2] http://www.biocentury.com.cn/english/ (accessed 13 September 2007).

the full quantity of fertilizer. The companies buy back the seed cotton and deduct the credit.

The cotton is harvested by hand in several pickings, cleaned roughly of bracts and other contamination and collected by the four VCC branch companies or joint stock companies. There are cotton processing factories (gins) at Hanoi, Nha Trang, Binh Thuan and Dong Nai, and two new gins were built recently in the Dak Lak and Gia Lai Provinces, each with capacity for 15,000 t/ year. All of the cotton fibre produced goes to supply the Vietnamese textile industry. Cottonseed, seed meal and seed oil is exported. Because the cotton is processed centrally, the farmers generally do not save seed for replanting.

2.3. Cotton Fibre Demand and the National Plan for Increasing Cotton Production in Vietnam

At present, only about a tenth of the cotton required annually by Vietnam's textile industry is produced in the country and the demand for cotton is continuing to grow as textile exports grow, stimulated by Vietnam's accession to the World Trade Organization in 2006, giving it tariff-free access to European and US markets. In 2005, the textile industry used 165,000 t of cotton fibre and the 2007 demand is expected to be at least 20% higher (USDA, 2007). In the past 4 years, Vietnam has imported cotton fibre from countries including the USA, Pakistan, West African countries, Australia, Mexico and India (USDA, 2007). In early 2005, the Vietnam government removed all import quotas on cotton fibre (Xinhuanet, 2005).

After increasing every year to a high point of nearly 34,000 t seed cotton in 2002/03, total cotton production has declined to under 15,000 t seed cotton (Figs 2.1 and 2.2). The international world cotton price index has been decreasing since the late 1990s and the cotton price is not expected to recover in the near future (Gruere, 2007). The Vietnamese government is concerned about providing sustainable supplies to the textile industry in the face of world price fluctuations and has published plans to increase cotton production greatly by 2010. In May 2006, the earlier cotton industry investment plan of 2001 was annulled,[3] but the separate decision of 2002[4] still promises government support for development and intensification of cotton production and aims to increase production of seed cotton to 68,000 t by 2010 and 135,000 t by 2015 (Figs 2.1 and 2.2), which is equivalent to 25,000 and 50,000 t of cotton fibre, respectively. Also planned is an irrigation infrastructure to increase dry season production greatly. The decision assigns responsibility for the development of cotton production in Vietnam to the Ministry of Agriculture and Rural Development (MARD) and mandates the establishment of cotton grower cooperatives. The plans also foresee a role for cotton production in rural development

---

[3] Decision 126/2006/QD-TTg issued on 30 May 2006, invalidating Decision 55/2001/QD-TTg dated 23 April 2001.
[4] Decision 17/2002/QD-TTg issued on 31 January 2002 on the orientations and solutions for the development of industrial cotton plant in the 2001–2010 period.

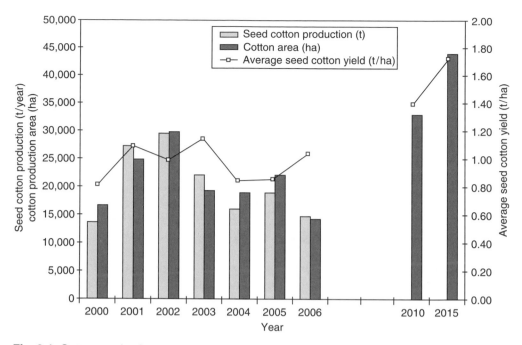

Fig. 2.1. Cotton production, area and yield in the rainy season 2000–2006 and planned increase for the seasons 2010 and 2015.

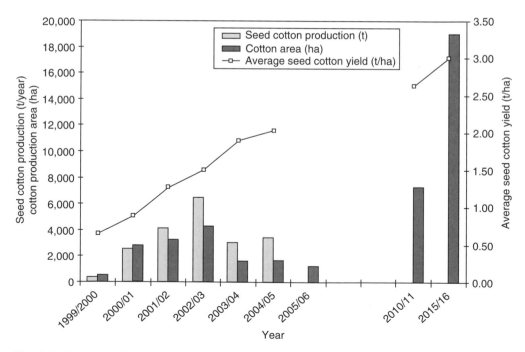

Fig. 2.2. Cotton production, area and yield in the dry season 2000–2006 and planned increase for the seasons 2010/11 and 2015/16.

(Vietnam News Agency, 2006). The Vietnamese government does not plan to provide subsidies for cotton growers.

The following section describes the current environmental and economic context of cotton growing in the different cotton cropping regions of Vietnam and discusses possibilities for expansion in each region.

2.4. Environmental and Economic Conditions for Cotton Cultivation in Vietnam

Cotton is grown in four of the agro-ecological regions of Vietnam: the central highlands and the coastal lowlands of central Vietnam, the south-eastern region and parts of the northern area (see Colour Plate 1). The largest area of cotton production is in the central highlands, followed by the coastal lowlands region, particularly the provinces of Binh Thuan and Ninh Thuan. The area of cotton production fluctuates from year to year as farmers choose to plant or not, based on the relative price of cotton compared to other annual crops, such as hybrid maize, soybean (*Glycine max* (L.) Merr.), groundnut (*Arachis hypogaea* L.) or mung bean (*Vigna radiata* (L.) R. Valiczek (1954)[5]), and the yield they obtained in the previous season. Because of the low cotton price, farmers regard cotton as a supplementary income crop rather than a crop to grow on a commercial scale. In the 2006 season, the total cotton production area was 15,620 ha, which represented a significant decrease from the period 2001–2005 (Figs 2.1 and 2.2).

Currently, 90% of the cotton area is rainfed and irrigated production is limited to less than 1500 ha in the coastal lowlands region. The area of rainfed production has declined to 15,000 ha after peaking in 2002/03 at nearly 30,000 ha and the average yield has fluctuated between 1.1 and 0.8 t/ha (Fig. 2.1). Dry season cotton production was restarted in Vietnam in the season 1999/2000. The area of production peaked in 2003/04 at 4300 ha and has also declined since, but yields have increased steadily and, at present, dry season cotton yields over 2 t/ha and makes up 20% of the total cotton production (Fig. 2.2).

The four regions differ considerably in climate, soils and cropping systems, so that cotton production in each region is faced with different challenges and opportunities. These are described for each region below.

Central highlands region (Gia Lai, Dak Lak and Dac Nong Provinces)

Cotton was introduced into the central highlands region in the 1990s and the region is now the biggest cotton production area in Vietnam, producing two-thirds of the rainfed cotton production. Dry season irrigated cotton was grown on a small area until 2003, but has since stopped completely, mainly because of water shortages.

[5] Previously known as *Phaseolus aureus* Roxb.

Rainfed cotton production

The 7-month rainy season from May to October/November can be used for two crops on better soils. The rainfed cotton crop is planted in mid-July to early August as the second crop, shortly after harvesting the hybrid maize crop, or is planted in between the maize rows 15–20 days before harvesting maize (relay cropping). Hybrid maize is often intercropped with dwarf green (French) beans (*Phaseolus vulgaris* L.), but these are harvested before the cotton is planted. The cotton is harvested from December to January, so the last month of the cropping period is in the dry season, which is suitable for boll maturation. Because of the lower average temperature, the growth period of cotton is longer than that in other regions.

The region has received a rapid increase in population through migration from other parts of Vietnam (WriteNet, 2006). Many immigrants have no previous experience of growing cotton, but IPM Farmer Field Schools in this region have improved cotton production (FAO-EU, 2002). Since 2000, the number of farming households growing rainy season cotton varied between 13,000 and nearly 42,000 in 2002 and the cotton area fluctuated between 8000 and nearly 20,000 ha.

However, rainfed cotton production is limited by pest and disease damage to bolls late in the season. Although cotton plays a role as a cash crop in the hybrid maize–cotton system, coffee and other perennial crops remain the main cash crops in the region, as well as rice (Müller and Zeller, 2002). During the 1990s, coffee occupied over half the arable land in Dak Lak (Ha and Shively, 2007). Annual crop alternatives to cotton, which have a shorter growing season, require less labour and bring higher income, are maize, soybean and mung bean. Hybrid maize is an increasingly significant source of cash for many households, especially those who cannot earn from rice or coffee, and is replacing local maize varieties and upland rice (Müller, 2004). Farmers are able to grow two hybrid maize crops a year, where they can grow only one cotton crop, and can combine maize with soybean, groundnut and mung bean.

This is the main region targeted for increase in rainfed cotton production. The cotton production plan aims to increase the area in this region massively from the 8250 ha planted in 2006 to 20,000 ha in 2008/09 and 28,000 ha in 2015/16, with an average yield increase from 1.15 t/ha currently to 1.45 t/ha. In order to increase the cotton yield and the income households earn from cotton, investments are needed to develop more suitable cotton varieties and cropping practices, and to support cotton growers.

Dry season irrigated cotton production

From 2001 to 2003, irrigated cotton was grown in the dry season in Ajunpa in Gia Lai, which gave a yield of almost 2 t/ha with good quality fibre. The cotton was sown at the end of the rainy season in November or December in rotation with rainfed rice, irrigated during December to February during flowering and yield development, and then harvested at the end of March. The cotton was irrigated by using the rice irrigation channels fed by river water. However, there are often droughts in the rainy season and, since 2003, there has been a severe drought every year (D'haeze *et al.*, 2005). During these droughts, the water flows in rivers in the central highlands have decreased markedly and most households have suffered water shortages (Vietnam News Agency, 2004; Cheesman and Bennett, 2005). The 2003 drought affected a fifth of rainy season cotton plantations in the central highlands (Vietnam News Agency, 2004) and the 2004 drought was worse. As a consequence of the water shortages, there has been no dry season cotton production in the region since 2003.

The plan to increase cotton production includes the reintroduction of dry season cotton production in Gia Lai and Dak Lak. No investments are being made currently, but these plans would require the construction of new irrigation systems, as the rice irrigation canals are not entirely suitable for cotton production.

Coastal lowlands region

The coastal lowlands region has a long tradition of planting cotton and is the main region for dry season cotton production. Rainfall varies considerably from year to year, with variations of over 1000 mm. It rains mostly at the end of May and in September, October and November. January to April are particularly dry months. The soils are mostly alluvial sediments and most of the agricultural area in the region would be suitable for cotton growing (Le, 1998a,b).

The region has the largest number of farmers growing cotton: at the peak of cotton production in 2003–2004, nearly 28,000 farming households grew cotton here and, in 2006, it was still over 11,100 farming households. Cropping systems in this region are very diverse and the amount of each kind of crop is based on topography, climate and the market price of farm produce of the previous year or the previous crop. We divide the region, on the basis of cotton cropping patterns and climate, into two parts; the southern coastal lowlands, where cotton is grown in both rainy and dry seasons, and the central coastal lowlands, where dry season cotton production was introduced in 2003 (Colour Plate 1).

Southern coastal lowlands (Binh Thuan and Ninh Thuan Provinces)

RAINFED COTTON PRODUCTION. In these provinces, cotton is grown in the rainy season from July, with the last month of maturation in the dry month of December. This subregion has maintained both dry and rainy season cotton production since 2000, but areas have declined; the average rainfed yield for the period since 2000 is 0.7 t/ha. Each year, between 2000 and 4000 farming households choose to cultivate cotton. On average, households have been growing over 1 ha of cotton in the rainy season. The cotton production plan foresees the stabilization and gradual increase of rainfed cotton production. However, the rainfall variability is a problem for farmers. The severe drought in 2004 also affected rainy season production in the coastal lowlands. The pest pressure in cotton is high and farmers sometimes prefer to grow rice and maize. In Binh Thuan, farmers try to fit the cotton crop in after a 2-month sesame crop that they plant at the start of the rainy season. Alternative rainfed crops that have shorter growing seasons, and sometimes have a higher overall value, are rice, hybrid maize, sugarcane, mung bean, soybean, cassava, sweet potato, or vegetables.

DRY SEASON IRRIGATED COTTON PRODUCTION. Dry season irrigated cotton is sown in the last month of the rainy season, November, immediately after a crop of rainfed wetland rice, and irrigated between December and February, using the rice irrigation channels. The cotton therefore replaces some of the irrigated rice crop, which is grown on about two-thirds of the farmed area, though a number of other crops are also grown under irrigation, such as hybrid maize, soybeans or tobacco.

The number of farming households growing irrigated cotton has fluctuated from over 10,000 in 2003 to 200 in 2006 and the average area of cotton grown by each household is less than 0.8 ha. In 2006, only 170 ha of dry season cotton was grown, down from 1000 ha in 2003, though the average raw cotton yield of dry season cotton was 1.6 t/ha and yields with a good irrigation system reached 2.4 t/ha. The cotton production plan targets this region as the main area for increase in dry season production, both in area and yield, up to 10,000 ha in 2015, though the goal for 2010 is a less dramatic increase to 3,000 ha but with a significant increase in the average yield to 2.70 t/ha. This will require substantial investments in improving irrigation and developing new cotton varieties and production systems (see dry season IPM later in the chapter).

Central coastal lowlands (Thua Thien-Hue, Quang Nam, Quang Ngai, Binh Dinh, Phu Yen and Khanh Hoa Provinces)

DRY SEASON IRRIGATED COTTON PRODUCTION. In this region, cotton has been grown since 2003 on between 1000 and 3000 ha. Cotton is grown in intercrop from December to May, with the first 3 months rainfed and the final 3 months with supplementary irrigation. For example, in Quang Nam, most cotton farmers irrigate after picking the first harvest so as to stimulate a second harvest 2 months later. Most of the cotton is intercropped with dwarf green (French) beans, mung bean, or groundnut, and sometimes with soybean, watermelon, Cassaba melon, tobacco, or sweet potato, to increase the overall value of the cropping system. In Phu Yen, the cotton replaces the rainfed rice crop from December to March, but most of the cotton areas in Quang Nam are on higher land and so replace other crops. No further crops are grown after cotton, until the next season. Cotton is planted after a rice, maize or bean crop, grown in the first part of the rainy season. Alternative rainfed crops are rice, or an intercrop of green beans and hybrid maize or capsicum, or intensive groundnut, soybean, or groundnut intercropped with cassava, sweet potato, tobacco, or hybrid maize, or mung beans followed by sesame. Beans and maize are also grown under irrigation between April and July/August.

This subregion currently produces two thirds of the dry season cotton production and the average yield for the subregion is increasing; in 2005, it was over 2 t/ha. The cotton production plan for this region is to stabilize the area of production at 3500 ha in 2010/11 and to increase average yield to 2.70 t/ha. However, in 2006 over 8000 farmers cultivated the 1000 ha of cotton, which means the average holding size (0.15 ha) is very small. This could make it difficult to optimize irrigation management, or to introduce new irrigation systems in order to increase production.

South-eastern region (Binh Phuoc, Dong Nai and Ba Ria-Vung Tau Provinces)
In the 1990s, the province of Dong Nai was the principal cotton producing region in Vietnam, with close to 5000 ha of rainfed cotton (Crozat *et al.*, 1997), but since then the area under cotton has decreased considerably, to just over 1000 ha in 2006. Irrigated cotton was grown in the province of Ba Ria-Vung Tau on a small area until 2002, but yields were very low and it was then stopped.

RAINFED COTTON PRODUCTION. Some aspects of the climate of the south-eastern region make it suitable for rainy season cotton production; the climate is moderate and the long rainy season from May to late October makes it possible to grow two crops a year without irrigation. Cotton is planted in early August as the second crop, so as to avoid rain during harvest in December/January (Le, 1998b). However, humidity is generally high at over 85% and high humidity during the open boll period causes boll rot, reducing yield and quality (Tran *et al.*, 1997). Recently, the rainy season has lasted for longer and has affected cotton productivity in this region; often, there is too little rain in October, when the cotton is at maximum growth, and too much in December, at harvest time.

Cotton production is decreasing, mainly because the fall in cotton prices has meant that other crops bring higher overall value, so farmers have switched to other crops, but also because damage from pests and diseases is high, farmers lack labour for weeding and harvesting and rainfall is unreliable. Currently, high value agricultural options in the region include fruit and vegetable production for the market of Ho Chi Minh City, hybrid maize, mung bean, soybean and cassava, as well as sugar, starch, and rubber (Minot and Goletti, 2000). However, the cotton production plan does include a substantial increase in rainfed cotton production in this region, to 6000 ha in 2010/11.

Northern region (Son La, Lai Chau, Hoa Binh and Dien Bien Provinces)
RAINFED COTTON PRODUCTION. Since 2001, between 1000 and 2000 ha of rainfed cotton has been grown in Son La, Lai Chau and Hoa Binh, with a small area in Dien Bien, accounting for 6% of the rainy season cotton area. The rainy season in northern Vietnam begins in mid-April and reaches a peak in August. Rainfall levels fall sharply from October to November and the dry season is from November to early April. Cotton is planted in late April to early May and harvested in August to September. The production plan is for stabilization and a gradual increase in production area and yield.

New regions for dry season irrigated cotton production

Northern coastal lowlands (Thanh Hoa and Nghe An Provinces)
In 2007, an 8.5 ha trial area of irrigated cotton in the province of Thanh Hoa was set up after the VCC and the Nhaho – RICOTAD held a workshop on cotton production. If the trial is successful, farmers may adopt dry season

production of cotton in this region in future. However, the low temperatures early in the season prolonged the cotton growing period to 7 months, from sowing in November to harvest in July.

In summary, increasing cotton production in Vietnam will require significant increases in dry season cotton yields and area, as well as stabilizing rainfed production. This will require investment in suitable irrigation systems and cotton grower cooperatives. Irrigation efficiency and investment in irrigation projects in Vietnam is at a low level and many irrigation systems are degraded because irrigation charges are not enough to cover operation and maintenance costs (Vietnam Agenda 21, 2004). It will also require improvements in crop management systems and pest and disease management. In the rest of the chapter, we describe the main diseases, arthropod pests and weeds of cotton and IPM for rainfed and dry season cotton.

2.5. Main Diseases, Arthropod Pests and Weeds of Cotton in Vietnam

Main cotton diseases in Vietnam

Cotton blue disease (CBD)

Cotton blue disease (CBD) is the most destructive disease of cotton in Vietnam (Nguyen, 1997–2001, 2003). The causal agent of CBD in Vietnam is presumed to be a virus, but it has not been isolated or identified. CBD reduces the quantity of harvested cotton and affects the quality of fibre. Cotton plants produce almost no yield if the symptoms appear before 50 days after emergence. The first CBD symptoms appear at the top of the plant on young leaves – the veins turn a light yellow-greenish and the edge of the leaf curls downwards. As the leaf ages, it gets a leathery texture and turns a dark blue-green colour. Affected plants are usually dwarfed or stunted and the stems have shortened internodes and a peculiar zigzag pattern of growth (see Colour Plate 2). CBD may appear on plants of various ages. The bolls are reduced in number and size.

HISTORY OF THE DISEASE IN VIETNAM AND CROP LOSSES. The disease was first noted in 1984–1985 at the Nhaho – RICOTAD in the Ninh Thuan province in the coastal lowlands (Nguyen, 1999b). At that time, its damage was negligible. The first CBD outbreak occurred in 1991 and, from 1994 onwards, the disease occurred early in the cotton season and caused heavy damage in Ninh Thuan and Binh Thuan. Many fields were highly susceptible and incidence was heavy. The disease has since spread to the other coastal lowland provinces and to the cotton area of Son La in the north. In the central highlands, CBD was first noted during the 1999 rainy season in the Chu Se District in the province of Gia Lai and it then became prevalent in Ajunpa, where irrigated cotton was grown. In the Dak Lak and Dac Nong provinces, CBD first occurred in 2000 and subsequently spread. In 2001, an outbreak occurred in the Cum Nga, Buon Don and Cu Jut districts of Dak Lak and the Dac Min district of Dak

Nong, and the disease has spread to Dong Nai and Binh Phuoc. Since 2002, CBD has been prevalent in all cotton regions, particularly in both rainy and dry season cotton in Ninh Thuan and Binh Thuan and in the central highlands and the south-eastern regions, and has become a key threat to cotton production in Vietnam. In these regions, CBD decreases yield by 10–15% (even 30%) a year on average. It is not yet such a problem in dry season production in the central coastal lowlands.

TRANSMISSION, VECTOR AND POSSIBLE CAUSAL AGENT.   CBD in Vietnam is transmitted by the cotton aphid (*Aphis gossypii* Glover), but not by leafhoppers (*A. devastans*) or thrips (*Thrips* sp.) (Nguyen, 1999b, 2002b). It is transmitted in a persistent, circulative manner, which means that once an insect vector has fed for the minimum acquisition period on a diseased plant, it is able to transmit the virus to other plants throughout most or all of its life and the virus moves inside the aphid's body into the salivary glands (Maramorosch, 1992). The longer the acquisition and transmission feeding time of aphids, the heavier will be the infection of cotton. In Vietnam, aphids can retain their transmission capacity for a long time after one feeding bout on infected cotton and only two aphids containing the causal agent are sufficient to transmit CBD to cotton seedlings. The latent period in the plant is about 20–25 days. CBD symptoms usually can be seen on cotton plants from 20 to 40 days after emergence and damage spreads quickly to a peak at 50–75 days after emergence. After that, new infections occur at a low rate. If farmers ratoon (prune) cotton and leave the cut stems and roots in the ground to grow back in the new season, new shoots will appear with the first rains bearing the typical symptoms of CBD. Cotton aphids colonize these shoots immediately in high densities and can transmit CBD to other new cotton plants nearby. Volunteer and feral cotton plants also act as disease reservoirs and *Hibiscus sabdariffa* L., *Sida rhombifolia* L., *S. acuta* Burm. F. and *Thespesia lampas* Cav. have been identified as alternative hosts (Nguyen, 1999b). The first three do not show CBD symptoms but act as reservoirs for the causal agent.

Since 1995, the Hanoi Agricultural University and the Nhaho – RICOTAD have cooperated in an attempt to identify the causal agent (Nguyen, 2002b). Electron microscope photographs showed the same virus-like structures in a cotton leaf and in a *S. acuta* plant, both infected with CBD. A seemingly similar disease in Brazil, also known as blue disease and spread by cotton aphids, has been identified as being caused by a luteovirus (Corrêa *et al.*, 2005). In Australia, a similar aphid-transmitted disease, called cotton bunchy top, affects cotton leaves and roots, but seems to be caused by an agent that is unrelated to the Brazilian luteovirus or other known viral structures (Ali *et al.*, 2007).

Other diseases of cotton
Several fungal diseases of cotton are a problem in Vietnam. In the rainy season, Rhizoctonia leaf spot and boll rot, caused by the fungus *Rhizoctonia solani*, appears from 60 days after emergence and develops rapidly in rainy and humid conditions. Currently, the disease is very prevalent and damaging in

rainfed cotton areas of the central highlands and the south-eastern region, but it does not cause significant damage in the coastal lowlands. Since the late 1990s, false mildew (also known as white mould or grey mildew) has been a problem in all the cotton growing regions. False mildew is often severe in the rainy season, causing defoliation of cotton plants, leading to early boll opening, reducing the yield and quality of fibre. The causal agent of false mildew is the fungus *Ramularia areola* G.F. Atkinson 1890 (synonym: *Ramulariopsis gossypii* (Speg.) U. Braun, previously known as grey or areolate mildew) (Hillocks, 1992).

Irrigated cotton in the central coastal lowland provinces is affected seriously by seedling damping off (also caused by the fungus *R. solani*), particularly where cotton is alternated with rice, and the disease can cause up to 50% loss of seedlings. False mildew is also prevalent in these dry season cropping systems. In contrast, in Ninh Thuan and Binh Thuan, dry season production fungal diseases are not a problem, though false mildew is present.

Bacterial blight, caused by the bacterium *Xanthomonas campestris* pv. *Malvacearum* (E.F. Smith) Dye, was a serious problem in cotton until the 1990s, occurring every season in many provinces, particularly in the Mekong Delta, but since 1995 the disease has virtually disappeared because resistant *G. hirsutum* varieties replaced the susceptible ones and cotton seeds are delinted with concentrated sulfuric acid.

Main insect and mite pests on cotton in Vietnam

More than 66 species of herbivorous insects and mites have been identified in cotton crops in Vietnam (Nguyen, 1996a). About 30 of these are abundant enough to be considered pests and eight are considered major pests requiring control measures: cotton aphid (*A. gossypii*), cotton bollworm (*H. armigera*), pink bollworm (*Pectinophora gossypiella*), armyworms (*Spodoptera* spp.), leafhopper (*A. devastans*), thrips (*T. palmi, Scirtothrips dorsalis*) and spider mites (*Tetranychus* spp.).

Cotton aphid (Aphis gossypii Glover) (Homoptera: Aphididae)

Before the 1980s, the cotton aphid was considered a minor pest on cotton, causing occasional problems and controlled probably as a side effect of the broad-spectrum insecticides used to control other pests (Leclant and Deguine, 1994). However, the importance of the cotton aphid as a pest is increasing worldwide and yield losses associated with cotton aphids have been documented from the USA (Andrews and Kitten, 1989; Harris *et al.*, 1992; Layton *et al.*, 1996). Cotton aphids are known to transmit over 50 plant viruses (Kennedy *et al.*, 1962). The cotton aphid is now a major pest in most of the cotton areas in Vietnam because they transmit the pathogen that causes CBD.

The first adults on seedling cotton are winged adults coming from other plant hosts. *A. gossypii* has a very broad range of host plants, including squashes, melon, citrus fruit trees, coffee, chilli, okra and many ornamental plants and weeds, some of which are alternative hosts of aphid-transmitted diseases (Leclant and Deguine, 1994; Nguyen and Sen, 2003). More than 25 different host plants of *A. gossypii* are found in and around cotton cropping systems in the Ninh Thuan and Binh Thuan provinces (Nguyen and Nguyen, 2003). However, not all these plants may be possible sources of aphid pests on cotton because *A. gossypii* probably consists of several genetically differentiated host races, or biotypes (Blackman and Eastop, 2000). There is some genetic evidence for a differentiated biotype on Cucurbitaceae host plants from Europe and South-east Asia (Vanlerberghe-Masutti and Chavigny, 1998) and for different biotypes on Malvaceae, Cucurbitaceae and Solanaceae in Cameroon (Brévault *et al.*, 2006). Aphids live in colonies and reproduce parthenogenetically. Duration of development, longevity and reproduction rate vary considerably with temperature and humidity, between winged (alate) and apterous aphids and between different host plants (Xia *et al.*, 1999). At optimum temperatures (27°C), the nymphal development period lasts for 4–7 days and average fecundity per adult is 38 nymphs (Nguyen and Nguyen, 2003).

In Vietnam, aphid population development differs on rainfed versus irrigated cotton and varies according to the climatic conditions in each cotton region. In the rainy season, aphids occur on cotton seedlings from 7 to 10 days after emergence. The population increases slowly and generally peaks at 40–50 days after sowing. In the central highlands, aphids rarely reach economic threshold levels because heavy rains in September and October wash the aphids off the cotton plants and reduce the populations, and a wide range of natural enemies are present (Nguyen, 2001). Rain also creates favourable conditions for the entomopathenogenic fungus *Neozygites fresenii* (Nowakowski) Remaudière & Keller to grow and parasitize aphids.

In contrast, during the dry season and in the coastal lowlands where rainfall is lower, aphids attack cotton as soon as it emerges and cause damage throughout the cotton season (Nguyen, 2001; Le and Tran, 2004). Observations in Ninh Thuan show that aphids appear on the cotton seedling immediately after emergence. When the seedling has two cotyledons, they can be present with about 10–50 individuals/100 cotyledons. Then the aphid population increases to a peak density of 500–5000 individuals/100 leaves at 35–40 days after emergence. After this, populations decrease and remain at a lower level until 75–80 days after emergence. The peak of CBD infection is related to one peak of aphid infestation in the crop season, but there is no apparent relationship between infection rate and aphid population size, possibly because the proportion of aphids carrying the disease is variable. Thus, we cannot use aphid density to forecast CBD infection. However, aphid population levels from emergence to 45 days after emergence determine whether CBD infection will be damaging, so aphids must be controlled early in the season to reduce CBD incidence.

Cotton bollworm (Helicoverpa armigera Hübner) (Lepidoptera: Noctuidae)
H. armigera is recorded from all the cotton growing areas in Vietnam and is considered a serious pest on maize, cotton, tobacco, tomato, soybean and mung bean in most areas where these crops are grown. *H. armigera* larvae feed on a

wide range of crops and wild plants (Ravi *et al.*, 2005). Experiments in Vietnam found that in choice tests with tobacco, maize and cotton, *H. armigera* larvae preferred to feed on tobacco, then cotton, with maize least preferred, and in field experiments, the highest egg and larval populations occurred on tobacco, followed by cotton, with the lowest populations on maize plots (Nguyen, 1996a). However, in general, maize and grain sorghum are preferred; cotton is not considered to be the most preferred host plant (Fitt, 1989). Female moths will move from numerous neighbouring crops on to cotton and also will undertake interregional migrations. In the most popular cotton cropping system in Vietnam, cotton is relayed after maize and intercropped with pulses and other crops, providing a continuous series of host plants suitable for *H. armigera* (but also ensuring a continuous population of natural enemies: see section on biological control). Maize is usually a suitable host for *H. armigera* for only one generation, but can produce large numbers of moths as the larvae are well protected in the ear. The crop can therefore act as a source of pests that move on to cotton, which is attractive for a much longer period (Fitt, 1989; Wu *et al.*, 2004).

A *H. armigera* female lays between 800 and 1800 eggs singly on cotton buds, young leaves and bracts. The first-instar larvae eat their empty eggshell before feeding on the young leaf or bud. Second- and third-instar larvae prefer feeding on buds or flowers and move about, partially injuring many fruiting structures (King, 1994). Fully-grown larvae are between 35 and 42 mm long and bore into the boll to eat inside, often leaving the hind part of their body exposed outside. In Vietnam, it was found that at 27–30°C, most larvae go through five instars on cotton (11.6% developed to sixth instar) (Nguyen, 1996a). The life cycle takes 31.4–33.0 days at 28.7–29.0°C (Nguyen, 2000).

Cotton in the rainy season is a suitable host for two to three generations of *H. armigera*. The first generation occurs from 20 to 25 days after sowing in Binh Thuan province in the coastal lowlands and 30 to 40 days after sowing in the central highlands, feeding on cotton growing tips and young leaves and buds. The larval population of this generation is kept in check, even though the egg density is very high. An important mortality factor in the central highlands is high rainfall during September and October, which washes eggs and first-instar larvae off young leaves. Another important factor is control by natural enemies, including the *Trichogramma* species (Hymenoptera: Trichogrammatidae), an egg parasitoid, which is present on the preceding maize crop, and egg and larval predators and pathogens (see section on natural biological control). The cotton plant also has a high capacity to compensate for damage during the vegetative period (e.g. Sadras, 1998; Wilson *et al.*, 2003), so the first generation does not cause significant damage. The second generation is found on cotton squares from 50 to 60 days after sowing. The population density of the second generation depends on the management of the cotton field; if insecticide is applied early in the season, this can cause outbreaks of the second and third generations, due to the depletion of the natural enemy population. The second and third generations cause economically significant loss by causing damage to cotton squares and bolls. *H. armigera* infestations occur as soon as the weather conditions become dry for a while.

In dry season cotton, *H. armigera* develops through three to four generations. The population density on cotton in the dry season is much higher than in the rainy season and the infestation is more severe (Nguyen, 2001). *H. armigera* damage in the dry season is estimated at 30–40% from the second and third generations and can sometimes reach to 70% of the cotton yield.

Pink bollworm (Pectinophora gossypiella Saunders) (Lepidoptera: Gelechiidae)

In Vietnam, *P. gossypiella* occurs mainly on cotton crops. Studies in Nha Ho in the southern coastal lowlands have shown that each female lays an average of 427 eggs that hatch after 3–5 days in protected places, such as in the calyx or tops of young bolls (Nguyen, 1985). The larvae penetrate the bolls to eat the seeds. A larva may destroy the entire contents of a young boll, while in older bolls it may complete its development on 3–4 seeds. Before bolls are available, eggs are laid in squares and flowers and the larvae eat the squares, preventing flower development. However, the economically significant damage is caused by boll feeding, which stops boll development, leading to premature or partial boll opening and reduced fibre quality. The feeding injury and faeces (frass) create favourable conditions for fungi and bacteria to develop, which cause boll rot, also affecting boll development and fibre quality, as well as aflatoxin contamination of seeds. Larvae pupate inside the boll capsules. *P. gossypiella* can diapause from 3 to 5 months inside cottonseeds or lint on or in the soil, or any other place where there is cotton debris, such as in the cotton gin, or in capsules of wild malvaceous hosts (Ingram, 1994). Adult lifespan lasts between 6 and 32 days, on average 17.8 days, at a temperature of 28.1°C (Nguyen, 1985). In Vietnam, there are usually two generations of *P. gossypiella* per cotton crop, from flowering to boll maturation (Nguyen, 1985).

P. gossypiella is a problem in the southern coastal lowlands (Binh Thuan and Ninh Thuan) and the central highlands and northern regions, particularly where farmers grow *G. arboreum* in their garden and where the wild host plant *T. lampas* occurs. In these regions, it is estimated to reduce cotton yield by 7–10%, in both the rainy and dry seasons. In the past, *P. gossypiella* caused serious damage to cotton fields in the highland areas where a few farmers traditionally grew perennial cotton through regeneration, but most have now abandoned this practice. Egg predation of *P. gossypiella* is relatively low because the eggs are protected within tissues. The larvae inside the buds, flowers and bolls are also protected against natural enemies and pesticides and are therefore hard to control.

Armyworms (Spodoptera exigua Hübner and Spodoptera litura Fabricius) (Lepidoptera: Noctuidae)

Both *Spodoptera* species are highly polyphagous on many crops and weeds. Both species lay eggs in clusters or 'rafts' on the underside of cotton leaves, often near flowers and stem tips, and cover them with fine hair-like scales from the moth's body. The hatched larvae remain near the egg raft, mainly on the underside of the leaves, gnawing the epidermis and veins, reducing photosynthesis (see Colour Plate 2). They are gregarious, feeding at night, and rarely move far until the fourth instar. Larvae occasionally may infest cotton seedlings heavily enough to cause defoliation, and resowing may be necessary. The solitary fourth and fifth instars eat both buds and flowers at night and rest during the day in the soil around the base of the plant. Mature larvae skeletonize whole leaves and are highly mobile; heavy infestations defoliate large areas, also destroying cotton squares and flowers. Under typical conditions in Vietnam, the larval period lasts between 7 and 13 days and the complete life cycle takes between 25 and 28 days, which means that one or two generations can occur on cotton. Pupation occurs in the soil close to the food plants. The species have very varied colouring but can be distinguished in the fourth instar. Mature *S. exigua* larvae have a white stripe along each side of the dorsal surface (see Colour Plate 2); mature *S. litura* larvae have a bright yellow stripe down the dorsal surface and are twice the size of *S. exigua* larvae of the same age. *Spodoptera* can be distinguished from *H. armigera* larvae by the smaller size and lack of hairs and spines. In addition, *H. armigera* larvae are always solitary.

S. exigua is an important pest of maize, tobacco, beans, soybean, groundnut, flowers and vegetables in southern Vietnam, although on maize it is not regarded to be as much a problem as *H. armigera*. It developed from a relatively minor pest to a major pest on dry season cotton during the 1980s (Vũ, 1978; Napompeth, 1987), when high infestation generally occurred from an early stage but, since 2002, abundance has been very low. During the rainy season, *S. exigua* is almost absent. In contrast, *S. litura* is a major pest on rainy season cotton and other crops such as groundnut, soybean and maize, but is not a problem on cotton during the dry season.

Cotton leafhopper (Amrasca devastans Distant) (Homoptera: Cicadellidae)[6]
Leafhoppers occur in all the cotton producing regions in Vietnam. *A. devastans* has a wide range of host plants, including okra (*Abelmoschus esculentus* (L.) Moench),[7] aubergine (*Solanum melongena* L.), melon, chilli, sunflower and potato, as well as cotton, though it has not been found on bean crops, citrus or coffee in Vietnam (Tran and Pham, 2005). Populations can build up on weeds, especially the castor oil plant (*Ricinus communis* L.), *Solanum* spp. and wild Malvaceae, and then colonize nearby cotton (Tran and Pham, 2005). In maturing cotton, heavy leafhopper injury to the leaves can reduce photosynthesis so severely that it causes boll drop and affects fibre and seed development, reducing fibre quality and yield by up to 30%. In the dry season, the leafhopper occurs from 10 to 15 days after sowing and they often grow better and cause damage to cotton earlier than in the rainy season (Nguyen, 2001, 2002a). On moderately hairy cotton varieties, leafhopper outbreaks occur from

---

[6] Synonyms include *A. biguttula biguttula* (Ishida), *Empoasca biguttula biguttula*, *E. devastans* Distant, *Chlorita biguttula* Ishida, *Sundapteryx biguttula biguttula* (Ishida). There is considerable confusion over the systematics of this species and close relatives. Source: Australian Pests and Diseases Image Library, http://www.padil.gov.au/viewPest.aspx?id=79 (NB: this species does not occur in Australia).
[7] Formerly known as *H. esculentus* L.

40 to 50 days after sowing in the dry season and 60 to 80 days after sowing in the rainy season (Tran and Pham, 2005).

Adults lay eggs into cotton leaf veins and the oviposition rate is lower on hairy leaf veins (Sharma and Singh, 2002). The eggs hatch after 4–8 days. Leafhoppers go through five nymphal stages in 6–12 days and adults have a lifespan of about 9 days. Both nymphs and adults reside on the cotton leaves on the lower two-thirds of the plant (see Colour Plate 2). During the day, they hide on the undersurface of the leaf, but can be found anywhere on the leaves during the night. Leafhoppers pierce and suck leaf mesophyll and leaf veins, preferentially at the base of the leaf, and inject toxic saliva. The plant response to the feeding wound and the saliva lead to leaf chlorosis, curling and stunting, called hopperburn (Backus *et al.*, 2005). If hopperburn is slight, cotton leaves turn yellow and the leaf edge curls, reducing photosynthesis. If hopperburn is severe, leaves will turn brown-yellow and then red, dry up and drop (see Colour Plate 2). Heavy injury can be caused by as few as one to two individuals per leaf.

Thrips (Scirtothrips dorsalis Hood and Thrips palmi Karny) (Thysanoptera: Thripidae)

Four species of thrips are found on cotton in Vietnam, though *S. dorsalis* Hood and *T. palmi* Karny are the most important species. Both species are highly polyphagous, occurring on many annual crops and weeds. They especially cause economic damage to vegetable crops and transmit tospovirus diseases of vegetables (Capinera, 2001). In rainy season cotton in the central highlands, thrips only appear on cotton late in the season and do not affect yield. In rainy season cotton in Binh Thuan and Ninh Thuan, they appear earlier, but their density remains too low to cause damage. However, in dry season irrigated cotton in Binh Thuan, thrips appear early and cause injury to cotton throughout crop growth, with most injury during the seedling period. No data are available on whether this damage affects yield.

At an average temperature of 27.8°C and humidity of 70%, the development time of *S. dorsalis* is 15 days and that of *T. palmi* is 14 days (Hoang, 2002), so in 6 months there can be 7–9 generations. A female *S. dorsalis* lays an average of 24 eggs and a female *T. palmi* lays 22 eggs on average (Hoang, 2002). Eggs are laid inside the leaf, flower or fruit tissue. Pupation occurs after falling to the ground and burrowing into the soil (Bournier, 1994). Young and mature thrips feed on plant parenchyma cells, lacerating the epidermis, injecting their saliva, which causes lysis of cell contents, and then sucking up the contents. Thrips are generally a pest of seedling cotton, making leaves curly and brittle, sometimes resulting in poor root development, leaf drop and delayed maturity. During the reproductive period, thrips usually cause little injury and are important predators of mite eggs. They are also attracted to flowers and feed on flower tissues and pollen, and they concentrate on the parts of the cotton plant that bear flowers (Parajulee *et al.*, 2006).

Spider mites (Tetranychus spp.) (Acarina: Tetranychidae)

There are several species of pest mites found on cotton in Vietnam and among them the Tetranychid spider mites are the most important, including *T. urticae.*

In Vietnam, the *Tetranychus* species occur on a wide range of host plants including fruit, cassava, chilli, beans, maize, squashes, groundnut and many weed species (Mai, 2006). Spider mites can crawl short distances along the plants and the ground and disperse over longer distances by being blown along on silk threads (Gutierrez, 1994), enabling them to invade cotton from nearby weeds and intercrops. During the rainy season, spider mites are minor pests on cotton and usually appear late in the season in most of the cotton growing areas in the central highlands and coastal lowlands. However, in dry season irrigated cotton, spider mites cause significant damage after heavy applications of insecticide to control thrips and *H. armigera*, because the pyrethroids do not affect spider mites but kill coccinellids and predatory mites, the main predators of *Tetranychus* species, and stimulate the spider mites to disperse (Penman and Chapman, 1988; Wilson *et al.*, 1998).

Spider mites have a high reproductive rate, particularly at high temperatures. At 30–32°C, which is the optimum temperature for development, the egg stage lasts 2–3 days, the nymphal stage 4–5 days and, with a pre-oviposition period of 1–2 days, the total life cycle takes only 7–12 days (Wilson, 1994a). Females lay an average of 90–110 eggs during a lifetime of about 30 days; therefore, mite populations can increase very rapidly under favourable conditions and there can be as many as ten successive generations on a cotton crop. Spider mites live on the undersides of cotton leaves and suck the mesophyll cells, which damages vascular tissue, reduces chlorophyll content and increases water loss, leading to reduced photosynthesis, carbon dioxide assimilation and transpiration (Reddall *et al.*, 2004).

Weeds in cotton in Vietnam

Cotton is very sensitive to weed competition in the first month of growth because, during this period, the cotton seedling mainly develops its root system and grows slowly above ground. If weeds are not controlled, cotton will grow poorly and productivity will decrease. Weeds can also contaminate the harvested cotton with bits of plant matter. The weed flora in cotton in Vietnam is diverse, with 33 key species of weed in 14 families (Table 2.2). *Ageratum conyzoides* L. is the most common weed in cotton and is a strong competitor (though it also has value as a medicinal plant; Ming, 1999). In Vietnam, weed control has become an urgent and important issue for raising productivity and fibre quality. Some cotton growers are using herbicides on rainfed cotton in the central highlands and the south-eastern regions because of lack of labour for weeding. Cotton is usually planted at a density of 5 plants/m^2, which is a moderate density, but some farmers also plant at lower densities.

2.6. IPM Strategy for Cotton in Vietnam

Integrated pest management (IPM) is a strategy for managing pests and producing a good crop in a healthy environment that pays attention to management throughout the whole year and in the whole cropping system. The objectives

1

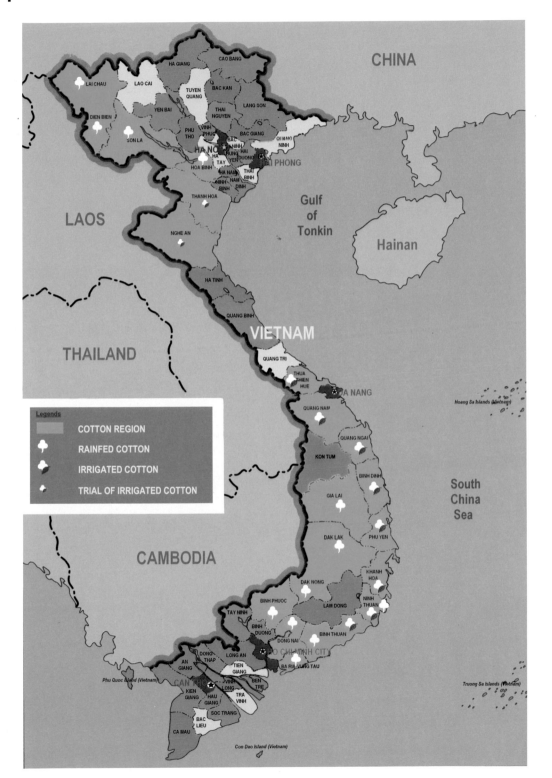

Plate 1. Map of cotton production regions.

2

Plate 2. (a) Cotton blue disease damage, (b) hairy cotton variety, (c) *Amrasca devastans* damage (late season), (d) *Amrasca devastans* on a cotton leaf, (e) young *Spodoptera litura* larvae, (f) mature *Spodoptera exigua* larva. (Photographs: a, c and d courtesy of L.J. Wilson, CSIRO; b, e and f courtesy of Nhaho Research Institute for Cotton and Agricultural Development)

Table 2.2. Main species of weeds in cotton in Vietnam and their relative abundance.

Vietnamese name	Scientific name	Plant family	Relative abundance in cotton
Cứt lợn	*Ageratum conyzoides* L. and *Ageratum* spp.	Asteraceae	+++
Bọ xít	*Synedrella nodiflora* G.	Asteraceae	++
Chân gà	*Dactylotenum aegytiacum* W.	Poaceae	++
Trứng rận	*Eragrostis amabills* Wight	Poaceae	++
Mần trầu	*Eleusine indica*, Guentin	Poaceae	++
Lét	*Mollugo pentaphylla* L.	Aizoaceae	++
Rêu rêu chỉ	*Cyanotis axilaris* Kunth	Commelinaceae	++
Dền gai	*Amaranthus spinosus* L.	Amaranthaceae	+
Đầu rìu	*Commelina bengalensis* K.	Commelinaceae	+
Cúc nút áo	*Tridax procumbens* L.	Asteraceae	+
Bạch đầu ong	*Vernonia cinerea* Lee	Asteraceae	+
Setaria	*Setaria barbata* Kunth	Poaceae	+
Đuôi chồn (Setaria)	*Setaria aurea* Hochst. ex A.Br.	Poaceae	+
Diếc không cuồng	*Alternanthera sesilis* (L.) DC.	Amaranthaceae	+
Màng màng	*Polanesia cholidonei* L.	Capparaceae	+
Sửa long	*Euphorbia hirta* L.	Euphorbiaceae	+
Chó đẻ	*Phyllanthus nirurii* L.	Euphorbiaceae	+
Tổ đỉa	*Aeschynomene aspera* L.	Fabaceae	+
Trinh nữ (mắc cỡ)	*Mimosa pudica* L.	Fabaceae	+
Mắc cỡ móc	*Mimosa invisa* Mart. ex Colla	Fabaceae	+
Rau dừa	*Jussioea linifolia* Vahl	Onagraceae	+
Bông tua	*Digitaria margitana* Lin	Poaceae	+
Đuôi phượng	*Leptochloa chinensis* Nee	Poaceae	+
Đuôi chồn	*Setaria aurea* A. Br.	Poaceae	+
Cỏ chỉ	*Cynodon dactylon* (L.) Pers.	Poaceae	+
Lác	*Cyperus radians* Nee & Mey	Cyperaceae	+
Cỏ cú	*Cyperus rotundus* L.	Cyperaceae	+
Cỏ chác	*Fimbristylis miliacea* (L.) Vahl.	Cyperaceae	+
Rau sam	*Portulaca deracea* L.	Portulacaceae	+
Thù lù cạnh	*Phasalis angulata* L.	Solanaceae	+
Bìm bìm	*Ipomoea chryseides* (Ker.)	Convolvulaceae	+
Cối xay	*Abutilon indicum* (L.) Sweet	Malvaceae	+
Chổi đực	*Sida acuta* Burm. F	Malvaceae	+

are to grow a healthy crop, optimize natural enemies, limit the use of chemical pesticides, observe fields weekly, farmers as expert and to prevent the development of resistance in pests. The benefits of the strategy are savings through less (or no) use of chemical pesticides, a better crop yield and quality, improved health of the farmer and his or her family and improved environmental quality.

We describe the main features of cotton IPM in Vietnam for both the rainy season and the dry season. Generally, extension workers help farmers to observe insects and advise them on when to apply insecticides. Some farmers have competence in IPM and observe pests according to action thresholds and decide themselves whether to spray. Farmers are aware of the negative consequences of the overuse of pesticides and cannot afford to buy many inputs, so they tend not to spray excessively. The FAO-EU programme has organized Farmer Field Schools on IPM every year since 1995 and, from 2003, facilitators were trained on IPM of cotton in the dry season (FAO-EU, 2002). As a result of this training, 207 technicians with university-level education and 63 farmers are qualified as cotton IPM facilitators.

Rainfed cotton IPM

One of the main problems for cotton production in Vietnam is CBD, transmitted by the cotton aphid, which has become prevalent in all cotton regions in both rainy and dry seasons. Although cotton aphid populations only reach economic thresholds for direct feeding damage in the coastal lowlands, they are controlled prophylactically in most regions because even small numbers of aphids are sufficient to transmit the disease.

On rainfed cotton, the fungal disease, false mildew, is prevalent in all regions, while Rhizoctonia leaf spot and boll rot cause yield losses in the central highlands and the south-eastern region. Leafhoppers cause some damage, but are limited by the use of hairy cotton varieties, and thrips can sometimes damage seedling cotton. *H. armigera* causes boll losses in all cotton growing regions. The pink bollworm is a problem in Binh Thuan and Ninh Thuan, the northern region and in the central highlands.

The principal IPM strategies on rainfed cotton are described in the following sections:

Use of hairy varieties with resistance to leafhoppers
Cotton varieties with hairy leaves are less attractive to leafhoppers than glabrous varieties with smooth leaves and little hair and can be used in Vietnam because cotton is hand harvested (whereas in countries with mechanical harvesting, smooth-leaved varieties are necessary to avoid contaminating lint). The widely grown hybrid cotton varieties, VN35 and VN01–2, have hairy leaves (see Colour Plate 2) and good resistance to *Amrasca*, as well as good quality fibre, vigour and yield (Table 2.1). These varieties can reduce by half the number of chemical insecticide applications necessary to control *A. devastans* compared to smooth-leaved varieties and, during the rainy season, using these varieties together with an insecticidal seed treatment can delay the timing of the first spray needed to control leafhoppers until 70–80 days after sowing (Nguyen and Ngo, 1996). For the central highlands, the moderately hairy varieties, VN20 and VN15, have been bred. These retain good leafhopper resistance but also have lower susceptibility to fungal diseases, such as false mildew, than the very hairy varieties grown in the coastal lowlands.

Destruction of crop residues after harvesting, destruction of infected crop material and destruction of alternative hosts of pests and diseases
Most cotton farmers in Vietnam remove cotton stalks and destroy crop residues after harvest (e.g. by burning) in order to reduce the incidence of pests and dis-

eases that carry over, such as *P. gossypiclla*, CBD, bacterial blight and the soil borne fungal diseases, and farmers also have learned to recognize and remove CBD-infected cotton plants during the growing season and burn them. They also remove feral cotton plants.

False mildew infection is carried over from one crop to another through conidia attached to cotton crop residues, cottonseeds, weeds, or on the soil surface, and spreads by wind and irrigation water or rain (Hillocks, 1992). Secondary sources of infection are the conidia produced on infected cotton plants. No host plants of the false mildew pathogen other than cotton have been found in Vietnam. *Rhizoctonia* fungal spores are transmitted to cotton from the soil and crop residues of maize, mung bean and soybean, which are grown in crop rotation with cotton. Many sclerotia can be found on the debris and the innoculum is also present in soil. The fungal spores are discharged to the air by wind and rainsplash to infect healthy leaves and bolls. Then, infected cotton leaves serve as the secondary source of infection. Most farmers in the south-eastern and central highlands regions spray 1–2 applications of fungicide 10 days after emergence and 1–2 applications around 50–60 days after emergence to control Rhizoctonia leaf spot, boll rot and false mildew. No increased tolerance to false mildew has been found in Vietnamese varieties so far, so it has not been possible to breed for false mildew tolerance.

Weeds can influence disease incidence critically by acting as reservoirs of diseases and their vectors, either as alternative hosts or as obligate alternate hosts (Wisler and Norris, 2005). *H. sabdariffa, S. rhombifolia, S. acuta* and *T. lampas* are alternative hosts of CBD and the cotton aphid and can carry the disease over between cotton seasons (Nguyen, 1997–2001). Destroying these weeds before sowing decreases the incidence of aphids and CBD in cotton, and many farmers do this. From 2002 to 2004, the VCC branch for the Ninh Thuan and Binh Thuan area paid farmers to destroy these weeds in an effort to control CBD. Similarly, *T. lampas* is a wild host of *P. gossypiella* and destroying it has been shown to decrease numbers of this pest. *A. devastans* has been shown to be common on *Solanum* species, weedy Malvaceae and the castor oil plant in the Ninh Thuan and Binh Thuan provinces (Tran and Pham, 2005). However, all these weeds are often abundant and cannot be controlled effectively by farmers. On the other hand, the populations of pests on these weeds will also ensure the continuous presence of natural enemies.

Enhancing natural biological control through diverse cropping systems
More than 136 species of natural enemies of cotton pests have been identified in cotton producing regions in Vietnam and they play an important role in controlling cotton pest populations (Pham, 1993). The diversity is high in the mixed Vietnamese cropping systems with an average field size of less than 0.2 ha. Cotton is intercropped, relay cropped and/or grown in mixed cropping systems. In an intercropped system, the crops grow in alternate rows in the field. In a relay cropped system, the second crop is planted between the first crop, shortly before the first crop is harvested. In mixed cropping systems, small areas of different crops are grown in close proximity.

An intercrop (e.g. beans) may act as a source of natural enemies for the more slow-growing primary crop if natural enemies colonize the intercrop before the primary crop develops, but the intercrop may act as a sink for natural enemies when crop and interplanted vegetation develop at the same time (Andow, 1991). Most predators on cotton are very mobile generalists. Investigations show that populations of predators (such as coccinellids, spiders, pentatomid bugs, predatory mites) and parasitoids (such as *Trichogramma* spp.) are very abundant on many crops that are not treated with insecticides in Vietnam, such as maize. After the maize is harvested, these natural enemies move on to the seedling cotton and help keep cotton pest insects below economic threshold levels (Nguyen and Nguyen, 1995). In the most popular cotton cropping system, cotton is relayed after hybrid maize and intercropped with pulses and other crops. Maize is sown in April, cotton is sown in mid-July in between the maize rows and the two crops overlap for 20–30 days in alternate rows.

NATURAL ENEMIES OF SUCKING PESTS.   The fungal pathogen *N. fresenii* is an important mortality factor of cotton aphid in the rainy season (possibly the most important; see Weathersbee and Hardee, 1994) and more than 52 species of arthropod natural enemies have been reported preying on *A. gossypii*, including the coccinellids *Menochilus sexmaculatus*, *Scymnus* sp., syrphid larvae, lacewings, *Geocoris* bugs and the parasitoids *Aphidius* sp. and *Aphelinus* sp. (Nguyen, 1996a, 2001). Coccinellid beetles are important and abundant aphid predators. Leafhoppers are mainly preyed on by spiders and lacewings, though leafhopper predators are not as abundant as aphid predators. Key predators of thrips are the minute pirate bug *Orius* sp. and predatory thrips. Spider mites are usually well controlled by natural enemies, especially egg predation by thrips and *Geocoris* and *Nabis* bugs (e.g. Wilson *et al.*, 2006).

NATURAL ENEMIES OF LEPIDOPTERA.   The main lepidopteran pests of cotton in Vietnam (except *P. gossypiella*) are highly mobile, generalist pests of a number of important crops such as maize, soybean, or other beans, as well as cotton, and can therefore infest cotton crops rapidly and lay their eggs. Unless natural enemies are present and well established in high numbers before the pests arrive, they cannot respond rapidly enough to control the pests. However, in the diverse cropping systems of Vietnam, natural enemy populations are usually already present in high numbers. For example, observations over 8 years have shown that 30–60% of *H. armigera* eggs are destroyed in early rainy season cotton in the central highlands, coastal lowlands and south-eastern regions (Nguyen Thi Hai, 1996a,b, 2001). Many eggs of *H. armigera* are lost because they dry out or are washed or blown from leaves (King, 1994), but also more than 70 species of parasitoids, predators and pathogens are recorded to attack *H. armigera* eggs or larvae.

Trichogramma egg parasitoids are important natural enemies of *H. armigera* and other Lepidoptera. During 5 years (1995–2000) in the central highlands rainy season, the rate of *Trichogramma* parasitism of *Anomis flava* eggs was found to be 25% or more and *Trichogramma* spp. parasitized over 10% of *H. armigera* eggs (Tran and Nguyen, 2003). Ants are significant predators of lepidopteran eggs. Other egg predators include coccinellids, lacewing larvae and *Geocoris* and *Nabis* bugs (see Chapter 7, this volume). Lepidopteran larvae are killed by pathogens, including a *Beauveria* species, a blue fungus (identity not clear), *Bacillus thuringiensis* and nuclear polyhedrosis viruses (NPVs). The rate of mortality of larvae from pathogens is higher during the rainy season than during the dry season. A study on *H. armigera* on cotton in the central highlands also found an unidentified nematode parasitizing 20% of larvae (Tran and Nguyen, 2003). Predators of lepidopteran larvae include pentatomid bugs (Heteroptera: Pentatomidae) and assassin bugs (Heteroptera: Reduviidae).

Seed treatment to control sucking pests (also increasing use of foliar insecticides)

An important component of the IPM strategy in Vietnam is seed treatment with the systemic neonicotinoid, imidacloprid (Crozat *et al.*, 1997). On rainfed cotton, it can provide control of aphids and *A. devastans* for the first 60–70 days after sowing (Nguyen and Ngo, 1996). Seed treatment is effective at low concentrations and, as it works systemically (i.e. it is taken up inside the plant), it has fewer effects on natural enemies than sprayed insecticide. Before the arrival of CBD, farmers in the Dak Lak and Dong Nai Provinces were encouraged to delay the first foliar insecticide application until 60–70 days after sowing, because the efficacy of the seed treatment was sufficient and natural enemies were protected and enhanced.

However, the damage caused by CBD is high even when aphid populations are low and, since 2003, farmers have applied 1–2 insecticide (acetamiprid or diafenthiuron) foliar sprays early on to rainfed cotton to control aphids and thereby reduce CBD transmission. Farmers are recommended to apply the first spray 7–10 days after cotton emergence if about 15–20% of cotton plants have aphid colonies. If the aphid population continues to be high, they apply a second spray in the following 10 days. After that, farmers are recommended not to spray against aphids.

Use of alternative control methods and selective insecticides to control Helicoverpa armigera *and* Amrasca devastans *late season*

From 70 days after sowing, farmers usually apply one or two sprays of imidacloprid, or other neonicotinoids (e.g. thiamethoxam), or dinotefuran, or occasionally buprofezin, to control leafhoppers. Farmers are recommended to spray against leafhoppers if at least one nymph per leaf is present from 70 days after sowing in the southern coastal lowlands, or from 80 days after sowing in the central highlands. Another neonicotinoid, dinotefuran, was introduced in 2006 to control aphids and leafhoppers. To control *H. armigera* from 70 days after sowing, farmers use mainly spinosad or abamectin, or chlorfenapyr or lufenuron. Abamectin and chlorfenapyr will also help suppress mite and mirid populations. Table 2.3 lists the synthetic insecticides used commonly on cotton in Vietnam. Various alternative control products, produced by the Nhaho – RICOTAD or small private companies, are also used to control Lepidoptera (Table 2.4). Two strains of nuclear polyhedrosis virus have been mass reared in the Nhaho – RICOTAD since the 1990s and control *H. armigera* (HaNPV) or *S. exigua* (SeNPV) effectively. Efficacy can be high and these products are very selective (Nguyen and Viet, 1996), but they are less reliable and cost the same

Table 2.3. Synthetic insecticides and acaricides used on cotton in Vietnam.

		Synthetic insecticides and acaricides			
Chemical group	Chemical	Commercial name	Target pests	Impact on arthropod natural enemies[a]	Frequency of use on cotton
Neonicotinoid	Imidacloprid	Gaucho (seed treatment)	*Aphis gossypii, Amrasca devastans*	Low (as seed treatment)	+++ (as seed treatment)
Neonicotinoid	Acetamiprid	Mospilan	*Aphis gossypii*	Moderate[b]	+++
Neonicotinoid	Imidacloprid	Admire, Confidor	*Thrips palmi, Aphis gossypii, Amrasca devastans*	Moderate (mite and *H. armigera* resurgence)	+++
Thiourea (site II electron transport inhibitor)	Diafenthiuron	Pegasus	*Aphis gossypii, Tetranychus* spp.	Low (mite resurgence)	++
Insect growth regulator (chitin)	Buprofezin	Applaud	*Amrasca devastans*	Low[c]	++
Insect growth regulator	Lufenuron	Match	*Helicoverpa armigera, Spodoptera* spp.	?	++
Pyrethroid	Cyfluthrin	Baythroid	*Aphis gossypii, Helicoverpa armigera, Spodoptera exigua*	Very high (pest resurgences)	++ (dry)
Pyrethroid	Cypermethrin	Sherpa	*Helicoverpa armigera, Spodoptera exigua*	Very high (pest resurgences)	++ (dry)
Pyrethroid	Deltamethrin	Decis	*Helicoverpa armigera, Spodoptera exigua*	Very high (pest resurgences)	+ (dry)
Pyrethroid	Beta-cyfluthrin	Bulldock	*Helicoverpa armigera*	Very high (pest resurgences)	+ (dry)
Pyrethroid	Lambda-Cyhalothrin	Karate	*Amrasca devastans, Helicoverpa armigera, Pectinophora gossypiella*	Very high (pest resurgences)	+ (dry)
Pyrethroid + organophosphate	Cypermethrin + Profenofos	Polythrin	*Aphis gossypii, Amrasca devastans, Helicoverpa armigera, Spodoptera exigua*	Very high (pest resurgences)	+

Neonicotinoid	Thiamethoxam	Actara 25 WG; 350FS	Aphis gossypii, Amrasca devastans	Moderate (mite and H. armigera resurgence)	+
Sulfite ester	Propargite	Comite	Tetranychus spp.	Moderate	+
Insect growth regulator	Hexythiazox	Nissorun	Tetranychus spp.	?	+
Benzoyl urea	Chlorfluazuron	Atabron	Helicoverpa armigera, Spodoptera exigua	?	+
Diacylzydrazine	Tebufenozide	Mimic	Helicoverpa armigera, Spodoptera exigua	?	+
Pyrazole/pyrrole	Chlorfenapyr	Secure 10EC	Helicoverpa armigera, Spodoptera litura, Tetranychus spp.	High (mite resurgence)	+
Neonicotinoid	Dinotefuran	Oshin	Amrasca devastans	Moderate	+ (recently introduced)
Carbamate	Methomyl	Lannate	Helicoverpa armigera, Spodoptera exigua	High (mite resurgence)	+ (use is restricted)
Organophosphate	Profenofos	Selecron	Aphis gossypii, Helicoverpa armigera	High (mite resurgence)	–
Organophosphate	Triazophos	Hostathion	Aphis gossypii, Helicoverpa armigera	High (mite resurgence)	–

Notes: [a]Unless otherwise marked, the references for impacts on arthropod natural enemies are Wilson et al. (2006) and Elzen (2001). Resurgence means that repeated applications of this product are likely to increase the risk of spider mite, aphid or H. armigera outbreaks. [b]Naranjo and Akey (2005). [c]Naranjo et al. (2004).

Table 2.4. Biopesticides used on cotton in Vietnam.

	Biopesticides				
Origin	Chemical/biological substance	Commercial name	Target pests	Impact on non-target species	Frequency of use on cotton
Streptomyces avermitilis	Abamectin	Tap ky, Vertimec	*Thrips palmi, Tetranychus* spp., *Helicoverpa armigera, Spodoptera exigua*	High to moderate[a]	+++
Saccharopolyspora spinosa	Spinosad	Success 25SC	*Helicoverpa armigera, Spodoptera exigua, Spodoptera litura*	Moderate to low (mite resurgence)[b]	++
Bacillus thuringiensis var. *kurstaki*	Cry exotoxins, Bt spores	Delfin WG (32 BIU), Dipel 3.2 WP, 6.4 DF, Aztron DF35000 DMBU, Xentari 15FC; 35WDG	*Helicoverpa armigera*	Low[c]	++
Virus	NPV (nuclear polyhedrosis virus)	HaNPV	*Helicoverpa armigera*	Negligible[d]	++
Virus	NPV (nuclear polyhedrosis virus)	SeNPV	*Spodoptera exigua*	Negligible[d]	+

Notes: [a]Youn *et al.* (2003), Biddinger and Hull (1995). [b]Cisneros *et al.* (2002), Medina *et al.* (2003) and references therein; Wilson *et al.* (2006). [c]Wilson *et al.* (2006). [d]Flexner *et al.* (1986).

as synthetic insecticide sprays (unless the efficacy of insecticide sprays has been reduced significantly by resistance). The Nhaho – RICOTAD also produces a product which is a combination of NPV and Bt. The VCC recommends the use of Bt sprays to control *H. armigera* during the period between 40 and 70 days after sowing, to avoid disruption of natural enemies, and many farmers apply two to three foliar Bt sprays on cotton and also use Bt on crops such as vegetables, beans and tobacco. Various Bt sprays are sold in all regions of Vietnam (Table 2.4), produced by small Vietnamese companies or imported.

Challenges for IPM of irrigated dry season cotton

The IPM system that has been implemented on rainfed cotton successfully does not perform as well on irrigated dry season cotton and Vietnam has an urgent need to develop effective strategies for IPM of irrigated cotton. Early season aphid populations are higher during the dry season and improved management of CBD is critical. In Binh Thuan and Ninh Thuan, *H. armigera* is estimated to cause 30–40% loss of bolls, and *S. exigua* can also cause significant loss of productivity in the dry season through damage to squares and flowers. Seedling damping-off disease and thrips are problems on seedling cotton and leafhoppers are much more abundant than in the rainy season. Spider mites can be a problem if a flare-up is induced by pesticide use (e.g. Wilson *et al.*, 1998).

On irrigated cotton, seed treatments are not as effective against sucking pests as during the rainy season (Nguyen *et al.*, 2004) and farmers apply about seven or eight pesticide sprays; half of them against sucking insects and half against *H. armigera*. They apply one or two sprays of acetamiprid or diafenthiuron early season (before 25–30 days after sowing) to control aphids and prevent CBD damage. If thrips populations are high, farmers may also spray abamectin, which also controls mites. From 50–70 days after sowing, two or more sprays are applied to control leafhoppers. However, application of chemical insecticides to control sucking pests early in the season destroys predators and parasitoids and leads to resurgences of *H. armigera* and mites. Studies of dry season cotton found that spraying considerably reduced the rate of egg parasitism of *H. armigera* (Nguyen *et al.*, 1990) and that natural enemy population densities were reduced by 70–80% directly after spraying, while 2 weeks later they were still 34% lower than in cotton that was not sprayed (Nguyen, 1996a). Table 2.3 lists the non-target effects on natural enemies of the synthetic insecticides used commonly on cotton in Vietnam. Farmers use abamectin, Bt sprays and pyrethroids to control *H. armigera* from 60 to 90 days after sowing, but it has not been possible to control *H. armigera* damage effectively during the dry season because of the disruptive early season spraying.

Based on this analysis, the priorities for developing IPM of dry season cotton are described in the following sections:

Collecting and breeding varieties resistant to cotton blue disease (CBD)
Host plant resistance to the most serious cotton disease, CBD, has not yet been found in Vietnam, though many cotton varieties have been tested. Under green-

house conditions, all the G. *hirsutum* varieties are 100% susceptible, including the cultivars that were introduced because of their resistance to CBD in other countries: SR1-F4, L1186 (cross HAR × BJA) from Africa, Reba P288 from South America and SSR-2 from Thailand (Bachelier, 2000). Under field conditions in Vietnam, the varieties show complicated responses – infection is variable in the same place on the same variety from year to year, but all varieties are never affected heavily at the same time. A strategy to avoid CBD infection therefore can be to grow three to four different varieties in one region. Most G. *arboreum* varieties are susceptible to CBD, but some appear to be tolerant and the variety Nghe An is resistant (Nguyen, 2002b). This provides hope of selecting and creating new varieties resistant or tolerant to CBD in the future. In Brazil, where a similar disease is prevalent, cottonseed companies identify their varieties as blue disease susceptible or tolerant (Lacape *et al.*, 2002) and researchers are working on the development of molecular markers linked to disease resistance as tools to facilitate breeding (Marc Giband, Brazil, September 2007, personal communication). In Thailand, a breeding programme is reporting successes in developing tolerance to CBD (Koshawatana, 2001).

Releasing varieties with resistance to Helicoverpa armigera and sucking pests
Looking for host plant resistance to pests in cotton often involves trade-offs – a resistance trait may make cotton more resistant to one pest but increases its attractiveness to other pests, and often affects yield potential. Experimental evidence from various countries is mixed as to whether hairy varieties are more or less resistant to H. *armigera*, though the hairy and medium hairy varieties in Vietnam do not show less resistance than the smooth-leaved varieties. Hairy cotton varieties are more attractive to oviposition by H. *armigera* females than glabrous varieties (Robinson *et al.*, 1980), but survival and mobility of young larvae may be lower because of the hairs (Ramalho *et al.*, 1984). The hairs have been shown to be a hindrance to predators and parasitoids of H. *armigera* eggs, slowing their search rate on the leaf surface and also trapping honeydew, which hinders progress and increases grooming time (Treacy *et al.*, 1986, 1987). The use of host plant resistance against H. *armigera* is a complex problem (King, 1994); it could contribute to improving dry season production in Vietnam but will not solve the problem of crop damage.

The okra leaf type could be a useful trait for Vietnamese cotton to reduce sucking pest damage. It reduces humidity in the canopy, which reduces the population increase of spider mites (Wilson, 1994b) and whitefly (Chu *et al.*, 2002). However, hairy varieties can be more susceptible to whitefly (Chu *et al.*, 1999), so combining the hairiness that counters leafhoppers with the okra leaf shape could diminish the value of the okra leaf trait to reduce whitefly damage. Investigations in Vietnam have shown that hairy cotton varieties are less affected by thrips than smooth-leaved varieties, although most of the current cotton varieties do not show marked resistance to thrips (Nguyen and Nguyen, 2003). Cotton aphid abundance has been found to be higher on less hairy varieties (Khan and Agarwal, 1990; Weathersbee and Hardee, 1994) and on varieties with high gossypol levels (Li *et al.*, 2004). The cotton plant has a high capacity to compensate for damage during the vegetative growth period (Sadras, 1998; Wilson *et al.*, 2003), although there is a lot of variability among varieties (Sadras and Fitt, 1997). Studies in the USA have shown that cotton can compensate strongly for early season thrips and aphid feeding damage with no yield loss (Rosenheim *et al.*, 1997; Sadras and Wilson, 1998). However, this recovery capacity cannot be taken advantage of when aphids transmit disease, as is the case in Vietnam. If CBD-resistant varieties could be found, then early season damage by thrips and aphids could be tolerated because of the compensation capacity of the cotton plant.

In 2005, the VCC entered into collaboration with the Chinese Biocentury Transgene Company to test new transgenic Chinese varieties that produce the Cry1Ab/Ac fused protein, with high toxicity to *H. armigera* (Wu and Guo, 2005). The Chinese varieties are hybrid, with high yield, large bolls and high fibre production, but are less hairy than the Vietnamese varieties and less resistant to sucking pests. According to testing so far, their disease susceptibility is about the same as the Vietnamese varieties. Researchers currently are working on breeding the Bt transgene into Vietnamese varieties.

Suitable cropping systems that include intercropping and rotation of cotton with other crops and suitable cultivation practices, including sowing time, that maintain a healthy crop and reduce the incidence of sucking pests and diseases

Cotton should be planted into warm, well-aerated soil (Le, 1998b) and, as it is sensitive to waterlogging, it should be planted on ridges (firm, high, well-shaped beds) and the soil surface hoed after rain or irrigation. This will prevent seedling damping-off. Farmers who grow irrigated cotton after wetland rice prepare the ridges into which the cotton is planted by hand and then, after the cotton harvest, they must smooth the soil again in preparation for wetland rice. To control seedling damping-off, farmers in the coastal lowlands apply tebuconazole and hexaconazole to cottonseed before planting.

The sowing time has a significant effect on the attack of cotton pests and diseases and farmers are recommended to plant during a defined period in each region, and not to plant earlier or later. In the central highlands, early planting usually leads to high infestation of false mildew, whereas late planting results in high infestation of bollworm, leafhoppers, aphids and CBD (Nguyen, 1996a). The sowing period for all farmers in a region should be as synchronous and compact as possible in order to avoid cross-colonization of pests from older to younger crops (Wilson *et al.*, 2006).

Cotton ridges can be mulched with maize crop residues, stubble or other organic material to maintain and improve soil structure, with better soil aeration and drainage. Mulching the soil surface also limits the population density and damage caused by leafhopper, aphids and thrips in cotton (Tran *et al.*, 2005). Farmers use plastic sheeting for watermelon production and, in Ninh Thuan province, it is used in the production of hybrid cottonseed, but the technique is too expensive for most cotton farmers.

Soil degradation from erosion, flooding, acidification and salinization is a problem on much of the agricultural land in Vietnam and soil fertility is a limiting factor for agricultural production (Vietnam Agenda 21, 2004). Most cotton farmers follow the recommendations of the extension workers and apply NPK fertilizer, which they buy from the cotton companies. Currently, the recommended cotton rotations for Ninh Thuan and Binh Thuan are cotton after maize or legumes, or cotton after sesame.

Good water management and control of vegetative growth

Cotton usually needs between 500 and 700 mm water, spread evenly over its whole 120-day growing season, to meet its water requirements for growth (FAO AGL, 2006). In the first month of growth, the cotton plant develops a strong root system if soil conditions are suitable and, by maturity, it has a rooting depth of about 1.5 m. Because of the slow above-ground growth of the cotton seedling during the first month, this phase needs only 10% of the total water requirements and excessive water will restrict root and crop development (Oosterhuis and Jernstedt, 1999).

During the vegetative growth period when squares (floral buds) are formed, adequate water supply is important to avoid loss of squares and delayed flowering, which affects yield. Once flowering has begun, the water supply should be managed carefully, both to restrict vegetative growth and maximize flower development and to avoid stress, and the water needs of the cotton plant must be met through the whole fruit set and maturation period in order to obtain high yield. Too much irrigation and nutrient availability will lead to excessive vegetative growth, which makes cotton very attractive to sucking pests such as aphids and leafhoppers. If the canopy is dense, it increases humidity and stimulates fungal infections. Conversely, although cotton can tolerate short drought periods because of its deep root system, high water deficits may restrict growth and cause shedding of squares and bolls (Oosterhuis and Jernstedt, 1999).

As the number of bolls on the plant increases, their demand for photosynthetic assimilate approaches the capacity of the cotton plant to supply the bolls with energy. This results in dramatically reduced growth and the cotton plant stops producing new squares, in a phase known as 'cut-out'. After cut-out, the water requirement decreases to 3–4 mm water/day and irrigation can be stopped if the available soil moisture is sufficient to meet the crops' daily evapotranspiration requirement until boll maturity. During this phase, low humidity is desirable and cotton fields must be airy and get enough light. High humidity, excessive irrigation or fertilization will delay boll opening and bacteria and insects will cause fruit loss and boll rot (Oosterhuis and Jernstedt, 1999).

Cotton farmers use mepiquat chloride (PIX) as a plant growth regulator in most cotton areas, according to the VCC specifications in the loan package. They use 2–3 applications per season from first pinhead square (30–40 days after sowing) onwards, with some variation according to the rate of plant growth in each region. The dosage of PIX should be adjusted to plant growth parameters, such as water status (Stewart, 2005). This reduces the amount of fresh regrowth and the attractiveness of the crop to pests, and so maximizes the plant's input of resources to the developing fruit. This is particularly important under the high humidity conditions in rainfed cotton.

Biological control: methods to conserve and augment natural enemies; alternative pest control methods; rearing and release of biological control agents

The Nhaho – RICOTAD is searching for alternative pest control methods for dry season cotton. Bt sprays provide little control of Lepidoptera on dry season cotton. The efficacy of the botanical biopesticide neem is poor, so it is not used for cotton production, nor are pheromones. The Nhaho – RICOTAD has a programme to study biology and mass rearing of biological control agents (*Eocanthecona furcellata*, *Orius* spp. and *Trichogramma* spp.) and has conducted trials with *Trichogramma* egg cards, but the cards did not bring any added advantage. This is probably partly because natural *Trichogramma* populations are already abundant in cropping systems where cotton is grown and partly because ant predation of *Helicoverpa* eggs limits the effectiveness of *Trichogramma* biocontrol. So, *Trichogramma* mass release is not practised. The Nhaho – RICOTAD is also searching for biological control agents against diseases, particularly seedling damping-off, and has tested *Trichoderma* using three different application methods (seedcoating, mixing with soil surface and pouring directly on to soil ridge surface), but no treatment was effective.

Care must be taken, before recommending mass releases of biological control agents, to establish that the approach is effective and particularly that the pest management system is conducive to the survival of the agents, e.g. that the released parasitoids are not killed by insecticide applications.

Insecticide resistance management strategy for sustainable use of neonicotinoids against aphids

Additionally, some measures should be taken to enhance the sustainability of both rainy and dry season IPM systems. The current pest management system relies heavily on neonicotinoid pesticides from early in the season to control aphids and leafhoppers. This creates a long window of selection for resistance in the aphids. Cotton aphids are known for their capacity to develop resistance to pesticides and a suitable insecticide rotation scheme, together with a monitoring system to check for resistance evolution in aphids, is needed. Until now, lack of resources has prevented monitoring of pests on cotton for insecticide resistance, but there is indirect evidence that certain pests have developed resistance because the efficacy of some insecticides has declined (Nguyen, 1999a; Nguyen and Sen, 2003). The efficacy of imidacloprid against aphids is decreasing and resistance of aphids to neonicotinoids is a current concern in many countries (Haviland, 2004; Denholm and Nauen, 2005). In other countries, cotton aphids also show widespread resistance to pyrethroids (Herron *et al.*, 2000; Ahmad *et al.*, 2003).

Anticipate potential future threats to cotton production

It is also important that cotton researchers anticipate potential future threats to cotton production in Vietnam, notably the possibility that whitefly-transmitted cotton diseases, such as the begomoviruses (cotton leaf curl disease; Briddon, 2007), enter Vietnam.

2.7. Conclusions and the Future of Cotton Production in Vietnam

Cotton can be grown in many regions in Vietnam during both the dry and rainy seasons, but the principal centre of production currently is concentrated in the central highlands and coastal lowlands during the rainy season. Dry season cotton suffers considerable insect attack and needs more insecticide applications than rainfed cotton. Pest problems are lower in rainfed cotton as compared to irrigated cotton, but productivity and quality are also lower. Due to the lack of an effective irrigated cotton IPM system for dry season production, currently more than 90% of the Vietnamese cotton area is grown under rainfed conditions. In the small-scale diverse cropping systems, natural biological control of cotton pests is high, although yields are often constrained by competition for water within the intercropping system. In recent years, a range of new agronomic technologies and new varieties with higher yield potential, as well as pest resistance, have been transferred to cotton growers. However, cotton production has declined in all regions due to the higher overall economic value and lower labour requirements of other crops such as hybrid maize, soybean, groundnut and cassava. Recently, the cooperation with China for transfer of the *cry1A* fused transgene into Vietnamese cotton varieties has brought new perspectives. Testing results show that these varieties have high resistance to bollworm (*H. armigera*), reduce insecticide applications and the cotton yield is higher than traditional cotton varieties, giving hope for increasing yield from irrigated cotton with Bt varieties. The Chinese Cry1Ac/Ab cotton may not provide control of *Spodoptera* species (Chapter 6, this volume) and other Bt transgenes that effectively control *Spodoptera* species as well as *H. armigera* could be more useful for Vietnam in the future (see Chapter 4, this volume).

However, it is very important that the gains made in IPM of cotton in Vietnam are not jeopardized. Bt cotton varieties are likely to control *H. armigera* and a small number of other lepidopteran species in Vietnam (Chapters 4 and 6, this volume), but will not provide control of any other pests, so IPM of these other pests will continue to be important. The main problem that has destabilized the IPM system of rainfed cotton in Vietnam is the presence of cotton blue disease (CBD) and, ideally, the Bt transgene could be used in CBD-resistant varieties. However, such varieties are not anticipated in the near future. It is therefore essential that the new cotton varieties with the Bt transgene have equally strong leafhopper resistance as the varieties currently used, so that pesticide applications against sucking pests can be kept to the minimum necessary to control CBD. It will also be key to maintain and enhance the action of naturally occurring biological control agents (predators, parasitoids, pathogens) (see Chapters 7 and 8, this volume) and to maintain soil fertility (see Chapter 10, this volume). It will be important to test season-long expression levels of the Cry toxins in the Vietnamese Bt varieties to assess whether expression levels decrease towards the end of the growing season, which could mean that some *H. armigera* damage still occurs on Bt cotton (Wan *et al.*, 2005). It will be necessary to implement a monitoring programme for target pests at risk of resistance, in order to sustain the effectiveness of the technology for the planned increase in cotton production in Vietnam (see Chapter 12, this volume), but it is also urgent to introduce monitoring and management of resistance to insecticides in other pests, such as aphids.

Acknowledgements

We express our thanks to Drs Lewis Wilson (CSIRO) and Pierre Silvie (CIRAD) for their helpful comments on earlier versions of the chapter.

References

Ahmad, M., Arif, M.I. and Denholm, I. (2003) High resistance of field populations of the cotton aphid *Aphis gossypii* Glover (Homoptera: Aphididae) to pyrethroid insecticides in Pakistan. *Journal of Economic Entomology* 96(3), 875–878.

Ali, A., Reddall, A., Roberts, J., Wilson, L.J. and Rezaian, M.A. (2007) Cytopathology, mode of aphid transmission and search for the causal agent of cotton bunchy top disease. *Journal of Phytopathology* 155(4), 220–227.

Andow, D.A. (1991) Vegetational diversity and arthropod population response. *Annual Review of Entomology* 36, 561–586.

Andrews, G. and Kitten, W. (1989) How cotton yields are affected by aphid populations which occur during boll set. *Proceedings of the Beltwide Cotton Conferences 1989*. National Cotton Council of America, Memphis, Tennessee, pp. 2–7.

Bachelier, B. (2000) Rapport de mission au Vietnam, Analyse des recherches cotonnières sur la maladie bleue: proposition d'une programmation de recherche pour l'amélioration de la résistance du cotonnier à la maladie bleue et de la connaissance de l'agent causal. CIRAD document. CIRAD, Montpellier, France.

Backus, E., Serrano, M.S. and Ranger, C.M. (2005) Mechanisms of hopperburn: an overview of insect taxonomy, behavior, and physiology. *Annual Review of Entomology* 50, 125–151.

Biddinger, D.J. and Hull, L.A. (1995) Effects of several types of insecticides on the mite predator, *Stethorus punctum* (Coleoptera: Coccinellidae), including insect growth regulators and abamectin. *Journal of Economic Entomology* 88(2), 358–366.

Blackman, R.L. and Eastop, V.F.E. (2000) *Aphids on the World's Crops: An Identification and Information Guide*. Wiley & Sons, Ltd, Chichester, UK.

Bournier, J.P. (1994) Thysanoptera. In: Matthews, G.A. and Tunstall, J.P. (eds) *Insect Pests of Cotton*. CAB International, Wallingford, UK, pp. 381–391.

Brévault, T., Carletto, J., Picault, S. and Vanlerberghe-Masutti, F. (2006) Cotton races in *Aphis gossypii* evidenced by microsatellite markers and life history traits. Poster abstract. In: *Beltwide Cotton Conferences January 3–6, 2006, San Antonio, Texas*. National Cotton Council of America. Memphis, Tennessee. http://ncc.confex.com/ncc/2006/techprogram/P4797.HTM (accessed 1 October 2007).

Briddon, R.W. (2007) The leaf curl epidemics: the situation with cotton leaf curl disease. In: ISAAA (eds) *Regional Consultation on Biotech Cotton for Risk Assessment and Opportunities for Small-Scale Cotton Growers*. National Institute for Biotechnology and Genetic Engineering, Faisalabad, Pakistan, March 6–8. Common Fund for Commodities, Report CFC/ICAC 34FT, pp. 53–61. http://www.icac.org/projects/CommonFund/cfc_icac_34/ (accessed 1 October 2007).

Capinera, J.L. (2001) *Handbook of Vegetable Pests*. Academic Press, San Diego, California.

Cheesman, J. and Bennett, J. (2005) Natural resources, institutions and livelihoods in Dak Lak, Vietnam. *Managing Groundwater Access in the Central Highlands (Tay Nguyen),*

Vietnam: Working Paper No 1. Australian Centre for International Agricultural Research (ACIAR). The Australian National University, Canberra. http://www.crawford.anu.edu.au/students/showphd.php?surname=Cheesman (last accessed 30 January 2008).

Chu, C.C., Cohen, A.C., Natwick, E.T., Simmons, G.S. and Henneberry, T.J. (1999) *Bemisia tabaci* (Hemiptera: Aleyrodidae) biotype B colonisation and leaf morphology relationships in upland cotton cultivars. *Australian Journal of Entomology* 38, 127–131.

Chu, C.C., Natwick, E.T. and Henneberry, T.J. (2002) *Bemisia tabaci* (Homoptera: Aleyrodidae) biotype B colonization on okra- and normal-leaf upland cotton strains and cultivars. *Journal of Economic Entomology* 95(4), 733–738.

Cisneros, J., Goulson, D., Derwent, L.C., Penagos, D.I., Hernández, O. and Williams, T. (2002) Toxic effects of spinosad on predatory insects. *Biological Control* 23, 156–163.

Corrêa, R.L., Silva, T.F., Simões-Araujo, J.L., Barroso, P.A.V., Vidal, M.S. and Vaslin, M. (2005) Molecular characterization of a virus from the family Luteoviridae associated with cotton blue disease. *Archives of Virology* 150, 1357–1367.

Crozat, Y., Renou, A., Nguyen Huu Binh and Nguyen Tho (1997) Diagnostic agronomique sur les systèmes de culture cotonniers de la province de Dông Nai. *CIRAD Cahiers Agricultures* 6, 493–500.

Denholm, I. and Nauen, R. (2005) Resistance of insect pests to neonicotinoid insecticides: current status and future prospects. *Archives of Insect Biochemistry and Physiology* 58(4), 200–215.

D'haeze, D., Deckers, J., Raes, D., Phong, T.A. and Loi, H.V. (2005) Environmental and socio-economic impacts of institutional reforms on the agricultural sector of Vietnam: land suitability assessment for Robusta coffee in the Dan Gan region. *Agriculture, Ecosystems and Environment* 105, 59–76.

Elzen, G.W. (2001) Lethal and sublethal effects of insecticide residues on *Orius insidiosus* (Hemiptera: Anthocoridae) and *Geocoris punctipes* (Hemiptera: Lygaeidae). *Journal of Economic Entomology* 94(1), 55–59.

FAO AGL (Food and Agriculture Organization of the United Nations, Land and Water Division) (2006) Crop Water Management – Cotton. AGLW Water Management Group. http://www.fao.org/ag/aGL/aglw/cropwater/cotton.stm (accessed 20 July 2007).

FAO-EU (2002) State of IPM in Vietnam (Vietnam Country Report, FAO-EU Cotton IPM Programme Steering Committee Meeting Chizhou, China, September 2002). Integrated Pest Management Programme for Cotton in Asia. Report GCP/RAS/164/EC. http://www.communityipm.org/Countries/vietnam.htm (accessed 10 June 2007).

Fitt, G. (1989) The ecology of *Heliothis* species in relation to agroecosystems. *Annual Review of Entomology* 34, 17–52.

Flexner, J.L., Lighthart, B. and Croft, B.A. (1986) The effects of microbial pesticides on non-target, beneficial arthropods. *Agriculture Ecosystems and Environment* 16(3–4), 203–254.

Gruere, A. (2007) *2007/08 World Cotton Outlook*. Paper by Secretariat Staff presented to the 9th International Cotton Conference, Gdansk, Poland, 6–7 September, 2007. International Cotton Advisory Committee, Washington, DC. http://www.icac.org/cotton_info/speeches/gruere/2007/gdynia_2007.pdf (accessed 20 October 2007).

Gutierrez, J. (1994) Acari – leaf feeding mites. In: Matthews, G.A. and Tunstall, J.P. (eds) *Insect Pests of Cotton*. CAB International, Wallingford, UK, pp. 407–424.

Ha Dang Thanh and Shively, G. (2007) Coffee boom, coffee bust: smallholder response in Vietnam's central highlands. *Review of Development Economics*, in press.

Harris, F., Andrews, G. and Caillavet, D. (1992) Cotton aphid effect on yield, quality and economics of cotton. *Proceedings of the Beltwide Cotton Conferences 1992*. National Cotton Council of America, Memphis, Tennessee, pp. 652–656.

Haviland, D. (2004) Neonicotinoid resistance management in cotton. *California Cotton Review* 71, 7–8.

Herron, G., Powis, K. and Rophail, J. (2000) Baseline studies and preliminary resistance survey of Australian populations of cotton aphid *Aphis gossypii* Glover (Hemiptera: Aphididae). *Australian Journal of Entomology* 39(1), 33–38.

Hillocks, R.J. (1992) Fungal diseases of the leaf. In: Hillocks, R.J. (ed.) *Cotton Diseases*. CAB International, Wallingford, UK, pp.191–238.

Hoang Anh Tuan (2002) Studies on cotton thrips and their management. Masters Degree thesis, Ha Noi Agricultural University, Hanoi.

Ingram, W.R. (1994) Pectinophora (Lepidoptera: Gelechiidae). In: Matthews, G.A. and Tunstall, J.P. (eds) *Insect Pests of Cotton*. CAB International, Wallingford, UK, pp. 107–150.

Khan, Z.R. and Agarwal, R.A. (1990) Mechanism of resistance to aphid (*Aphis gossypii*) in cotton. *Indian Journal of Entomology* 52(2), 236–240.

Kennedy, J.S., Day, M.F. and Eastop, V.F. (1962) *A Conspectus of Aphids as Vectors of Plant Viruses*. Commonwealth Institute of Entomology, London.

King, A.B.S. (1994) Heliothis/Helicoverpa (Lepidoptera: Noctuidae). In: Matthews, G.A. and Tunstall, J.P. (eds) *Insect Pests of Cotton*. CAB International, Wallingford, UK, pp. 39–106.

Koshawatana, P. (2001) Cotton selection for leaf roll disease in Thailand. In: Michel, B. (ed.) *Sustainable Development of Cotton Production in South-East Asia: Constraints and Perspectives*. Proceedings of the 2nd South-east Asian Cotton Research Consortium Meeting, 20–22 November 2001, Ho Chi Minh City, Vietnam.

Lacape, M., Vaissayre, M., Silvie, P., Michel, B. and Hau, B. (2002) The cotton blue disease: subject review and elements of a strategy to improve genetic resistance. In: Michel, B. (ed.) *Sustainable Development of Cotton Production in South-east Asia: Constraints and Perspectives*. Proceedings of the 2nd South-east Asian Cotton Research Consortium Meeting, 20–22 November 2001, Ho Chi Minh City, Vietnam, pp. 155–163.

Layton, M., Smith, H. and Andrews, G. (1996) Cotton aphid infestations in Mississippi: efficacy of selected insecticides and impact on yield. *Proceedings, Beltwide Cotton Conference 1996*. National Cotton Council, Memphis, Tennessee, pp. 892–893.

Le Cong Nong (1998a) Nutritional need of cotton plant and cotton cultivation techniques. In: Vietnam Cotton Company (ed.) *Cotton Production Technology for High Productivity*. Agriculture Publishing House, Hanoi, pp. 151–163 (in Vietnamese).

Le Quang Quyen and Tran The Lam (2004) Studies on cotton insect pests in dry season in Coastal Central. *Plant Protection Bulletin* 3(195), 3–7 (in Vietnamese).

Le Xuan Dinh (1998b) Natural condition in some cotton regions. In: Vietnam Cotton Company (ed.) *Cotton Production Technology for High Productivity*. Agriculture Publishing House, Hanoi, pp. 11–38 (in Vietnamese).

Leclant, F. and Deguine, J.P. (1994) Aphids (Hemiptera: Aphididae). In: Matthews, G.A. and Tunstall, J.P. (eds) *Insect Pests of Cotton*. CAB International, Wallingford, UK, pp. 285–324.

Li, D., Ge, F., Zhu, S. and Parajulee, M.N. (2004) Effect of cotton cultivar on development and reproduction of *Aphis gossypii* (Homoptera: Aphididae) and its predator *Propylaea japonica* (Coleoptera: Coccinellidae). *Journal of Economic Entomology* 97(4), 1278–1283.

Mai Van Hao (2006) Studies on management of cotton red mites. In: *Annual Report of Nhaho Research Institute for Cotton and Agricultural Development*. Nhaho Research Institute for Cotton and Agricultural Development, Nha Ho, Vietnam, 69 p. (in Vietnamese).

Maramorosch, K. (1992) *Plant Diseases of Viral, Viroid, Mycoplasma and Uncertain Etiology*. Oxford & IBH Publishing, New Delhi, India.

Medina, P., Budia, F., Estal, P. del and Viñuela, E. (2003) Effects of three modern insecticides, pyriproxyfen, spinosad and tebufenozide, on survival and reproduction of *Chrysoperla carnea* adults. *Annals of Applied Biology* 142, 55–61.

Ming, L.C. (1999) *Ageratum conyzoides*: a tropical source of medicinal and agricultural products. In: Janick, J. (ed.) *Perspectives on New Crops and New Uses*. ASHS Press, Alexandria,

Virginia, pp. 469–473. http://www.hort.purdue.edu/newcrop/proceedings1999/v4-469. html (accessed 24 September 2007).

Minot, N. and Goletti, F. (2000) Rice market liberalization and poverty in Vietnam. Research report 114. International Food Policy Research Institute, Washington, DC. http://pdf.dec. org/pdf_docs/Pnack940.pdf (accessed 10 June 2007).

Müller, D. (2004) From agricultural expansion to intensification: rural development and determinants of land-use change in the central highlands of Vietnam. Report F-VI/6e. Tropical Ecology Support Programme (TOEB). GTZ, Eschborn, Germany.

Müller, D. and Zeller, M. (2002) Land use dynamics in the central highlands of Vietnam: a spatial model combining village survey data with satellite imagery interpretation. *Agricultural Economics* 27, 333–354.

Napompeth, B. (1987) FAO Consultancy report on cotton integrated pest management in Vietnam: second mission report (April 2 to May 28, 1987). Project VIE/84/001 – Cotton improvement and extension in Vietnam. UNDP/FAO, Hanoi.

Naranjo, S.E. and Akey, D.H. (2005) Conservation of natural enemies in cotton: comparative selectivity of acetamiprid in the management of *Bemisia tabaci*. *Pest Management Science* 61, 555–566.

Naranjo, S.E., Ellsworth, P.C. and Hagler, J.R. (2004) Conservation of natural enemies in cotton: role of insect growth regulators in management of *Bemisia tabaci*. *Biological Control* 30, 52–72.

Nguyen Huu Binh and Nguyen Thi Hai (1995) The effect of intercropping cotton on cotton insect pests and their natural enemies. *Plant Protection Bulletin* 4, 9–10 (in Vietnamese).

Nguyen Minh Tuyen (2000) Population dynamic of cotton bollworm (*Heliothis armigera*) and its management. PhD thesis, Vietnam Academy of Agricultural Sciences, Hanoi, (in Vietnamese).

Nguyen Thi Hai (1996a) Biology and ecology of some key insect pests and natural enemies on cotton in Ninh Thuan and Dong Nai provinces. PhD thesis, Vietnam Academy of Agricultural Sciences, Hanoi (in Vietnamese).

Nguyen Thi Hai (1996b) Cotton pests and their natural enemies. In: *Selected Study Reports on Cotton During 1976–1996*. Nhaho Research Institute for Cotton and Agricultural Development. Agriculture Publishing House, Ho Chi Minh City, Vietnam, pp. 108–120 (in Vietnamese).

Nguyen Thi Hai (1999a) *Insecticide Resistance as a Part of IPM Strategy on Cotton in Vietnam*. ICAC Publishing, Vietnam (in English).

Nguyen Thi Hai (2001) *Cotton Insect Pests and Their Natural Enemies*. Annual report. Institute for Cotton and Fiber Crops, Nha Ho, Vietnam (in Vietnamese).

Nguyen Thi Hai (2002a) *Study Report on Cotton Insects and Mites in Central Coastal Region*. Institute for Cotton and Fiber Crops, Nha Ho, Vietnam (in Vietnamese).

Nguyen Thi Hai and Nguyen Thi Thanh Binh (2003) *Study on Cotton Insect Pests and Diseases and Their Management*. In: Annual Report of Ministry of Industry 2003. Ministry of Industry, Hanoi, 51 p. (in Vietnamese).

Nguyen Thi Hai, Mai Van Hao, Phan Cong Kien and Tran The Lam (2004) *Studies on Management of Sucking Insects and Seedling Disease for Irrigated Cotton*. In: Annual Report of Ministry of Industry 2004. Ministry of Industry, Hanoi, 48 p. (in Vietnamese).

Nguyen Thi Thanh Binh (1997–2001) *Annual Reports on Cotton Diseases*. Institute for Cotton and Fiber Crops, Nha Ho, Vietnam (in Vietnamese).

Nguyen Thi Thanh Binh (1999b) Study on cotton blue disease and its management. PhD thesis, Vietnam Agricultural Sciences Institute, Hanoi (in Vietnamese).

Nguyen Thi Thanh Binh (2002b) Cotton diseases in Vietnam. In: Michel, B. (ed.) *Sustainable Development of Cotton Production in South-east Asia: Constraints and Perspectives*. Proceedings of the 2nd South-east Asian Cotton Research Consortium Meeting, 20–22 November 2001, Ho Chi Minh City, Vietnam.

Nguyen Thi Thanh Binh (2003) Study result of cotton diseases. In: *Annual Reports from 1989–2003*. Institute for Cotton and Fiber Crops, Nha Ho, Vietnam, pp. 102–120 (in Vietnamese).

Nguyen Thi Toan (1985) Some results of studies on pink bollworm (*Pectinophora*). In: *Annual Report 1985*. Institute for Cotton and Fiber Crops, Nha Ho, Vietnam.

Nguyen Tho and Ngo Trung Son (1996) Twenty years of research and application of IPM on cotton in Vietnam. In: Cotton Research Centre (eds) *Results of Research on Cotton 1976–1996*. Agriculture Publishing House, Hanoi, pp. 16–24 (in Vietnamese).

Nguyen Tho and Nguyen Duc Quang (1990) Studies on cotton IPM. In: *Annual Reports*. Institute for Cotton and Fiber Crops, Nha Ho, Vietnam, 43 p.

Nguyen Van Cam and Hoang Thi Viet (1996) Influence of some factors on the formulation process and application of NPV-Ha for controlling cotton bollworm (*Helicoverpa armigera*) on tobacco. In: Nguyen Van Cam and Pham Van Lam (eds) *Selected Scientific Reports on Biological Control of Pests and Weeds (1990–1995)*. Agriculture Publishing House, Hanoi, pp. 24–33 (in Vietnamese).

Nguyen Van Huynh and Sen, L.T. (2003) *Textbook of Agricultural Entomology*. Can Tho University, Can Tho, Vietnam (in Vietnamese).

Oosterhuis, D.M. and Jernstedt, J. (1999) Morphology and anatomy of the cotton plant. In: Smith, C.W. and Cothren, J.T. (eds) *Cotton: Origin, History, Technology and Production*. Wiley & Sons, New York, pp. 175–206.

Parajulee, M.N., Shrestha, R.B. and Leser, J.F. (2006) Influence of tillage, planting date, and Bt cultivar on seasonal abundance and within-plant distribution patterns of thrips and cotton fleahoppers in cotton. *International Journal of Pest Management* 52(3), 249–260.

Penman, D.R. and Chapman, R.B. (1988) Pesticide-induced mite outbreaks: pyrethroids and spider mites. *Experimental and Applied Acarology* 4(3), 265–276.

Pham Van Lam (1993) Preliminary results of identification of natural enemies of cotton insect pests. *Plant Protection Journal* 5, 2–5 (in Vietnamese).

Ramalho, F.S., Parrott, W.L., Jenkins, J.N. and McCarty, J.C. (1984) Effects of leaf trichomes on the mobility of newly hatched tobacco budworms (Lepidoptera: Noctuidae). *Journal of Economic Entomology* 77, 619–621.

Ravi, K.C., Mohan, K.S., Manjunath, T.M., Head, G.P., Patil, B.V., Angeline Greba, D.P., Premalatha, K., Peter, J. and Rao, N.G.V. (2005) Relative abundance of *Helicoverpa armigera* (Lepidoptera: Noctuidae) on different host crops in India and the role of these crops as natural refuge for *Bacillus thuringiensis* cotton. *Environmental Entomology* 34(1), 59–69.

Reddall, A., Sadras, V.O., Wilson, L.J. and Gregg, P.C. (2004) Physiological responses of cotton to two-spotted spider mite damage. *Crop Science* 44(3), 835–846.

Robinson, S.H., Wolfenbarger, D.A. and Dilday, R.H. (1980) Antixenosis of smooth leaf cotton to the ovipositional response of tobacco budworm. *Crop Science* 20, 646–649.

Rosenheim, J.A., Wilhoit, L.R., Goodell, P.B., Grafton-Cardwell, E.E. and Leigh, T.F. (1997) Plant compensation, natural biological control, and herbivory by *Aphis gossypii* on pre-reproductive cotton: the anatomy of a non-pest. *Entomologia Experimentalis et Applicata* 85(1), 45–63.

Sadras, V.O. (1998) Herbivory tolerance of cotton expressing insecticidal proteins from *Bacillus thuringiensis*: responses to damage caused by *Helicoverpa* spp. and to manual bud removal. *Field Crops Research* 56(3), 287–299.

Sadras, V.O. and Fitt, G.P. (1997) Resistance to insect herbivory of cotton lines: quantification of recovery capacity after damage. *Field Crops Research* 52(1–2), 127–134.

Sadras, V.O. and Wilson, L.J. (1998) Recovery of cotton crops after early season damage by thrips (Thysanoptera). *Crop Science* 38(2), 399–409.

Sharma, A. and Singh, R. (2002) Oviposition preference of cotton leafhopper in relation to leaf-vein morphology. *Journal of Applied Entomology* 126(10), 538–544.

Stewart, A.M. (2005) Suggested guidelines for plant growth regulator use on Louisiana cotton. Louisiana State University AgCenter Publication 2918. Louisiana State University Baton Rouge, Louisiana. www.lsuagcenter.com/NR/rdonlyres/8E3A2145-FCFD-43EB-9C3F-0B7D4F1540E4/12012/pub2918cotton1.pdf (accessed 20 June 2007).

Tran Duc Hanh, Doan Van Diem, and Nguyen Van Viet (1997) *Theory of the Proper Exploitation of Agricultural Climate Resource*. Vietnam Academy of Agricultural Sciences. Agriculture Publishing House, Hanoi (in Vietnamese).

Tran The Lam and Nguyen Thi Hai (2003) Biological control study in support of cotton IPM in Vietnam. In: *The 6th International Conference on Plant Protection in the Tropics*, 11–14 August 2003, Kuala Lumpur, p. 57 (in English).

Tran The Lam and Pham Van Lam (2005) Some biological characteristics of cotton jassid *Amrasca devastans* Distant (Homop.: Cicadellidae) feeding on cotton in Vietnam. *Proceedings of the 5th Vietnam National Conference on Entomology*, Hanoi, 11–12 April 2005, pp. 415–418.

Tran The Lam, Nguyen Huu Binh and Le Quang Quyen (2005) Research on influence of mulching materials to sucking insect feeding on cotton in dry season. In: *Proceedings of the 5th Vietnam National Conference on Entomology*, Hanoi, 11–12 April 2005, pp. 419–422 (in Vietnamese).

Treacy, M.F., Benedict, J.H., Segers, J.C., Morrison, R.K. and Lopez, J.D. (1986) Role of cotton trichome density in bollworm (Lepidoptera: Noctuidae) egg parasitism. *Environmental Entomology* 15, 365–368.

Treacy, M.F., Benedict, J.H., Lopez, J.D. and Morrison, R.K. (1987) Functional response of a predator (Neuroptera: Chrysopidae) to bollworm (Lepidoptera: Noctuidae) eggs on smooth-leaf, hirsute and pilose cottons. *Journal of Economic Entomology* 80, 376–379.

USDA (2007) *Vietnam Cotton and Products Annual 2007*. Global Agriculture Information Network Report No. VM7035. USDA Foreign Agricultural Service. http://www.fas.usda.gov/gainfiles/200706/146291286.pdf (accessed 22 February 2007).

Vanlerberghe-Masutti, F. and Chavigny, P. (1998) Host-based genetic differentiation in the aphid *Aphis gossypii* Glover, evidenced from RAPD fingerprints. *Molecular Ecology* 7(7), 905–914.

Vietnam Agenda 21 (2004) Decision by the Prime Minister on Promulgation of the Strategic Orientation for Sustainable Development in Vietnam (Vietnam Agenda 21). 17 August 2004. http://www.va21.org/eng/va21/va21_de.htm (accessed 13 September 2007).

Vietnam News Agency (2004) Drought will continue to plague central highlands. Thursday, 25 March 2004. http://vietnamnews.vnanet.vn/2004–03/24/Stories/12.htm (accessed 1 March 2007).

Vietnam News Agency (2006) Vietnam Pictorial No. 5. For Bumper Cotton Harvests. Story by Van Quy. http://vietnam.vnagency.com.vn/VNP-Website/Print/Print.asp?ID_cat=19&ID_NEWS=3933&language=EN&number=5&year=2006 (accessed 25 February 2006).

Vu Cong Hau (1971) *Development of Cotton Planting in Vietnam and Cotton Varieties*. Scientific and Technical Publishing House, Hanoi (in Vietnamese).

Vu Cong Hau (1978) *Cotton Production Techniques*. Agriculture Publishing House, Ho Chi Minh City, Vietnam (in Vietnamese).

Wan, P., Zhang, Y.-J., Wu, K.-M. and Huang, M. (2005) Seasonal expression profiles of insecticidal protein and control efficacy against *Helicoverpa armigera* for Bt cotton in the Yangtze River valley of China. *Journal of Economic Entomology* 98(1), 195–201.

Weathersbee, A.A. and Hardee, D.D. (1994) Abundance of cotton aphids (Homoptera: Aphididae) and associated biological control agents on six cotton cultivars. *Journal of Economic Entomology* 87(1), 258–265.

Wilson, L.J. (1994a) Plant-quality effect on life-history parameters of the two-spotted spider mite (Acari: Tetranychidae) on cotton. *Journal of Economic Entomology* 87(6), 1665–1673.

Wilson, L.J. (1994b) Resistance of okra-leaf cotton genotypes to two-spotted spider mites (Acari: Tetranychidae). *Journal of Economic Entomology* 87, 1726–1735.

Wilson, L.J., Bauer, L.R. and Lally, D.A. (1998) Effect of early season insecticide use on predators and outbreaks of spider mites (Acari: Tetranychidae) on cotton. *Bulletin of Entomological Research* 88, 477–488.

Wilson, L.J., Sadras, V.O., Heimoana, S.C. and Gibb, D. (2003) How to succeed by doing nothing: cotton compensation after simulated early season pest damage. *Crop Science* 43, 2125–2134.

Wilson, L.J., Deutscher, S., Mensah, R. and Johnson, A. (2006) Integrated Pest Management (IPM) guidelines for Australian cotton II. In: Farrell, T. (ed.) *Cotton Pest Management Guidelines 2006/2007*. New South Wales Department of Primary Industries, Narrabri, New South Wales, Australia, pp. 18–31.

Wisler, G.C. and Norris, R.F. (2005) Interactions between weeds and cultivated plants as related to management of plant pathogens. *Weed Science* 53(6), 914–917.

WriteNet (2006) *Vietnam: Situation of Indigenous Minority Groups in the Central Highlands*. WriteNet Report commissioned by United Nations High Commissioner on Refugees, Status Determination and Protection Information Section (UNHCR). http://www.unhcr.org/home/RSDCOI/44c0f55a4.pdf (accessed 13 August 2007).

Wu, K., Feng, H. and Guo, Y. (2004) Evaluation of maize as a refuge for management of resistance to Bt cotton by *Helicoverpa armigera* (Hübner) in the Yellow River cotton-farming region of China. *Crop Protection* 23(6), 523–530.

Wu, K.M. and Guo, Y.Y. (2005) The evolution of pest management practices in China. *Annual Review of Entomology* 50, 31–52.

Youn, Y.N., Seo, M.J., Shin, J.G., Jang, C. and Yu, Y.M. (2003) Toxicity of greenhouse pesticides to multicolored Asian lady beetles, *Harmonia axyridis* (Coleoptera: Coccinellidae). *Biological Control* 28, 164–170.

Xia, J.X., van der Werf, W. and Rabbinge, R. (1999) Influence of temperature on bionomics of cotton aphid, *Aphis gossypii*, on cotton. *Entomologia Experimentalis et Applicata* 90, 25–35.

Xinhuanet (2005) Vietnam allows free import of cotton, milk materials, maize. http://old.usvtc.org/News/Mar%2005/news%20brief%2022.htm (accessed 25 February 2007).

3 Consideration of Problem Formulation and Option Assessment (PFOA) for Environmental Risk Assessment: Bt Cotton in Vietnam

Nguyễn Văn Uyển, Phan Văn Chi, Nguyễn Văn Bộ, Hoàng Thanh Nhàn, Lê Quang Quyến, Nguyễn Xuân Hồng, Lê Minh Sắt, Arjen Wals, Deise M.F. Capalbo and Kristen C. Nelson

Throughout the world, countries are discussing the role genetically modified organisms (GMOs) will have in their future. Each country begins the discussion at a different starting point, depending on distinct historical, economic, social and environmental factors. For some, GMOs are a new technology that should be used based on market principles – if it is a viable product, it will survive and contribute to economic growth. For others, it is a question of considering long-term risks and uncertainties before making short-term decisions. Precaution is the guiding principle for these decision makers. Still others are caught in the conflict between these points of view as they make decisions regarding the introduction of GMOs to their countries. Decision makers and citizens have the right and responsibility to design their own policy and regulatory systems to address GMOs.

Vietnam is an agriculture-based developing country and, as such, the export of agricultural commodities is and will play an important role in its economy. Despite the fact that Vietnam joined the Cartagena Protocol officially on 19 January 2004, the conflict over GMOs in the world, as reflected by mass media reporting, has created considerable hesitation among policy makers as they formulate and adopt Vietnam's National Biosafety Guidelines. Several meetings and seminars were organized to stimulate discussion between scientists and decision makers, including well-known foreign scientists. These events were designed to encourage deliberation and reflection about the official point of view of the Vietnam Government on the GMO issue.

This chapter is a result of reflection by the authors, begun in sessions held in Ho Chi Minh City, Vietnam, 1–5 April 2004 and continued over a 2-year period. One critical component focused on deliberation as a core element of environmental risk assessment. In this case, it is clear that Vietnam must create a responsive system to facilitate social recognition of the risks and participate in the selection of acceptable choices. At its core, the discussion focuses on the critical societal need that will be addressed by the GMO, i.e. what needs will be satisfied and at what risk? Societal risk requires societal reflection. A deliberative process with multi-stakeholder participation allows members of society to participate in the evaluation of critical needs and risks. A cross-section of society – farmers, consumer groups, industry, environmental representatives, policy makers, etc. – must have a vehicle to express their concerns and evaluate the future alternatives for addressing basic needs. Finally, this deliberative process will be increasingly important for resource-scarce nations if public investment is involved, because a comparative reflection by a cross-section of society may be beneficial in the prioritizing and targeting of resources. To meet these requirements, the Problem Formulation and Options Assessment (PFOA) (Nelson *et al.*, 2004; Capalbo *et al.*, 2006; Nelson and Banker, 2007) concept is presented in this chapter.

3.1. What is the PFOA?

The PFOA is a science-based multi-stakeholder process to formulate problems and assess options as a basis for environmental risk assessment when a country is considering the introduction of a genetically modified organism (GMO) into a specific environment (see Nelson *et al.*, 2004; Capalbo *et al.*, 2006; Nelson and Banker, 2007).

The goals of this process are to help multiple stakeholders to assess their needs, evaluate the risks related to multiple future options and to make recommendations to decision makers about policies to reduce societal risks and enhance the benefits provided by adoption of the GMOs. To fulfil these objectives, the authors suggest that a PFOA conducted in Vietnam should meet the following requirements:

1. *All stakeholders' input* should be emphasized
The PFOA should have all stakeholders' input in the identification of priorities, assessment of possible harms, formulating of options and recommendations for a decision by government authorities. All stakeholders, decision makers, environment representatives, farmers and consumer groups have the right to express their concerns about the use of GMOs and to contribute to the formulation of appropriate GMO policies and decisions for the biosafety of their country.

2. The PFOA should be *legitimate to the public*
The process should not be considered as the private work of one person or one group of stakeholders. It should be legitimate, so public contributions are possible and citizens know what is happening in the environmental risk assessment, management and communication.

3. The PFOA should be *transparent*
Transparency in the PFOA means it should be conducted to facilitate public awareness of the problems and benefits, while encouraging the public's eagerness to contribute to formulating options and evaluating risks.

4. The PFOA should be *sanctioned formally*
Vietnamese authorities should recognize the contribution the PFOA makes to environmental risk assessment and use the recommendations from the stakeholder discussions to inform their decisions.

5. PFOA data and information should be driven with *professional expertise*
In order for the decision to be well informed, the discussion is best served when driven by sound, scientifically guided assessment and review. A robust environmental risk assessment clearly delineates when scientific knowledge, information and analysis can respond to key questions effectively.

6. *The PFOA should be country specific*
The PFOA is country specific and should be conducted with strong references to the local social and natural conditions in Vietnam. Not only do countries have different starting points with regards to the introduction of new technologies such as GMOs, they also differ in the way the discussions about the GMOs' introduction are conducted and in the way the outcomes of such discussions influence the decision making process.

In many cases, the previously mentioned characteristics are the ideal requirements and some may face constraints and limitations during implementation. The PFOA organizers may find there is limited information, financial support, or political will to meet all these characteristics. Every effort can be made to recognize the ideal characteristics while working with what is available or feasible at the time, always planning for improvements in the future.

The PFOA is started with the *Initiation of a Proposal* to the competent authority (CA) of the relevant ministry for risk assessment of a specific introduction of GMO(s) to the environment. For example, the relevant ministry will depend on whether the GMO is a crop, fish, tree, or a medicine. The PFOA could be conducted by the relevant ministry, when the CA decides to proceed with an evaluation of a GMO. The following section presents a brief overview of the PFOA methodology.

3.2. Relation of the PFOA to Environmental Risk Assessment[1]

Practitioners and scholars have tested numerous techniques that serve as a methodological foundation for the PFOA in environmental risk assessment (Grimble and Wellard, 1997; Kessler and Van Dorp, 1998; Loevinsohn *et al.*, 2002, to name a few). Two crucial steps in risk assessment are addressed by many of these techniques and the PFOA is designed specifically to address

---

[1] We provide a few definitions to support the discussion of risk assessment and PFOA in Vietnam (Box 3.1).

them. The first critical step in risk assessment is *problem identification* (NRC, 1983, 1996). What is the problem that the GMO technology is going to address? What is the scope of the problem, how is it defined? Problem identification frames the entire risk assessment. A second critical step is the *identification of potential alternative solutions* to the problem and their possible risks (NRC, 1983, 1996). The proposed action, in this case, the use of Bt cotton in Vietnam, is never the only possible way to address the problem. Risk assessment depends entirely on an appropriate specification of alternatives (including taking no action and doing nothing), so that comparative risk can be assessed and appropriate controls for risk assessment science can be defined and used.

The PFOA is comprised of specific brainstorming, discussion and analytical components. First, formulating the problem serves as the core foundation. The problem is defined as an *unmet need that requires change*. This is the identified problem and its effect, which results in an unmet need that requires change. Basic human needs are identified most commonly as food, shelter and safety. For example, a particular agricultural pest may reduce yields in a crop that is an important staple of a nation's population. If pest damage results in extreme food shortages for a large per cent of the population, this unmet need threatens food security and requires change. Once the needs for food, shelter and safety are met, individuals can expand their interests to include numerous options for well-being. These interests will differ from one individual to another and from one group to another.

After a problem is defined, the PFOA requires a comparative approach to risk assessment. The participants clarify the relative importance of this problem as compared to other problems or issues. Once the group agrees the problem is sufficiently important to merit an analysis, the range of future alternatives for solving the problem are compared in relation to their attributes, potential ability to address the problem, changes required to implement the option and potential adverse effects. The PFOA is assessing alternative future options, not for the current conditions against one option, but rather making a comparative assessment of options that exist and are in use now, that exist but are not used due to identifiable barriers, or new options that could exist in the future, such as the GMO. After a complete analysis by a multi-stakeholder group, a recommendation is made to decision makers to continue research and development (in some cases, risk assessment research) with the technology or to halt the development of the technology.

A science-based PFOA must be a deliberative process (Forester, 1999) designed to provide for social reflection and discussion about transgenic organisms. A sound deliberative process is transparent, equitable, legitimate and data-driven when possible (Susskind *et al.*, 1999). Transparency allows for the open communication of information between all parties and easily accessible reporting of decisions to the public (Hemmati, 2002). Providing an equitable PFOA process means that information from the broadest spectrum of society must be included, with all stakeholders having the possibility to contribute. Civic society must perceive that there are sufficient avenues for input and consideration of diverse viewpoints and concerns. When transparency and equitable input are central to the process, the PFOA gains legitimacy in the public eye. This public legitimacy must be matched by traditional legitimacy or sanctioning by a formal political body that embeds the deliberative process. The deliberative process can be tied to a regulatory authority or legislative authority, but it must provide a means by which results from the PFOA inform government decision making and action. Finally, the foundation of the PFOA is a science-based inquiry. Questions are answered with data, impacts are assessed with valid indicators and the limits of our understanding are delineated clearly by a research agenda or procedures for taking uncertainty into account.

Again, each country will need to develop a country-specific deliberative process that fits the particular structure and authority of the relevant decision making bodies and implementing agencies. For many political systems in the world, the legitimating authority exists to incorporate the PFOA in a legislative or regulatory context, but there are debates about necessary modifications of policies and regulation for transgenic organisms (Munson, 1993; Miller, 1994; Hallerman and Kapuscinski, 1995; NRC, 2002). Depending on the legislative or regulatory situations, a PFOA can be incorporated into a public consultative process that is authorized by regulation, or it may be added as an alternative process, supported by civic society, that informs the debate in traditional decision making bodies, or it may be incorporated in existing decision making processes in order to make that process more inclusive, transparent and more science-based.

3.3. Steps in Conducting a PFOA

The Vietnam National Biosafety Regulation (BSR)[2] was adopted officially by the Vietnam Government in August 2005 (Box 3.1). The National Action Plan on Biodiversity, approved in May 2007,[3] identifies as a major solution the '...active participation of people in biodiversity protection and biosafety management' as described in objective 3(b). 'To ensure the community's right and participation in the process of appraising investment policies, strategies, master plans, plans, programmes and projects concerning natural reserves and the biosafety decision making process'. Implementation of the PFOA is therefore an immediate, necessary action if these regulations are to be applied properly.

A PFOA normally consists of several steps:[4]

- Step 1: Problem Formulation
- Step 2: Prioritization and Scale of Problem

[2] Decision No. 212/2005/QD-TTg promulgating the Regulation on Management of Biological Safety of Genetically Modified Organisms, Products and Goods originating from Genetically Modified Organisms, signed by Prime Minister Phan Van Khai, effective date 26 August 2005 (National Biosafety Regulations).
[3] Decision No. 79/2007/QD-TTg approving the National Action Plan on Biodiversity up to 2010 and Orientations towards 2020 for Implementation of the Convention on Biological Diversity and the Cartagena Protocol on Biosafety, signed by Prime Minister Nguyen Tan Dzung, effective date 31 May 2007.
[4] For a discussion of the PFOA Steps and Questions refer to Nelson *et al.*, 2004; Capalbo *et al.*, 2006; Nelson and Banker, 2007.

Box 3.1. Terminology for risk assessment and PFOA in Vietnam

Terminology used in the Vietnam National Biosafety Regulations:

1. **Biological safety** means measures to manage safety in scientific research, technological development and assay; production, trading and use; import, export, storage and transportation; risk evaluation and management and grant of biological safety certificates for genetically modified organisms; products, goods originating from genetically modified organisms.

2. **Gene** means a unit of heredity, a segment of genetic material of an organism determining the particular characteristics of the organism.

3. **DNA** (deoxyribonucleic acid) means genetic material of an organism, shaped like a double helix and composed of many genes (units of heredity).

4. **Gene transfer technology** means the transfer of a gene of one organism to another, forcing the DNA helix of the target organism to accept the foreign gene.

5. **Genetically modified organisms** mean animals, plants or microorganisms whose genetic structure has been altered by gene transfer technology.

6. **Products or goods originating from genetically modified organisms** mean products or goods created wholly or partly from genetically modified organisms.

7. **Release of genetically modified organisms** means the deliberate introduction into the environment of genetically modified organisms.

8. **Risk assessment** means the determination of the potential hazard and the extent of damage which has been caused or might be caused to human health, the environment and biodiversity in activities related to genetically modified organisms, particularly the use and release of genetically modified organisms; and to products and goods originating from genetically modified organisms.

9. **Risk management** means the application of safety measures to prevent, deal with and overcome risks to human health, the environment and biodiversity in activities related to genetically modified organisms, products and goods originating from genetically modified organisms.

Terminology used in the Problem Formulation and Options Assessment Handbook (Nelson and Banker, 2007):

1. **Adverse effects:** an undesired effect.

2. **Deliberation:** deliberation is a means by which a group of participants representing diverse interests in a governance process can work together to consider all relevant sides of an issue carefully in order to reach or move closer to a shared conclusion. It is characterized by an open sharing of ideas, listening to others, acknowledgement of diverse views and a spirit of collaboration.

3. **Future alternative:** any available option that could be implemented in place of what presently exists. This can include options that exist currently, options that will exist in the future and options that may exist in the future, whether or not they have yet been thought of.

4. **Problem formulation:** identifying the societal problem that the technology will address. Discussion focuses on whose problem is being addressed, whose problem should be addressed and what needs of the people identified are not being met by the present situation. The group assesses whether a problem truly exists based on extent, severity and relative importance compared to other problems.

Continued

Box 3.1. *Continued*

---

5. Problem Formulation and Options Assessment (PFOA): methodology for conducting deliberative formulations of a problem and comparative assessments of future alternatives for addressing the problem relative to the biosafety evaluation of GMOs. A PFOA process helps stakeholders analyse collaboratively and advise on the identification of possible harms and the enhancement of potential benefits within the specific contexts for which a GMO is being considered. To this end, a PFOA relies on being transparent, inclusive of all appropriate stakeholders and rationally informed by the best available science.

6. Stakeholder representative: individuals that participate directly in the core deliberation of the PFOA on behalf of the interests of a particular stakeholder sector or grouping of sectors with shared interests. Stakeholder sectors must have their interests represented in a PFOA by a representative because it is not practical or effective to involve directly every individual member in the process.

- Step 3: Problem Statement
- *Step 4: Authority decision to analyse options*
- Step 5: Options
- Step 6: Attributes for Solving Problem
- Step 7: Changes Required and Anticipated for a Solution Option
- Step 8: Impact to the System
- *Step 9: Authority decision about an option.*

Vietnam depends on imported cotton for its textile industry (90% of its raw material is imported). With the application of new varieties and modern integrated pest management (IPM) approaches, the area under cotton reached 30,000 ha, but has since declined to half this area. Obstacles to the development of cotton production are low prices and lack of an effective dry season production system, including suitable irrigation, varieties and IPM. Disease and pest damage is high, including lepidopteran pests.

We use Vietnam cotton as a case study for evaluating the merit, applicability and benefits of using the PFOA in environmental risk assessment. The participants in the trial run were the chapter authors. The authors do not represent the full diversity of stakeholders who may be involved in a PFOA, but they do represent a diversity of agency representatives and one farmers' union representative. An essential element of the PFOA process is the involvement of a broad spectrum of stakeholders whose representatives are allowed to contribute to the deliberative process. The identification and selection of the relevant stakeholders is particularly important for maintaining the public legitimacy of the proposed PFOA process. Representatives of all interested and affected stakeholders, both powerful and marginalized, need to be included in the deliberation process (see Nelson and Banker, 2007, for further explanation).

In this case study, Step 1: Problem Formulation, was done through brainstorming and debate. The group discussed that lepidopteran insects attacking cotton cause high yield losses (25–30%) and that some lepidopteran species (*Spodoptera exigua*, *Helicoverpa armigera*) have become highly resistant to

> **Box 3.2.** Problem Formulation and Option Assessment (PFOA): example responses to Step 2 for Bt cotton in Vietnam, from the group discussion 2004
>
> **Prioritization and scale considerations**
>
> - Most farmers growing cotton are affected, especially poor farmers with small-holdings who lack proper application equipment, money to buy pesticides and knowledge of pest control. In particular, farmers in the central coastal region of Vietnam are affected heavily.
> - Cotton companies who sign contracts with farmers (the company provides the means for growing cotton in return for the yields) lose investment.
> - Cotton yield usually is reduced by 25–30%.
> - To protect cotton fields from pests, farmers have to use more pesticides, leading to many health problems. In many areas, people suffer from allergies and many other diseases.
> - The quality of life for agricultural workers, farmers and their families could be reduced.
> - Soil, water and air in the cotton fields and surrounding areas have been polluted.
> - The ecosystem of the whole region can also be affected.

most pesticides. Several members pointed out that most farmers growing cotton are using higher doses of pesticides, some of which have little or no affect on insect control and that, as a result, farmers were hesitant to switch to cotton because of the high risk involved. Some researchers have found that early season insecticide applications to control the spread of cotton blue disease by aphids induce outbreaks of *H. armigera* (see Chapter 2, this volume).

The case study discussion continued with Step 2: Prioritization and Scale of Problem (Box 3.2) and Step 3: Problem Statement; all designed to answer questions such as: Who is affected by the problem? And at what scale? What losses have occurred?

In Step 5: Options, different options to alleviate the problem(s) have to be identified and discussed based on scientific data and field test results. In our case study discussion, we identified several options for the control of Lepidoptera in cotton fields:

- Biological control (including the use of transgenic Bt cotton varieties)
- Chemical control
- Cultivating system
- IPM package = biological, chemical and cultivation system.

Of these, we selected two options for the process of evaluating how option assessment would work: Option A – Use of insect-resistant transgenic cotton varieties; and Option B – Generic IPM package including biological control + chemical control + cultivation management. Steps 6, 7 and 8: Option Assessment, are designed for a multidisciplinary assessment of options regarding different aspects. Again, in our case study of transgenic cotton introduction to Vietnam, we identified examples of assessment responses regarding the two options (Box 3.3). Questions in Step 6 provide a discussion of the technology attributes and barriers to adoption

Box 3.3. Problem Formulation and Option Assessment (PFOA): example responses to Steps 6, 7 and 8 for Bt cotton in Vietnam, from the group discussion 2004

Option A: Use of insect-resistant GM cotton varieties

Option B: IPM package Biological, chemical and cultivation management

Attributes of the option

Characteristics: transgenic
Regions for use: all cotton growing areas
Barriers to technology adoption and efficacy:
- Seed cost (?) and source
- Adaptability of new varieties to local conditions
- Government authorization and intellectual property issues (risk assessment, permission for distribution and commercialization)
- Knowledge of farmers and acceptance
- Trade barriers (consumer concern)

Characteristics: integrated management system
Regions for use: central coastal region of Vietnam and similar regions
Barriers to technology adoption and efficacy:
- Difficulty in finding a good biocontrol measure
- Low acceptance by farmers to apply IPM
- Farmers' knowledge for applying IPM is limited
- Coordination of stakeholders across a region (farmers, local authorities, extension workers, companies...) is weak
- Applying biocontrol measures is costly and effect is slow (bioproduct is expensive)
- IPM is rather complex – some farmers may not apply it correctly

Needed or anticipated changes to the system if using the option

- Less pesticide use and pest-control cost
- Larger cotton growing area, especially in dry season
- More monoculture of some varieties

- Reduce use of pesticide
- Need for labour is increased
- Need IPM training for farmers and improved coordination of stakeholders
- Farmers are more independent from foreign input (seed, biopesticide)

Possible effects of the technology option

- Higher dependence of farmers on foreign seeds
- Biodiversity loss
- Dramatic change for non-target pests
- Breakdown of resistance
- Unforeseen other consequences (human health...)

- Working condition of farmers and environment is improved
- Sustainable practice
- Cost for production might be increased
- Knowledge of IPM is increased

and/or efficacy. Questions in Step 7 allow the PFOA participants to consider larger system changes from farm to societal scales. For example, participants evaluate anticipated or needed changes if a technology is implemented. Finally, questions in Step 8 allow the PFOA group to discuss possible adverse effects and benefits of the technology, with a special focus on environmental risks.

At the end of the PFOA deliberation, participants make a recommendation to the CA in the relevant ministry, who is responsible for deciding whether to proceed with the GMO proposal. There are several ways to agree on a final PFOA recommendation. Some countries suggest that everyone has to agree on the recommendations (consensus); other countries say most of the people have to agree (two-thirds voting in support); and others say a simple majority is fine. In Vietnam, the appropriate approach is recommendation by consensus. It is best to continue discussion and work for clarification, while focusing on matters where there are disagreements. This approach has been accepted as a working principle in meetings of the Association of South-east Asian Nations (ASEAN) by member countries on many topics. For PFOA recommendations, it is best to achieve a consensus opinion.

3.4. Suggested Sources of Information and Scientific Data

There are many sources of information and scientific data that would be helpful in answering the PFOA questions and assisting in discussion. General national level data on agriculture can be found in the Vietnamese General Office for Statistics and the Ministry of Agriculture and Rural Development (MARD). Provincial Departments of Agriculture and Rural Development could provide substantial farm-scale information. The Vietnamese Ministry of Natural Resources and Environment (MONRE) could provide basic information about biodiversity and ecosystem studies.

In relation to cotton, the Vietnam Cotton Company is a good source for information on production, cropping systems, processing and agricultural technology. The Vietnamese Textile and Garment Company manages data on fibre demand, import and export markets, as well as local markets.

The Plant Protection Department of MARD can provide pest and disease statistics and losses due to pesticide poisonings. The Nhaho Research Institute for Cotton and Agricultural Development and the Plant Protection Research Institute are another good source for scientific studies on pests, diseases and GMO testing. Information related to general biotechnology issues can be found in the Biotechnology Institute, the Agricultural Genetics Institute and the Cuulong Delta Rice Research Institute.

In general, information necessary for risk assessment can be found in a variety of ministries, depending on the nature of the questions.

3.5. Challenges and Recommendations for Implementing PFOA for Environmental Risk Assessment in Vietnam

After discussion of the PFOA process, using the case study on Bt cotton and learning from the experiences of Kenya and Brazil, we identified the challenges/ questions and recommendations for implementing PFOA in Vietnam that are listed in Table 3.1.

Table 3.1. Challenges/questions and recommendations for implementing PFOA in Vietnam.

Challenges	Recommendations
Should the PFOA play a role?	The PFOA should be considered and applied in the whole process of making a decision.
Who is responsible for organizing the PFOA in Vietnam?	See 'Decision making process' scheme including: information, consultation, decision.
How can we increase public awareness about GMOs?	Education and training on GMO risk assessment, PFOA for policy makers, decision makers, researchers and built-in communication with mass media in order to reach the general public.
How can we coordinate the process involvement of different actors?	Establish a National Biosafety Committee that includes representatives of all relevant ministries and social organizations.
How can we get groups actively involved?	Enhance people's awareness of their options to influence the decision making process and provide incentives for public participation.
Time, money and energy?	Applicant should assist in paying for the PFOA consultation.
How can we get data, information and expertise to improve the process?	The Competent Authority (CA) of the relevant ministry will consider the PFOA questions and consult with related organizations.
How can we get stakeholder cost–benefit analysis?	Information from stakeholders on data should be compiled and reported by the CA.
How can we establish criteria for choosing the best option and who does it?	The CA Technical Group develops criteria based on health, environmental, economic and social concerns.
How can we deal with diversity in the way different groups think and value? Are all ways equal or are some better than others?	Raise awareness within the public on a comprehensive understanding of all aspects of GMOs.
Who decides?	The CA.
When does the PFOA stop?	No conclusive recommendation.

3.6. Decision Making Scheme

The final step of the PFOA (Step 9) is the CA decision about the GMO proposal. This step is presented schematically in Fig. 3.1. The CA uses information gathering and sharing based on environmental risk assessment research and the PFOA deliberation to make their decision, as well as campaigns to inform the public about GMOs and the particular decision. Then the CA proceeds with consultation by reviewing the PFOA deliberations and recommendations, as well as other legal requirements for public consultation. Based on the foundation of information and consultation, the CA proceeds with a decision to recommend or decline the GMO proposal to the National Biosafety Committee (NBC). As of October 2007, no decision had been made about the

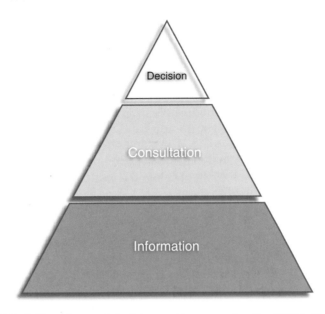

Fig. 3.1. PFOA deliberation and decision making regarding the GMO proposal as part of Vietnam's governance system.

make-up of the NBC, but Decision No. 102/2007/QD-TTg,[5] effective July 2007, requires the CAs to clarify roles and responsibilities for biosafety, as well as develop biosafety regulations. CAs include the Ministry of Natural Resources and Environment (MONRE), the Ministry of Agriculture and Rural Development (MARD), the Ministry of Science and Technology (MOST), the Ministry of Health (MoH) and the Ministry of Industry and Trade (MIT).[6]

The CA of the relevant ministry will conduct the PFOA, consider the PFOA recommendations and consult with related organizations. For example, the CA of MARD will be responsible for the risk assessment of projects related to agriculture. Since most of the primary ministries had representatives involved in our discussions during 2005, these suggestions were made with an understanding of the potential responsibilities and relationships of different agencies. Overall, there should be three phases to assessing a GMO.

The first phase of the environmental risk assessment (ERA) process would focus on information (Fig. 3.1). The GMO applicant would prepare documentation, including:

- Request for planting/rearing/commercializing a GMO;
- Risk assessment and risk management report;

[5] Decision No. 102/2007/QD-TTg promulgating Comprehensive Plan to Strengthen Management Capacity in Biosafety of GMOs, Goods and Products originating from GMOs until 2010, as well as Implementation of Cartagena Protocol on Biosafety, signed by Prime Minister Nguyen Tan Dzung, effective date 11 July 2007. http://antoansinhhoc.vn (Vietnamese and English translation).
[6] Decision No. 102/2007/QD-TTg, see above.

- Relevant documents supporting their plan for planting/rearing or commercialization.

All these documents would then be submitted to the CA of the relevant ministry.

The second phase focuses on consultation, which ranges from a simple comment period to a more active process of deliberation represented by the PFOA. The CA would establish a Technical Group to review and evaluate all the information and data provided by the applicant. At the same time, the CA would establish a Ministerial Biosafety Committee and seek comment from relevant ministry representatives. The Ministerial Biosafety Committee might include MONRE, MOST, MARD, MIT and MoH. During the consultation, the Ministerial Biosafety Committee would consider the PFOA questions/recommendations and consult with the relevant organizations. Finally, the Ministerial Biosafety Committee would review all findings from the Technical Group, the PFOA recommendations and conduct a general public consultation. From this, they would make a recommendation to the Minister of the Competent Authority regarding whether or not to permit development, planting, or commercialization of the GMO.

Finally, in the third phase, the CA Minister would decide whether to recommend supporting or declining the petition and pass along their recommendation to the National Biosafety Committee. After the National Biosafety Committee makes a decision, the CA would then inform the Applicant and the Biosafety Clearing House (BCH) of MONRE about the decision.

3.7. Conclusions

Vietnam is entering a new, historic era in its development with basic reforms to its economy policy. As a result, foreign investment is expected to reach US$15 billion in 2007 and, over the next 5 years, Gross Domestic Product (GDP) is forecast to increase by more than 8% annually. With the growth in the agricultural sector, within a global context, Vietnam is considering accelerating the introduction of GMOs to its agricultural system and environment.

Industrialization and modernization of rural agriculture have been given priority by government planners, with science and technology envisioned as the driving force. In January 2006, the Prime Minister issued a decision approving a critical programme of biotechnology development and implementation as a tool for agricultural and rural development through 2020.[7] State authorities view biotechnology, the key element of which is the introduction of GMOs in agricultural sectors, as promising important niche solutions to many problems in the future.

At the same time, Vietnamese decision makers are aware of the global debate about GMOs. Many nations and scholars critique the argument that biotechnology is a necessary component of modern agriculture. They suggest it is still unclear how important this technology will prove to be. For example,

---

[7] Decision No. 11/2006/QD-TTg approving the key programme on development and application of biotechnology in the domain of agriculture and rural development up to 2020, signed by Prime Minister Phan Van Khai, effective date 12 January 2006.

the European Union countries have varying acceptance of the production and use of GMO products. Considering the rapid changes in points of view towards GMOs and the pressure for Vietnam to modernize its agriculture, the mass introduction of some GMOs into Vietnamese agriculture is likely to be only a matter of time, most likely in the very near future. But the Vietnamese government wants to introduce GMOs in a safe manner, so that the environment does not sustain irreversible damage and the products can be used safely.

Several events suggest the government is preparing the policy and regulatory foundation for GMO technologies. Vietnam signed the Cartagena Protocol in 2004 and issued the Vietnam National Biosafety Regulation[8] in August 2005. Decision No. 79[9] and Decision No. 102[10] in 2007 go on to require the development of biosafety regulations with coordination among ministries and a clarification of roles and responsibilities. On the economic front, some in Vietnam argue that the introduction of GMOs is an economic necessity for keeping up with regional and international trends in the modernization of agriculture, as well as the protection of the environment. In general, the introduction of GMOs into Vietnamese agriculture is considered a key approach for establishing a modern, sustainable agriculture, by 'Developing biotechnology for the benefit of sustainable development of agriculture, forestry and fishery, as well as the protection of human health and the living environment'.[11]

As a guiding principle for decision makers worldwide, the application of Problem Formulation and Option Assessment (PFOA) should be considered as a critical step that will assist Vietnamese authorities in making any decision, from denying to giving permission for a GMO release. PFOA benefits Vietnamese decision making by providing a science-based, multi-stakeholder process to formulate problems and assess options as a basis for ERA when a country is considering the introduction of a GMO. Public awareness and multi-stakeholder participation in the whole process are the main benefits of a PFOA approach. Before giving permission for the introduction of any GMO to Vietnam, the government authorities would have to be sure that all environmental risks have been assessed fully and clearly on a scientific basis with the full participation of all the stakeholders that would be affected, namely the companies, the scientists, the farmers and the representatives of the general public in the case of Bt cotton. The decision is made only after getting full information on risk identification and ample consultation on management alternatives (options assessment) from the concerned scientific and management institutions.

Though there are many debates in international and national meetings about risk assessment and risk management concerning GMOs, Vietnam clearly benefits from the experiences of other countries where PFOA has been considered. This will help to reduce conflict associated with misunderstandings and increase the effectiveness of GMO management, assuring transparency and public acceptance.

---

[8] Decision No. 212/2005/QD-TTg, see above.
[9] Decision No. 79/2007/QD-TTg, see above.
[10] Decision No. 102/2007/QD-TTg, see above.
[11] Resolution 18/CP on Development of Biotechnology in Vietnam to 2010, signed 11 March 1994.

References

Capalbo, D.M.F., Simon, M.F., Nodari, R.O., Valle, S., Dos Santos, R.F., Coradin, L., Duarte, J. de O., Miranda, J.E., Dias, E.P.F., Le Quang Quyen, Underwood, E. and Nelson, K.C. (2006) Consideration of problem formulation and option assessment for Bt cotton in Brazil. In: Hilbeck, A., Andow, D.A. and Fontes, E.M.G. (eds) *Environmental Risk Assessment of Genetically Modified Organisms, Volume 2: Methodologies for Assessing Bt Cotton in Brazil*. CAB International, Wallingford, UK, pp. 67–92.

Forester, J. (1999) *The Deliberative Practitioner: Encouraging Participatory Planning Processes*. MIT Press, Cambridge, Massachusetts.

Grimble, R. and Wellard, K. (1997) Stakeholder methodologies in natural resource management: a review of principles, contexts, experiences and opportunities. *Agricultural Systems* 55, 173–193.

Hallerman, E.M. and Kapuscinski, A.R. (1995) Incorporating risk assessment and risk management into public policies on genetically modified finfish and shellfish. *Aquaculture* 137(1–4), 9–17.

Hemmati, M. (2002) *Multi-stakeholder Processes for Governance and Sustainability: Beyond the Deadlock and Conflict*. Earthscan Publications Inc., London.

Kessler, J. and Van Dorp, M. (1998) Structural adjustment and the environment: The need for an analytical methodology. *Ecological Economics* 27, 267–281.

Loevinsohn, M., Berdegué, J. and Guijt, I. (2002) Deepening the basis of rural resource management: learning processes and decision support. *Agricultural Systems* 73, 3–22.

Miller, H.I. (1994) A need to reinvent biotechnology regulation at the EPA. *Science* 266, 1815–1818.

Munson, A. (1993) Genetically manipulated organisms: international policy-making and implications. *International Affairs* 69, 497–517.

NRC (National Research Council) (1983) *Risk Assessment in the Federal Government: Managing the Process*. National Academy Press, Washington, DC.

NRC (National Research Council) (1996) *Understanding Risk: Informing Decisions in a Democratic Society*. National Academy Press, Washington, DC.

NRC (National Research Council) (2002) *Environmental Effects of Transgenic Plants*. Committee on Environmental Impacts Associated with Commercialization of Transgenic Plants. Board on Agriculture and Natural Resources: Division on Earth and Life Studies: National Research Council. National Academy Press, Washington, DC.

Nelson, K.C. and Banker, M. (2007) *Problem Formulation and Options Assessment Handbook: A Guide to the PFOA Process and How to Integrate it into Environmental Risk Assessment (ERA) of Genetically Modified Organisms (GMOs)*. University of Minnesota, St Paul, Minnesota and GMO ERA Project. http://www.gmoera.umn.edu (accessed 6 February 2007).

Nelson, K.C., Kibata, G., Lutta, M., Okuro, J.O., Muyekho, F., Odindo, M., Ely, A. and Waquil, J. (2004) Problem formulation and options assessment (PFOA) for genetically modified organisms: the Kenya case study. In: Hilbeck, A. and Andow, D.A. (eds) *Risk Assessment of Transgenic Crops: A Case Study of Bt Maize in Kenya*. CAB International, Wallingford, UK, pp. 57–82.

Susskind, L., McKearnan, S. and Thomas-Larmer, J. (1999) *The Consensus Building Handbook: A Comprehensive Guide to Reaching Agreement*. Sage Publications, Thousand Oaks, California.

4 Transgene Locus Structure and Expression

Trần Thị Cúc Hoà, Anna Depicker, Kakoli Ghosh,
Nelson Amugune, Phan Đình Pháp, Thương Nam Hải
and David A. Andow

A transgene locus is the physical location where a delivered transgene is integrated into a chromosome. It includes the various elements of the delivered DNA, such as the target gene and its promoter, introns, exons, terminators and spacer regions; and possibly a selectable marker gene and its promoter and the backbone of the plasmid carrying these transgene elements.

The integration of a transgene into the host genome is complex because of various patterns of rearrangements that can occur during a transgene insertion event, and because of the interactions between elements of the transgene locus and the adjacent plant genomic DNA. Such types of rearrangements can cause erratic expression of the target gene (referred to as ectopic expression) and lead to the formation of expressed, unpredicted open reading frames (ORFs). Transgene integration can also cause endogenous gene disruption and silencing. In addition, interactions between the transgene products and endogenous gene products in the plant are possible (referred to as pleiotropy or epistasis) and these may be different in different environmental conditions.

The potential consequences of these phenomena on human health or the environment have caused particular concerns about the use of transgenic plants. Therefore, a characterization of the structure of a transgene locus is necessary for anticipating the risks of transgene insertion. Molecular characterization allows and ensures a precise description of the nature and extent of integrated DNA sequences and their connection with the neighbouring plant sequences, and therefore will provide information necessary to perform a valid risk assessment.

Molecular characterization of the transgene locus entails: (i) sequence determination of the transgene components; (ii) sequence characterization of the border regions between the inserted DNA and the plant host genomic DNA; and (iii) characterization of the transgene related rearrangements of the plant host genome. Method development will prove crucial for the implementation of GMO management, including risk management and the related regulatory provisions, such as labelling and tracing. Figure 4.1 shows the interaction

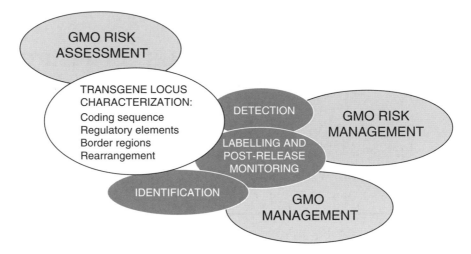

Fig. 4.1. A scheme to show the interaction of molecular characterization of the transgene locus with risk assessment, development of methods for detection, identification of the transgene, labelling and post-release monitoring and risk management.

between molecular characterization of a transgene locus, the detection of transgenic proteins and management of transgenic crops and their products.

In addition to the characterization of the structure of the transgene locus, information on its inheritance and expression is crucial for assessing the environmental and human health risks of transgenic crop plants destined for commercial production. Transgene design and production, as well as the transformation method, are major determinants of the subsequent nature of the transgene locus, allowing determination of the necessary components of risk assessment and pre-emptive reduction of risk or the need for risk assessment. Hence, this chapter covers three major themes: (i) transgene design and production; (ii) transgene locus structure and inheritance (genotypic analysis), and (iii) transgene expression (phenotypic analysis). The staging of these analyses in the risk assessment process will also be addressed, along with recommendations for each theme as a guide for minimizing risks and improving gene expression. In addition, the description of some selected Bt cotton events is given in view of using these events for field-testing in Vietnam in the near future.

4.1. Transgene Design and Production

Design elements that reduce risk or the need for some risk assessment

The literature on the characterization of transgene expression and locus structure indicates that simple, single copy inserts of the transgene sequence have the greatest probability to provide robust transgene expression and are less likely to become silenced (Butaye *et al.*, 2005). It has also been noted that simple non-rearranged transgene locus structures are less likely to result in ectopic expression or silencing of host plant genes or transgenes, or in the generation of spurious ORFs (European Commission, 2002; SBB, 2003). Thus, a number of strategies have been emerging recently to eliminate unnecessary transgene DNA and repeated copies from being incorporated into the transgene locus. Also, removal of intact antibiotic resistance genes, often used as selectable markers, is highly desirable so as not to propagate resistance genes for useful and valuable antibiotics in the environment. The absence of a marker gene results in a simpler transgene locus structure, which simplifies the risk assessment substantially.

The aim of transgene design and production is to produce transgenic events (plants) with a well-characterized nuclear transgene locus segregating in the next generation with Mendelian ratios. An 'ideal' transgenic plant may have the locus structure of the transgene possessing the following features:

- One transgene locus per plant
- Only one copy of the transgene at the locus
- No extra delivered DNA in the transgene locus
- No disruption of an existing ORF or expression of other plant genes
- No spurious ORFs
- Minimal rearrangements of genomic DNA flanking the integrated transgene.

Production details essential for subsequent assessment

Knowledge of the sequence of the DNA delivered into the recipient genome is essential to characterize the structure of the transgene locus in the plant, for screening for mutations and for determining the absence or presence of parts of the original transgene in the transgene locus. The transformation method used provides some insight into the probable nature and extent of rearrangements that may be created in the transgene locus and therefore provides useful information concerning the locus structure.

Sequencing information
DNA sequencing should be done for both the transgene construct and the integrated DNA (NRC, 2002; UK ACRE, 2002). This information helps to confirm that transgene integration proceeded as anticipated and to reveal the presence of any unintended expression sequences (e.g. bacterial selectable marker genes, plasmid backbones) at the transgene locus after transgene integration.

Transformation method information
Each transformation method has both disadvantages and advantages relating to transgene locus structure and expression, so knowledge of the transformation method used allows anticipation of possible problems and determines the screening strategy for eliminating potentially problematic transgene events.

For the particle bombardment method, transgene insertion is a complex process resulting in a high number of transgene copies per locus and a high frequency of transgene DNA rearrangements (Pawlowski and Somers, 1996; Makarevitch *et al.*, 2003). So if the particle bombardment method is used, procedures should be developed to reduce the transgene number, rearrangements of the transgene and target site and insertion of superfluous DNA.

For the *Agrobacterium*-mediated transformation method, plant transformation techniques are needed to eliminate or reduce the probability of insertion of plasmid and other superfluous DNA and limit the rearrangement of the T-DNA and the target-site DNA. This may be possible by using different *Agrobacterium* strains, redesigning the plasmids used for transformation, or making other modifications to limit rearrangements.

In either method, it is essential to use innocuous (harmless) or characterized selectable markers or to employ strategies to remove them. Production of marker-free plants is recommended to reduce the need for downstream characterization, especially if the transgenic plants or derivatives are to be used for food or feed. Removal of the selectable marker is also often desirable when the transgene locus will be dispersed widely via pollen. For targeted insertion of transgenes, homologous recombination or some other techniques of targeted transgene insertion should be developed to reduce the chance of disrupting functional sequences by random transgene insertion (Britt and May, 2003).

Recommendations for transgene design and production

Strategy to produce marker-free transgenic plants

The marker gene enables identification of plant cells in which the transgene is being expressed during the initial laboratory stage of development of transgenic plants. The use of either antibiotics or herbicides as selection agents to identify transgenic events has been applied widely (Christou, 1997; Hei *et al.*, 1997). However, the use of these marker genes in transformation procedures has caused concerns about the biosafety of transgenic plants (Kleter *et al.*, 2005). Thus, there is a need to develop marker-free transgenic plants and several techniques to eliminate a marker gene have been developed (Yoder and Goldsbrough, 1994; Puchta, 2003). Some marker genes are now considered to be unlikely to cause significant impacts on human or animal health (e.g. *nptII* and *hph*), but others remain controversial (e.g. vancomycin). In any event, elimination of a marker gene will reduce the requirements and costs of risk assessment.

USE OF CO-TRANSFORMATION SYSTEMS.   Co-transformation systems appear to be the simplest method to eliminate marker genes and have been used widely in direct transformation methods (Schocher *et al.*, 1986). In these systems, the marker gene and gene of interest are placed on separate DNA vectors, such that the frequency with which the genes are integrated in the plant genome as unlinked fragments is high. Subsequently, the gene of interest can segregate from the selectable marker gene in the progeny. Another advantage of the co-transformation systems is that construction of the separate molecules for marker genes and genes of interest is less tedious than creation of large DNA fragments in which all genes are linked.

Many isolates of *A. tumefaciens* contained more than one T-DNA, and crown gall tumours were often co-transformed with multiple T-DNAs (Hooykaas and Schilperoort, 1992). Thus, *A. tumefaciens* is a naturally occurring co-transformation vector for higher plants. Co-transformation mediated by *Agrobacterium* has been examined in several laboratories (Depicker *et al.*, 1985; Framond *et al.*, 1986; Simpson *et al.*, 1986; McKnight *et al.*, 1987; De Block and Debrouwer, 1991). Multiple T-DNAs were delivered to plant cells either from a mixture of strains ('mixture methods') or from a single strain ('single-strain methods'). Segregation of one T-DNA from others was observed on various occasions (Framond *et al.*, 1986; McKnight *et al.*, 1987; De Block and Debrouwer, 1991). These studies indicated that the *Agrobacterium* system potentially was useful for the production of marker-free transformants. Two important parameters must be considered in the evaluation of co-transformation methods: (i) efficiency of co-transformation must be high; and (ii) frequency of unlinked integration of T-DNAs must also be high. De Block and Debrouwer (1991) evaluated a large number of *Brassica napus* plants transformed by a mixture method and reported that the frequency of co-transformation was high, but introduction of genetically linked T-DNAs was also favoured. Although Depicker *et al.* (1985) described a single-strain method with a higher co-transformation frequency than a mixture method, single-strain methods have not been tested on a large scale.

Komari *et al.* (1986) developed a 'super-binary' vector system suitable for a single-strain method of co-transformation. These novel vectors carried two separate T-DNAs. One T-DNA contained a drug-resistant, selectable marker gene and the other contained a gene for β-glucuronidase (GUS). A large number of tobacco (*Nicotiana tabacum* L.) and rice (*Oryza sativa* L.) transformants were produced by *A. tumefaciens* LBA4404, which carried the vectors. Frequency of co-transformation with the two T-DNAs was greater than 47%. GUS-positive, drug-sensitive progeny were obtained from more than half of the co-transformants. Molecular analyses by Southern hybridization and polymerase chain reactions confirmed integration and segregation of the T-DNAs. Thus, the non-selectable T-DNA segregated independently from the selectable marker in more than a quarter of the initial, drug-resistant transformants. Since various DNA fragments may be inserted into the non-selectable T-DNA by a simple procedure, these vectors will likely be very useful for the production of marker-free transformants of diverse plant species. Delivery of two T-DNAs to plants from mixtures of *A. tumefaciens* was also tested, but the frequency of co-transformation was relatively low. The double right border system of Lu *et al.* (2001) may be another way to the double T-DNA system.

USE OF SITE-SPECIFIC RECOMBINATION SYSTEMS. The Cre/*lox* site-specific recombination system from *Escherichia coli* phage P1 (Dale and Ow, 1990; Stuurman *et al.*, 1996) consists of a specific target DNA sequence (*lox*, 34 bp) and a recombinase (Cre) that is necessary and sufficient to induce recombination between the two target sites. The Cre/*lox* system has been characterized most highly in plants. With this system, site-specific recombination is directed by the Cre protein (recombinase), which recognizes and interacts with the two *lox* sites to direct excision, integration, or inversion of the intervening DNA sequences, depending on the orientation of *lox* sites (Abremski *et al.*, 1983;

Hoess *et al.*, 1986). The Cre/lox system was applied to excise a selectable marker from the host genome of tobacco (Dale and Ow, 1991; Bayley *et al.*, 1992) and of *Arabidopsis* (Russell *et al.*, 1992).

Hoa *et al.* (2002) demonstrated the excision of the selectable marker *hpt* from the host genome of rice by crossing plants containing *lox/hpt/lox/gusA* with those containing *cre/hpt* (or *cre/barB*). In the hybrid, the *cre* gene and the *lox* sites were brought together into the same plant, and the Cre recombinase interacted with the *lox* sites directing the excision of the *hpt* gene, which was flanked by the two *lox* sites. The excision of the *hpt* gene led to the fusion of the *hpt* promoter with the promoterless, distally located *gusA* gene and, consequently, the *gusA* gene was activated and its expression could be observed.

USE OF NOVEL SELECTION SYSTEMS TO REPLACE ANTIBIOTIC SELECTION SYSTEMS.   The mannose selection system is based on using the selectable marker gene, *pmi*, a gene derived from *E. coli* encoding phosphomannose isomerase (Miles and Guest, 1984). In a culture medium containing mannose, mannose was converted by an endogenous hexokinase to mannose-6-phosphate, an unusable carbon source for most plant cells, which accumulated in the cells, resulting in severe growth inhibition of the cultured cells. In contrast, the transgenic *pmi*-expressing cells converted mannose-6-phosphate into fructose-6-phosphate, which is a carbohydrate source that can be used by the plant tissue. Only the transgenic plant cells expressing the *pmi* gene could metabolize mannose into a usable source of carbon, resulting in normal growth on a mannose medium, while non-transformed cells either stopped growing or died due to starvation (Hansen and Wright, 1999). The successful use of the mannose selection system has been reported in developing different transgenic crop lines, including transgenic sugarbeet (Joersbo *et al.*, 1998), maize and wheat (Wright *et al.*, 2001), japonica rice (Lucca *et al.*, 2001) and indica rice (Hoa *et al.*, 2003), although it has not been effective in cotton.

Production of transgenic plants free of plasmid backbone
Martineau *et al.* (1994) reported the transfer of binary vector sequences into transgenic plant DNA by *A. tumefaciens*-mediated transformation. Later, Wenck *et al.* (1997) and Kononov *et al.* (1997) found that the entire binary vector, including backbone sequences as well as T-DNA sequences, frequently could be transferred to tobacco and *Arabidopsis*. Zhang *et al.* (2008) found that a third of *Agrobacterium*-transformed cotton plants contained the vector backbone.

Hoa *et al.* (2003) designed a strategy for rice to monitor the presence or absence of beyond-border transfer by using PCR with specific primers (Fig. 4.2). Two PCR reactions were performed for each side (A, B on the left border and C, D on the right border). PCR reactions A and C confirmed the transformation event and acted as a positive control for checking the quality of the isolated genomic DNA and the PCR conditions. These two PCR reactions should yield a product with all the transgenes. In contrast, PCR reactions B and D should yield a product only when there is transfer of vector DNA beyond the

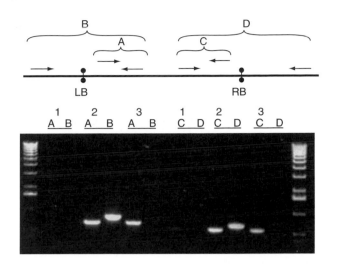

Fig. 4.2. PCR analysis to screen for beyond T-DNA border transfer. The scheme gives the primer combination used for the inside border (reactions A,C) and outside border amplicons (B,D). 1: Wild type, 2: Plasmid, 3: Transgenic event showing no beyond-border transfer (bands in lane A and C but no bands in B and D), LB: left border, RB: right border (adapted from Hoa *et al.*, 2003).

T-DNA border. Reactions A/B and C/D share a primer that binds within the borders. As shown in the scheme (Fig. 4.2), the transgenic event (designated No. 3) had no beyond-transfer because it did not have bands in lanes B and D. There are also counter selection strategies to reduce beyond-border transfers (e.g. by using a toxic gene outside of the left border region) that minimize multiple copy integrations and vector backbone in *Agrobacterium* transformation (Eamens *et al.*, 2004).

USE OF PURIFIED GENE CASSETTES. The insertion of plasmid DNA sequences can be avoided by using purified gene cassettes instead of entire plasmids during particle bombardment (Fu *et al.*, 2000; Breitler *et al.*, 2002; Loc *et al.*, 2002), although this does not solve the entire problem of multiple copy inserts.

Avoidance of promoter duplication in construct design
Transcriptional gene silencing (TGS) is known to be associated with methylation of homologous promoter sequences of the transgene (Mette *et al.*, 1999). Therefore, duplications of promoters in the construct design should be avoided. Likewise, promoter sequences should not duplicate gene sequences in the recipient genome to avoid transgene-specific suppression of plant gene expression. Zakharov *et al.* (2004) engineered seed-specific promoters from *Phaseolus vulgaris* and *Vicia faba* into related species and found that the reporter gene was expressed in non-seed tissues, such as pollen. To overcome this, it is advised to select promoters isolated from the genome of the same species for transformation of a particular species. For example, in a study of 15 promoters isolated from genes expressed in the rice seed, Qu and Takaiwa (2004)

demonstrated that many of these sequences directed GUS reporter expression in specific cell types of the rice seed, such as endosperm.

Avoidance of duplicated transgene sequences

Probably the most common phenomenon causing transgene suppression in plants is post-transcriptional gene silencing (PTGS) or RNA silencing, assumed to result from sequence homology in transcribed regions (Bender, 2004). This type of silencing is also known as repeat-induced gene silencing (RIGS). One of the clearest examples comes from *Arabidopsis*, where Assaad *et al.* (1993) demonstrated that tandemly- and directly-repeated transgenes either were not expressed or were expressed poorly. These genes are methylated and, as measured by nuclease sensitivity assays, assemble into a dense chromatin structure (Ye and Signer, 1996). Reducing the number of gene copies restores gene expression (De Buck *et al.*, 2001).

4.2. Characterization of the Structure of a Transgene Locus and Inheritance (Genotypic Analysis)

Characterization of the structure of a transgene locus is essential to: (i) assess the result of transgene integration; (ii) identify the integration site; (iii) predict the expression and stability of the locus; and (iv) provide a strategy to track the locus. Locus characterization allows assessment of the possibilities of: (i) altered recipient plant gene function or expression via insertional mutagenesis; (ii) ectopic expression of the transgene caused by the juxtaposition of novel enhancer elements to a plant gene; (iii) creation of a spurious ORF (open reading frame) that could produce unintended gene products, which could result in environmental and human health risks; (iv) repeated sequences in the transgene locus affecting the probability of homologous recombination, resulting in intra-locus instability; (v) genomic interspersions that may lead to ectopic recombination and possible movement of the transgene to other locations in the plant genome; and (vi) incorporation of a transgene into a known transposable element, thereby allowing it to move independently.

Southern analyses combined with segregation analyses are sufficient to characterize many aspects of transgene locus structure and copy number (UK ACRE, 2002). Southern analyses conducted over several generations are required to assure that the transgene locus is fixed and stably inherited, and to determine how many copies are integrated and whether multiple transgene inserts are linked or independent. Thus, Southern analyses should be set up to determine transgene copy number and to screen for the presence or absence of DNA sequences, such as bombarded plasmid sequences or binary T-DNA vector backbone sequences. Transgene locus sequencing is recommended to address the detailed structure of transgene rearrangements and changes to flanking genomic DNA (e.g. Ohba *et al.*, 1995). The transgene locus sequences of the transgene plant DNA junction fragments are especially relevant as they allow design of PCR primers for specific event detection, for further analysis of the locus inheritance and stability, and for tracing the transgene locus (Windels *et al.*, 2001). Newer methods continue to be developed.

Number of transgene loci in the plant

Transformation can lead to multiple transgene integration events in a single transformed plant and these can be at one or more genetic loci and either active or inactive (Pawlowski and Somers, 1996; Kohli *et al.*, 1998). The structural analysis should include information on non-expressing loci, because non-expressing loci tend to cause silencing of the active transgene. Transgene silencing is stochastic, creating uncertainties that would need to be addressed in risk assessment. Hence, this analysis should be based on DNA analyses rather than gene product analyses. Information on the linkage of multiple transgene loci also will be helpful in assessing the usefulness of a specific transgenic plant. Sexually fertile plants, such as cotton, maize and rice, can be examined by combining both segregation and molecular analysis. The sensitivity of these methods varies and should be specified. For sterile, vegetatively propagated plants like potato or banana, segregation analysis is not possible and the number of integration events could be characterized by molecular means using, for example, Southern analysis and fluorescent *in situ* hybridization (FISH) (Svitashev and Somers, 2002).

Number of copies of the transgene integrated at each locus

Multiple copies of linked whole or rearranged transgenes can lead to transgene silencing (Matzke and Matzke, 1995; Pawlowski *et al.*, 1998). In many cases, multiple copies of transgenes are linked directly in direct or inverted configuration and therefore are impossible to unlink via breeding methods. De Buck *et al.* (2004) found that single-copy T-DNAs in 21 different *A. thaliana* transformants were all integrated at different positions in the *Arabidopsis* genome, but displayed uniform and comparable expression measured in two subsequent generations. It is proposed that selecting single-copy T-DNA transformants should yield more transformants with stable and high transgene expression, and that integration position is not a major determinant of transgene expression variability.

Cytoplasmic genome transformed

Transgenes integrated into the cytoplasmic genomes will be inherited differently in different plant species. The most common mode is maternal inheritance. Localization can be determined by segregation analysis. If the transgene is cytoplasmic, is there positive selection for it to be retained? There is evidence that genes integrated into cytoplasmic genomes may be transferred to the nuclear genome at some low rate (Stegemann *et al.*, 2003; NRC, 2004; Ruf *et al.*, 2007). The consequences to such movement should be considered. Chloroplast transformation is not yet commercially available.

Sequencing of the transgene locus, including the genomic DNA flanking the transgene sequences

This analysis allows the most comprehensive assessment of the transgene locus structure. The purposes include:

1. Identification of all promoter, selectable marker and gene of interest sequences to compare these sequences with those from the original transferred construct (Makarevitch *et al.*, 2003). Any mutation or DNA sequence rearrangement present can be detected. A number of documented structures are of concern for both human and animal health and environmental risks. Sequencing information assists in assessing these factors by allowing determination in the transgene locus of:

- Rearrangements in the transgene, e.g. direct and indirect repeats, truncated portions of the delivered DNA and randomly recombined fragments of the delivered and genomic DNAs, such that the transgene or flanking genes in the recipient genome are likely to be expressed ectopically (UK ACRE, 2002).
- Occurrence of promoters in the flanking regions that might cause different expression of the transgenes. If there are any additional partial copies of the transgene, the possibility that they could be activated by flanking promoters in the plant could be assessed. Also, additional copies enhance the chance to trigger RNA silencing, which may be assessed, especially when the plant genome contains homologous sequences. For example, a number of the commercially released Bt cotton transgenes have one full-length gene and a truncated copy (OGTR, 2002; AgBios, 2005). The orientation of such truncated transgenes will influence whether they are likely to be expressed from flanking promoters.
- Any repeated sequences in the transgene construct or in the resultant transgene locus that would lead to locus instability via homologous recombination or other mechanism(s). Southern analysis in successive generations should enable identification and elimination of unstable events.
- Genomic interspersions in the transgene locus (e.g. Svitashev and Somers, 2001; Svitashev *et al.*, 2002). These structures represent substantial rearrangements of the integration site and have the potential for translocation of part of the transgene within the genome, if the interspersions are of ectopic origin.
- Spurious ORFs generated by the insertion or rearrangements. Spurious ORFs could give rise to transgene products with uncharacterized risk potential (e.g. RNA variants were found to be transcribed from the Roundup Ready soybean transgene; Rang *et al.*, 2005).

2. Determination of whether the genomic regions flanking the transgene insert exhibit sequence homology to:

- Known or predicted genes, suggesting that a recipient gene had been mutated via transgene integration (Jeong *et al.*, 2002; Forsbach *et al.*, 2003).
- Transposable element sequences in the genomic DNA flanking the transgene locus. Putative DNA transposable elements and retrotransposon elements comprise a large fraction of the non-coding sequences of plant genomes. Proximity of a transgene to a known or putative transposable element sequence raises the possibility that the transgene may become mobile if the element is active. This kind of theoretical possibility has not yet been demonstrated but sequence information, coupled with locus stability analysis using molecular methods that are specific to the transgene locus over several generations, will allay this concern.

Stable inheritance of the transgene locus

To determine the stable inheritance of the transgene locus, it is necessary to characterize it over several segregating generations, preferably up to F_4 or F_5. For example, in the case of the Bollgard cotton event 531, inheritance stability was monitored to F_4 (AgBios, 2005) and, in Bollgard II cotton event 15985, it was monitored to F_5 (US APHIS, 2000).

Molecular monitoring tools

Molecular monitoring tools need to be available, such as PCR primers based on the transgene locus sequence or Southern probes. Such tools can be used to determine transgene structure in commercial varieties and their derivatives for the purposes of monitoring locus stability and of allowing detection or traceability of the transgene locus (e.g. Hernández *et al.*, 2003). The molecular information is important as a tag to identify the particular insertion and allow tracking of the insertion event in the environment (or in contamination of non-transgenic crops, etc.).

Analysis of commercial transgenic plants

While the analyses described above are likely conducted on the original transgenic event, some data concerning these questions should be provided for each commercial transgenic variety. For example, once the information is generated for the original event, follow-up analysis on each commercial variety could be done with PCR primers to confirm that the transgene is still intact and in its expected location, i.e. that there was no breeding or other mix-up during variety production.

Recommendations for characterization of transgene locus structure and inheritance

For transgenic crop plants produced using the currently available plant transformation methods, we do not know what will be the typical or expected arrangement(s) of a transgene insertion event (Somers and Makarevitch, 2004; Latham *et al.*, 2006). This information would be useful for anticipating the risks of transgene insertion which arise from endogenous gene disruption and transgene rearrangement. To understand this generally, large-scale studies need to be carried out using various transformation procedures and in different crops. These studies should be carried out on large numbers of loci classified as single transgene inserts by Southern blot analysis, as single transgene inserts are the kind most often used for commercial purposes. DNA sequence analysis needs to be carried out on numerous transgene loci, as well as sufficiently large regions of flanking genomic DNA, to determine the full extent of target-site disruption. Such large-scale studies of random single-insert loci will help to clarify the nature and extent of genomic and transgene DNA rearrangement during transgene insertion. They will also help clarify the extent and nature of superfluous DNA insertion in the vicinity of the desired transgene(s).

In the meantime, it will be necessary to characterize transgene locus structure on a case-by-case basis. Wilson *et al.* (2004) suggested a rigorous set of transgene rejection criteria for the development of events in crops, including:

– Rejection of transgenic events associated with superfluous transgene, marker or plasmid DNA.
– Rejection of transgenic insertion events associated with large deletions and rearrangement of genomic DNA.
– Rejection of transgenic events inserted into or near functional plant gene sequences.

4.3. Transgene Expression (Phenotypic Analysis)

Target transgene expression

The characterization of transgene expression (phenotype) should address what is essential for evaluating efficacy, non-target effects, resistance management and gene flow. Transgene expression can be measured in terms of the concentration of transgene product, as well as the whole-organism phenotype. For insect resistance genes, the appropriate whole-organism phenotype is the efficacy of control of the target pest.

Methods for measuring accurately the products of transgenes in plants have been improving, but additional efforts are needed to establish accurate, repeatable measures for the various plant tissues and crops. Extraction artefacts, such as incomplete extraction from plant cells and product modification during processing (e.g. proteolytic degradation), have often produced misleading results, such that the stability or variability of the transgene product concentration could not be established conclusively.

An important limitation to the analysis of target gene expression lies in the fact that transgene products may interact with other gene products and chemicals in the plant, thereby altering plant physiology and chemical composition (e.g. secondary plant metabolites in cotton; Olsen and Daly, 2000). Possible interactions and resulting complex expression patterns will be difficult to predict and to measure, except that any significant negative effect can be monitored at the whole-organism level.

Target transgene promoter

The type of promoter used to drive the transgene obviously affects the transgene expression pattern. Based on the promoter characteristics, one can predict where and when the transgene is likely to be expressed. These predictions can be used to determine which plant tissues are relevant to be evaluated for target gene expression. The effects of transgene promoters on endogenous plant genes should also be assessed to determine whether transgene promoters influence expression of genes on either side of the inserted transgene. This is especially relevant when strong viral promoters are used to regulate transgenes, as such promoters are known to have the ability to affect nearby genes. Experiments carried out in transgenic *A. thaliana* and rice plants indicated that strong transgene enhancers or promoters could influence endogenous gene expression, even at a distance of several kilobase pairs (Wilson *et al.*, 1996; Weigel *et al.*, 2000; Jeong *et al.*, 2002).

Target transgene products

It is essential to analyse the active protein products produced by the transgene using appropriate techniques (e.g. Western blots). These products are compared with the ones predicted from the original transgene construct or the naturally occurring product of the gene. It is necessary to be sure that the analysis methods do not create artefacts and that the extraction efficiency can be measured (e.g. ELISA). The sampling method of plant materials for analysis should take into consideration: (i) the pattern of expression of the target gene among plant tissues in relation to potential non-target and resistance risks, for example, some potential non-target species interact only with certain plant tissues, such as pollen, leaf-litter, or leaf mesophyll tissue (see Chapter 5, this volume); and (ii) the changing expression of active product during plant development – the number of time periods that need to be sampled should be determined in relation to non-target and resistance risk, for example, resistance risk can be affected strongly by variation in or declines in the concentration of gene product during the period when the target pest is associated with the plant (Kranthi *et al.*, 2005).

Whole-plant phenotype

The relevant whole-plant phenotype is usually related to the intended use of the transgenic plant. For example, for *cry* genes that code for Cry toxins, efficacy of control of the target pest or pests is a relevant phenotype (Greenplate *et al.*,

2001; Haile *et al.*, 2004). It is important to measure this because it influences resistance risk in critical ways through direct exposure and indirect pathways, such as through gene flow. For each relevant phenotype (trait), appropriate methods of evaluation should be used.

Marker gene expression

In addition to the expression of the target gene(s), it is essential to characterize the expression of the marker gene (if present). Steps for characterizing marker gene expression are as follows:

1. Collect information on the marker gene contained in the original construct (its coding sequence, promoter, history of concerns of each marker/promoter).
2. Identify the presence of the marker gene at the transgene locus for each copy at each locus (its sequence, promoter).
3. Identify the effect of the promoter (active/partially active or inactive in the plants).
4. In the case that the marker gene is present and of potential concern, but not expressed, it is essential to identify the reason of non-expression (i.e. blocked at the DNA, RNA or protein level?). Some questions may need to be clarified, for example:

- Is the marker gene transcribed? If not, what would need to happen to allow the marker gene to be transcribed?
- Is the transcription product translated? If not, what would need to happen to allow the marker gene to be translated?
- Is the protein inactive? What evidence demonstrates this? What is the expression pattern of the inactive protein? Would the inactive protein have some other possible function in the plant or a potential environmental or health risk?

Marker gene product analysis

The marker gene product should be characterized in sufficient detail to confirm that it is identical to the product from the naturally occurring gene from which the marker is derived. If a difference is found, its effect or lack of effect should be documented.

Factors to be considered for marker gene expression analysis include: (i) analysis method and possible artefacts created; (ii) extraction efficiency; (iii) expression of the marker gene among plant tissues; and (iv) expression consistency among environments. Of course, none of this is needed if the marker has been eliminated from the transgenic plant.

Effects on other traits

If a transgene (or its associated marker gene) affects a plant trait in addition to the trait it codes for, it is said to have pleiotropic effects. If it interacts with the regulation and expression of other gene loci to affect a trait in addition to the trait it codes for, it is said to have epistatic effects. If it has either pleiotropic or epistatic effects, it could affect agronomic performance, end-product quality or result in the production of metabolites harmful to humans, animals and non-target organisms that consume the transgenic plants. For example, Hashimoto *et al.* (1999a,b) found increased glycoalkaloid content in potatoes that expressed transgenic soybean glycinin. Shewmaker *et al.* (1999) found multiple changes in metabolic pathways linked to a transgenically modified pathway in canola. Ye *et al.* (2000) found that modification of the carotenoid biosynthetic pathway resulted in formation of unexpected carotenoid derivatives.

　　Agronomic performance and crop quality evaluations should be conducted over multiple years and locations to provide assessment of these possible effects (Verhalen *et al.*, 2003). Compositional analysis should be conducted on the consumed portions of the plant to identify changes in known metabolites related to the transgenic trait (Cellini *et al.*, 2004). However, in many cases, it will be necessary to evaluate non-target performance on whole plants (Andow and Hilbeck, 2004).

Genotype by environment interaction

Tests as described above should be conducted during multiple years and on multiple locations (Rochester, 2006) to address how expression of the trans-gene product or any pleiotropic effect may vary among environments and under stress, such as stem splitting of Roundup Ready soybean under heat stress (Gertz *et al.*, 1999). For example, the fitness associated with herbicide tolerance genes can vary in different environments and when expressed in dif-ferent genetic backgrounds (Mercer *et al.*, 2006). Efficacy of Bt cotton against target insects can vary under different field conditions (Dong and Li, 2007), such as temperature stress (Olsen *et al.*, 2005), and genetic background can affect Cry1Ac expression in Bt cotton (Adamczyk and Meredith, 2004).

Transformation-induced mutations

The present plant transformation methods typically cause some genomic dis-ruption during transgene integration. As demonstrated by Wilson *et al.* (2004), each transformed plant genome contains a unique spectrum of mutations in addition to the transgene, resulting from (i) tissue culture procedures, (ii) gene transfer methods such as *Agrobacterium*-mediated or particle bombardment transfer, (iii) transgene insertion; and (iv) superfluous DNA.

　　These transformation-induced mutations can be separated into two types: inserted-site mutations, referring to those introduced at the site of transgene insertion, and genome-wide mutations, referring to those introduced at other sites in the recipient genome. There is still relatively little known about the mutations created in crop plants at the site of transgene insertion. Forsbach *et al.* (2003) studied 112 single-copy T-DNA insertion events in *A. thaliana* and found that exact T-DNA integration almost never occurred and that most of the insertions resulted in small (1–100 base pair) deletions of plant genomic sequences at the insertion site. In addition, a significant number of events (24/112) showed evidence of large-scale rearrangement of plant genomic DNA at the insertion site.

For genome-wide mutations, it was reported that many hundreds or thousands of such genome-wide mutations were likely to be present in transformed plants using typical transformation methods, especially those involving the use of plant tissue culture techniques (Sala *et al.*, 2000). Labra *et al.* (2001) estimated that the 'genomic similarity value' of the control plants was 100%, but only 96–98% for the transgenic plants.

Significance of transformation-induced mutations

Insertion-site and genome-wide mutations may be hazardous if they occur in a functional region of plant DNA. Mutations in functional plant DNA, including gene coding sequences or regulatory sequences, may have implications for agronomic performance or environmental interactions, or for animal or human health. For example, a transformation-induced mutation might disrupt a gene producing a product involved in nutrient biosynthesis, resulting in altered nutrient levels, or it might disrupt or alter a gene involved in the regulation or synthesis of compounds toxic to humans. Disruption of a gene encoding a regulatory protein, such as a transcription factor, could result in the altered expression of numerous other genes. Such biochemical changes would be unpredictable and difficult to identify, even with extensive biochemical testing (Kuiper *et al.*, 2001).

Wilson *et al.* (2004) recommended the following regulatory guidelines:

– Transformed sexually reproducing plants intended for field-scale trials for commercial release should be subjected to an extensive backcrossing programme, followed by testing for effective removal of transformed-induced mutations.
– Transgene insertion events into plant sequences, which are, or may be, functional DNA sequences, should be rejected from the breeding programme.
– Transgenic lines containing genomic alterations at the site of transgene insertion should be rejected.
– Both the transgene insertion event (including all transferred DNA and a large stretch of flanking DNA) and the original target site should be sequenced and compared, because this is the only known way to determine definitively whether gene sequences have been disrupted.
– Besides DNA sequencing, other techniques such as FISH, fibre-FISH, Southern blot analysis and PCR should be used to detect rearranged transgenes and superfluous DNA at or near the transgene insertion event.

Interactions with other transgenes

Breeders will be 'stacking/pyramiding' transgenic traits in the future. In cases where some transgene elements are identical between the two transgenic parents, combining these parental transgene loci in a single plant may silence one or more transgenes. This has not yet been a problem in transgenic cotton, but these interactions will best be assessed by monitoring transgene expression, and agronomic performance should be evaluated over multiple years and locations.

4.4. Transgene Inheritance

In general, it is important to evaluate the stability of inheritance of the transgene locus and phenotype over multiple generations and in different genetic backgrounds. A transgene can be integrated stably into the genome, but its phenotype can be altered because the genetic background changes or because the gene is silenced or enhanced. Under gene flow, the transgene may be expressed in crosses with other plants (including other varieties, landraces, wild escapees and feral subpopulations). For some crop plants, this may include interspecific crosses, such as to wild relatives or other sexually compatible species. If these crop relatives have the same mating system and chromosome structure as the transgenic plant, it would be expected that transmission and expression of the transgene in crosses with these relatives should follow predicted Mendelian ratios. When there is a difference in chromosome structure, the transmission and expression of the transgene may not follow predicted Mendelian ratios and repeated backcrosses into the relevant relatives should be evaluated (Oard *et al.*, 2000; Ammitzboll *et al.*, 2005).

Transmission in different genetic backgrounds

The transgene phenotype should be transmitted to progeny according to predicted Mendelian ratios. If true, this provides assurance that there are no major instabilities in the transgene locus or transgene expression, which would signal a need for additional risk assessment investigations. The following questions should be addressed by studying the inheritance of the transgene phenotype according to Mendelian ratios and the consistency of inheritance pattern over multiple generations:

1. When crossed with relatives, does the transgene locus segregate like a normal plant gene? If not, then molecular investigations may be needed to determine the reason for the difference.
2. When crossed with relatives, does the transgene phenotype segregate according to predicted Mendelian ratios? If not, then molecular investigations may be needed to determine the reason for the difference.

Gene expression in different genetic backgrounds

Transgene expression may depend on the recipient population to which the transgenic plant is crossed, as well as the environment in which the plant grows.

Even if crossing occurs, it does not imply necessarily that there is any environmental or human health risk, but the possibility should be investigated. Expression of the transgene should be assessed in:

1. Relevant environments when crossed to recipient populations with a mating system or chromosome structure that differs from the transgenic crop. Relevant environments would be those in which the hybrids are likely to be found. In addition, expression in backcrosses to the recipient population should be assessed. Some have argued that multiple hybrid generations need to be assessed (Hauser *et al.*, 1998).

2. Environments where the transgene may confer a selective advantage to the hybrid or the backcross (Snow *et al.*, 2003; Mercer *et al.*, 2006).

4.5. Transgenic Cotton Plants Expressing Insecticidal Genes that may be Transferred to Vietnamese Cotton Cultivars

The following section characterizes the Bt cotton events that include the *vip3A* gene and the *cry1Fa* and *cry1Ac* genes in transgenic cotton lines. These transgenes may be transferred into elite Vietnamese cotton cultivars via conventional breeding or by transformation to develop Bt transgenic cotton for use in Vietnam. Two other Bt cotton events that might be introduced in Vietnam – Bollgard I (expressing Cry1Ac) and Bollgard II (expressing Cry1Ac and Cry2Ab) – were described in a previous volume in this series (Grossi-de-Sa *et al.*, 2006). A detailed review of expression patterns of these and other Bt cotton events is provided in Tables 4.1, 4.2 and 4.3.

Bt cottons expressing the *vip3A* gene

The VIP3A cotton lines contain one of three different transformation events inserted into the US cultivar Coker 312. The COT102 Bt cotton line (COT102 line) contains insect resistance (*vip3A*) and antibiotic resistance (*aph4*) genes. The COT202 and COT203 lines contain only the *vip3A* gene and do not contain an antibiotic resistance gene. The Bt cotton lines produce a vegetative insecticidal protein (VIP3A) that is expected to be toxic to a range of lepidopteran species, including *Helicoverpa armigera* (Liao *et al.*, 2002), which is the key lepidopteran pest of cotton in Vietnam (Chapter 2, this volume). Much of the following information was derived from FSANZ (2004) and OGTR (2003b).

vip3A *transgene event*
The *vip3A* gene in these events is derived from *Bacillus thuringiensis* variety *kurstaki* (Bt) and was resynthesized with a codon bias suitable to optimum expression in cotton cells. The promoters for the *vip3A* gene are constitutive plant promoters. COT102 contains the *A. thaliana* actin-2 promoter (An *et al.*, 1996); COT202 and COT203 have different, unspecified plant promoters. Use of a plant promoter should provide strong expression of the bacterial

Table 4.1. Reported levels of Bt proteins in leaves, roots, root exudates and stems of cotton. All measurements are in µg/g fresh weight (= mg/kg = parts per million = ng/mg, 1000 ng/g), unless marked DW = dry weight. Numbers are means, unless otherwise indicated. If two or more measurements are indicated, they are means either of different sampling times or from different studies (different cultivars, locations or years). Standard deviations are given in superscript retaining all significant digits from the publications. DAP = days after planting. ND = not detectable, below limit of detection.

Transgenic cotton	Bt protein	Leaves	Terminal leaf	Roots	Stem	Reference[a]
COT102	VIP3a	3.0–22.0 DW 4.0–11.9		< 2.5; DW 7.5		8[b]
COT 102	VIP3a	3–22; DW 5–118		< 0.2–2; DW < 0.4–7		13
Event 531	Cry1Ac	1.40, 1.49, 3.55, 1.30 and 5.12, 3.21, 0.13, 0.23[c]		DW 1–43 (27–63 DAP)	DW 0.2–3 (27–63 DAP)	7[d]
Event 531	Cry1Ac	2.04				16
Event 531	Cry1Ac	Range 0.3–5.0[e]				10
Event 531	Cry1Ac		DW $38^{\pm1}$–$18^{\pm1f}$			3
Event 531	Cry1Ac			Root exudates: ND (in soil 55 DAP, in solution 40 DAP)[g]		11
Event 531	Cry1Ac	DW $3.5^{\pm0.3701}$, $10.6^{\pm4.9859}$ (63 DAP)	DW $11.4^{\pm2.1045}$, $27.4^{\pm2.7533}$ (63 DAP)	Taproot: DW $3.5^{\pm0.4450}$ (63 DAP); $0.5^{\pm0.0656}$ (119 DAP) Secondary root: DW $2.2^{\pm0.2690}$, $3.3^{\pm0.1040}$ (63 DAP); $0.8^{\pm0.0167}$, $0.2^{\pm0.1433}$ (119 DAP)[h] Fine root: DW $15^{\pm4.4374}$, $5.3^{\pm0.7900}$ (63 DAP); $0.4^{\pm0.2393}$, $1.3^{\pm0.2091}$ (119 DAP) Root exudates: DW 0.2–0.4 (in solution 14 and 56 DAP)	DW $0.8^{\pm0.2900}$, $10.6^{\pm1.2789}$ (63 DAP)	4
Chinese Bt cotton[i]	Cry1A + CpTI	2.29 (seedling)		1.33 (seedling)	1.07 (seedling)	5
Chinese Bt cotton (99B)	Cry1A	1.40 (seedling)		1.12 (seedling)	0.88 (seedling)	5
Chinese Bt cottons	Cry1A[j]c/Cry1ab fused gene Cry1Ac[l]		Range $0.19^{\pm0.0416}$–$0.77^{\pm0.04993x}$ Range $0.06^{\pm0.00195}$–$0.72^{\pm0.03092}$			17
Indian Bt cotton	Cry1Ac[m]	UC: $1.21^{\pm0.58}$–$1.99^{\pm0.94n}$ MC: $1.13^{\pm0.15}$–$1.94^{\pm0.30}$ LC: $1.31^{\pm0.28}$–$2.49^{\pm0.38}$				6

Continued

Table 4.1. *Continued*

Transgenic cotton	Bt protein	Leaves	Terminal leaf	Roots	Stem	Reference[a]
Bollgard II[o]	Cry2Ab	23.8 (range 10.1–33.3)	21.0, 40.1, 19.7, 16.7 (28, 55, 85, 108 DAP)			12
Bollgard II[p]	Cry1Ac	3.6±0.35 (lower canopy leaf)	2.4±0.15			1
	Cry2Ab	21.0±1.72, 14.4±0.34q (lower canopy leaf)	13.5±2.07, 5.5±0.80			
Cry1Ac synpro cotton[r]	Cry1Ac	DW 1.92±0.7 (young leaf 21 DAP to 42 DAP) (2.9, 2.4)	DW 1.44±0.5 (after 63 DAP) (0.77–2.5)	DW 0.20±0.1 (seedling stage) DW 0.10±0.07 (pollination) DW (0.05±0.04)s (defoliation)		15 and 9
Cry1Fa cotton[t]	Cry1Fa	DW 6.48±3.3 (young leaf 21 DAP to 42 DAP) (8.7, 11)	DW 7.67±5.3 (after 63 DAP) (5.4–28)	DW 0.72±0.6 (seedling stage) DW 0.36±0.1 (pollination) DW 0.61±0.5 (defoliation)		9
Cry1F cotton[u]	Cry1Fa	5.3	18.1	0.5 DW 1.6	11.9 DW 40.5	14[v]
WideStrike[w]	Cry1Ac	DW 1.82±0.6 (young leaf up to 56 DAP) (2.6, 1.8)	DW 1.31±0.4 (after 63 DAP) (1.1–2.0)	DW 0.17±0.05 (seedling stage) DW (0.07±0.06)x (pollination) ND (defoliation)		9[y] and 2
	Cry1Fa	DW 6.81±3.6 (young leaf up to 56 DAP) (7.3, 9.4)	DW 8.19±3.5 (after 63 DAP) (6.7–41)	DW 0.88±0.7 (seedling stage) DW 0.54±0.4 (pollination) DW 0.51±0.2 (defoliation)		

Notes: [a]References for Tables 4.1 to 4.3: 1 = Akin *et al.* (2002); 2 = EU (2004); 3 = Greenplate *et al.* (2001); 4 = Gupta and Watson (2004); 5 = Jiang *et al.* (2006); 6 = Kranthi *et al.* (2005); 7 = OGTR (2002); 8 = OGTR (2003a); 9 = OGTR (2003b); 10 = Rochester (2006); 11 = Saxena *et al.* (2004); 12 = US APHIS (2000); 13 = US APHIS (2003); 14 = US APHIS (2004a); 15 = US APHIS (2004b); 16 = US EPA (2001); 17 = Wan *et al.* (2005). [b]Data are not corrected for extraction efficiency. Limit of quantification: 0.02–0.0270 µg/g fresh weight of plant tissue, 0.3 µg/g dry weight. [c]Measured four times during season, in two separate studies in the US Bollgard Coker variety. [d]Limit of detection: 1.6 ng/g fresh weight of plant tissue. Data for roots and stems based on Gupta *et al.* (2002). Data for leaves based on US research. [e]Measured Cry1Ac levels in leaves at nodes 5, 10 and 15 under different environmental conditions (stress factors) and at mid-flowering, mid-boll fill and 20% open bolls periods. Measured eight cultivars (Sicot series). [f]Means for all 35 varieties at 5 different times during the season (2, 4, 6, 8 and 10 weeks after pinhead square). [g]Measured with EnviroLogix Lateral Flow Quickstix either in rhizosphere soil supernatant or directly in hydroponic solution. [h]When the means of the two varieties (Sicot 298i and Sicot 289Rri) are very different, they are listed separately. [i]Cultivar GK9708-41. [j]Cultivar GK19. [k]Measured 5 to 7 times during season, over 2 years. [l]Cultivar BG1560. [m]All Indian Bt cotton hybrids contain the Cry1Ac transgene from Coker 312. [n]UC = upper canopy leaves, MC = mid-canopy leaves, LC = lower canopy leaves. The range of seasonal means for the eight hybrids measured (MECH and RCH hybrids). [o]Line MON15985, containing Event 531 from parent variety DP50B and the Cry2Ab event. [p]Line MON15985, containing Event 531 from parent variety DP50B, and the Cry2Ab event. Variety NuCOTN 33BII. [q]The two means are from 2 different years, as transgene expression varied markedly between the 2 years. [r]Event 3006-210-23, cotton line MXB-7. [s]Calculated concentration is less than LOQ of the method. [t]Event 281-24-236, cotton line MXB-9. [u]Event 281-24-236, cotton line MXB-9. [v]Describe as 'high end estimates'. [w]Dow WideStrike cotton: line 281-24-236/3006-210-23. [x]Below the validated limit of quantification. [y]Averages of data from three cotton lines in 1 year at six sites in the USA.

Table 4.2. Reported levels of Bt proteins in squares, flower tissues and boll tissues of cotton. All measurements are in µg/g fresh weight, unless marked DW = dry weight. Numbers are means, unless indicated. If two or more measurements are indicated, they come either from different sampling times or from different studies (different cultivars, locations or years). DAP = days after planting. ND = not detectable, below limit of detection. WP = white petals, PP = pink petals, WS = white stamens, PS = pink stamens. Key to references in Table 4.1.

Transgenic cotton	Bt protein	Square	Flower	Boll	Reference
COT102	VIP3a	< 4.0; DW 17.0		0.3–1.5; DW < 0.3–9.0	8[a]
COT102	VIP3a			Declined from c.1 to ND at pre-harvest (DW c.7–9)	13
Event 531	Cry1Ac			Primary fruiting structures: DW 259 (46 DAP)–43 (116 DAP)	7[b]
Event 531	Cry1Ac	DW 13±1.5–27±2.5c		DW 17±1d	3
Chinese Bt cottons	Cry1Ae c/Cry1Ab fused gene	0.2±0.00301–0.9±0.15739	Petals 0.2±0.05567–0.7±0.02753; Stamens 0.4±0.0585c–0.8±0.00440; Ovule 0.05±0.04981–0.3±0.05756	0.1±0.04344–0.4±0.01480	17
	Cry1Ac[f]	0.2±0.00835–0.6±0.01188	Petals 0.5±0.05463–0.8±0.01371; Stamens 0.5±0.0581c–0.8±0.03304; Ovule 0.1±0.00439–0.2±0.02515	0.05±0.00305–0.2±0.03593	
Indian Bt cotton	Cry1Ac[g]	Square bracts: 0.06–0.63; Square buds: 0.05–0.08 (3 hybrids); 0.25–0.51 (5 hybrids)	Sepals 0.4–1.6; Petals 0.25–0.80; Anthers 0.05–0.6; Ovary ND–0.07 in 4 hybrids, 0.15–0.27 in 4 hybrids	Boll bracts: 0.19–1.17; Boll rind: 0.01–0.05 in 5 hybrids, 0.25–0.37 in 3 hybrids; Loculi wall: 0.38–1.98; Raw seed cotton: 0.65–2.02	6
Bollgard II[h]	Cry1Ac		WP 3.1±0.23; PP 2.4±0.11; WS 2.7±0.13; PS 2.2±0.13	Young boll: 2.1±0.15; 10–14 day old boll wall: 2.9±0.21; 10–14 day old boll–internal contents: 1.5±0.08	1
	Cry2Ab		WP 20.1±2.77, 8.4±0.60; PP 17.5±0.24, 5.4±0.48; WS 26.2±0.69, 8.8±0.70; PS 18.8±2.17, 5.6±0.46	Young boll: 22.9±1.43, 8.0±1.76; 10–14 day old boll wall: 10.0±2.27, 9.0±0.99; 10–14 day old boll–internal contents: 22.0±0.85, 6.4±0.32	

Continued

Table 4.2. _Continued_

Transgenic cotton	Bt protein	Square	Flower	Boll	Reference
Cry1Ac synpro cotton[i]	Cry1Ac	DW $1.84^{\pm0.5}$	DW $1.92^{\pm0.3}$ (2.1–2.2 at first flower; 1.9–1.7 at 2–3 weeks; 1.3–2.0 at 4–6 weeks)	DW $0.77^{\pm0.2}$ (0.75–0.58 early boll; 0.66–0.41 at 2–3 weeks; 0.46–0.43 at 4–6 weeks)	9 and 15[j]
Cry1Fa cotton[k]	Cry1Fa	DW $5.04^{\pm1.8}$	DW $5.71^{\pm2.1}$ (4.3–5.3 at first flower; 2.7–6.5 at 2–3 weeks; 4.4–1.5 at 4–6 weeks)	DW $4.02^{\pm2.0}$ (4.4–1.5 early boll; 6.7–5.1 at 2–3 weeks; 7.3–4.6 at 4–6 weeks)	9[l]
WideStrike[m]	Cry1Ac	DW $1.82^{\pm0.5}$	DW $1.83^{\pm0.4}$ (2.1–2.2 at first flower; 2.0–1.8 at 2–3 weeks; 0.91–1.9 at 4–6 weeks)	DW $0.64^{\pm0.2}$ (0.5–0.47 early boll; 0.44–0.33 at 2–3 weeks; 0.37–0.40 at 4–6 weeks)	9[n] and 2
	Cry1Fa	DW $4.88^{\pm1.8}$	DW $5.44^{\pm1.8}$ (4.8–5.6 at first flower; 4.4–6.5 at 2–3 weeks; 2.6–5.7 at 4–6 weeks)	DW $3.52^{\pm1.7}$ (3.5–1.4 early boll; 6.3–7.6 at 2–3 weeks; 5.0–4.8 at 4–6 weeks)	

Notes: [a]Data are not corrected for extraction efficiency. Limit of quantification: 0.02–0.0270 µg/g fresh weight of plant tissue, 0.3 µg/g dry weight. [b]Data for fruiting structures based on averages of several US field trials with Bollgard. [c]Range of season means for 35 pre-commercial varieties. [d]Season mean for all 35 pre-commercial varieties (no significant difference between short-season and long-season varieties). [e]Cultivar GK19. [f]Cotton line BG15960. [g]All Indian Bt cotton hybrids contain the Cry1Ac transgene from Coker 312. [h]Line MON15985, containing Event 531 from parent variety DP50B, and the Cry2Ab event. Variety NuCOTN 33BII. [i]Event 3006-210-23, cotton line MXB-7. [j]Ranges of mean values of nine replicate samples from two different US sites. [k]Event 281-24-236, cotton line MXB-9. [l]Ranges of mean values of nine replicate samples from two different US sites. [m]Dow WideStrike cotton: line 281-24-236/3006-210-23. [n]Ranges of mean values of nine replicate samples from two different US sites.

Table 4.3. Reported levels of Bt protoins in seed, pollen and nectar of cotton. All measurements are in µg/g fresh weight, unless marked DW = dry weight. Numbers are means, unless indicated. If two or more measurements are indicated, they come either from different sampling times during the season or from different studies (different cultivars, locations or years). DAP = days after planting. ND = not detectable, below limit of detection. Numbers in parentheses are ranges of means.

Transgenic cotton	Bt protein	Seed	Pollen	Nectar	Reference
COT102	VIP3a	2.0–3.0; DW 2.0–4.0	DW 1.1	ND	8[a]
COT102	VIP3a	c.3	1.1 (air-dried)	ND[b]	13
Event 531	Cry1Ac	0.86 (0.49–1.62), 2.18 (1.13–3.41) in Coker; 4.30$^{\pm0.86}$ in DP5415		ND[c]	7[d]
Event 531	Cry1Ac	1.62	0.0115[e]		16
Bollgard II[f]	Cry2Ab	43.2 (31.8–50.7)	ND[g]		12
Cry1Ac synpro cotton[h]	Cry1Ac	0.57$^{\pm0.09}$	1.44$^{\pm0.5}$	ND[i]	9 and 15
Cry1Fa cotton[j]	Cry1Fa	5.13$^{\pm1.1}$	-0.09$^{\pm0.30}$[k]	ND[l]	9
Cry1F cotton[m]	Cry1Fa	7.5[n]	0.7	< 0.05 ng/µl	14
WideStrike[o]	Cry1Ac	0.55$^{\pm0.07}$	1.45$^{\pm0.5}$	ND[p]	9 and 2
	Cry1Fa	4.13$^{\pm1.1}$	-0.06$^{\pm0.15}$[q]	ND[r]	

Notes: [a]Data are not corrected for extraction efficiency. Limit of quantification: 0.02–0.0270 µg/g fresh weight of plant tissue, 0.3 µg/g dry weight. [b]'VIP3a was not detectable in the pooled nectar sample and no protein of any kind was detectable in nectar by the Bio-Rad method.' [c]No detection of protein of any kind in nectar. [d]Limit of detection: 1.6 ng/g fresh weight of plant tissue. Data from US field trials with Bollgard varieties. [e]11.5 ng/g. Cited in US EPA (2001) and OGTR (2002). [f]Line MON15985, containing Event 531 from parent variety DP50B and the Cry2Ab event. Variety NuCOTN 33BII. [g]Not detected above limit of detection (0.25 µg/g) at either location. [h]Event 3006-210-23, cotton line MXB-7. [i]Below limit of detection, which was 0.001–0.4 ng/mg. [j]Event 281-24-236, cotton line MXB-9. [k]Estimated to be below detection limit. [l]Below limit of detection, which was 0.001–0.4 ng/mg. [m]Event 281-24-236, cotton line MXB-9. [n]Described as 'high end estimates'. [o]Dow WideStrike cotton: line 281-24-236/3006-210-23. [p]Below limit of detection, which was 0.001–0.4 ng/mg. [q]Estimated to be below detection limit. [r]Below limit of detection, which was 0.001–0.4 ng/mg.

gene in plant cells and a constitutive promoter is expected to express VIP3A protein in nearly all cotton tissues at all stages of growth. The termination and polyadenylation signals are provided by the 3' end of the A. tumefaciens nopaline synthetase (nos) gene (Depicker et al., 1982).

The COT102 line also contains the selectable marker gene aph4 from the bacterium E. coli that confers resistance to the antibiotic hygromycin B. The expression of the aph4 gene is also driven by a constitutive plant promoter and the 3' end of the A. tumefaciens nos gene provides the termination and polyadenylation signals. The COT202 and COT203 lines have had the selectable marker gene aph4 removed through segregation.

Short regulatory sequences that control expression of the transgenes are also present in all of these Bt cotton lines. Some of these are derived from a plant and some from A. tumefaciens.

Method of gene transfer in Bt cottons expressing the vip3A *gene*
The Bt cotton event COT102 was generated by *Agrobacterium*-mediated transformation using protocols similar to those described by Murray *et al.* (1999), but with some modifications. The disarmed *Agrobacterium* strains were constructed specifically for plant transformation. The disarmed strains did not contain the genes (*iaaM, iaaH* and *ipt*) responsible for the overproduction of auxin and cytokinin required for tumour induction and are therefore not, in themselves, capable of causing disease. A useful feature of the Ti plasmid is the flexibility of the *vir* region to act in either *cis* or *trans* configuration to the T-DNA. This has allowed the development of two types of T-DNA vectors for transformation: (i) co-integrating vectors, which integrate into the resident plasmid of *Agrobacterium* containing the *vir* region; and (ii) binary vectors, which coexist in *Agrobacterium* together with the plasmid containing the *vir* regions.

For developing the COT102 Bt cotton, a conventional, disarmed, binary vector pCOT1 was used for transformation. For COT202 and COT203 Bt cotton, the insecticidal and antibiotic resistance genes were contained within the same vector, pNOV103, but in separate T-DNAs, so that the genes inserted independently at different sites in the cotton genome, allowing segregation in subsequent generations of transgenic plants. No part of the *Agrobacterium* vector is present in the COT102 genetic modification event as confirmed by Southern blotting and PCR techniques.

Transgene characterization of Bt cottons expressing the vip3A *gene*
In the COT102 event, single copies of the *vip3A* and of the *aph4* selectable marker gene were inserted as determined by Southern blot analysis. The *vip3A* and *aph4* genes were inherited as a single dominant Mendelian trait over five generations of backcrossing and selfing. The expression of the inserted gene in the Bt cotton was stable. The COT202 and COT203 lines also contain a single copy of *vip3A* and Mendelian inheritance was stable over the three generations evaluated.

Transgene expression of the VIP3A protein
B. thuringiensis produces a range of insecticidal proteins (Bt toxins), each with specific toxicity to certain groups of insects. The biological role of the insecticidal proteins is unclear, but they provide these free-living bacteria with an ability to immobilize and colonize a ready source of nutrients in the insects they kill. The *vip3A* gene encodes a synthetic form of a protein, the VIP3A protein, which is secreted into the extracellular environment by *B. thuringiensis* during vegetative growth (hence, the name vegetative insecticidal protein), as well as during the stationary (sporulation) phase, unlike the *cry* genes, which are expressed only during sporulation (Estruch *et al.*, 1996; Donovan *et al.*, 2001). The VIP3A protein differs from the Cry insecticidal proteins that are present in other types of insecticidal Bt cottons, having no amino acid homology to the Cry1 and Cry2 proteins (Estruch *et al.*, 1996), and it binds to specific receptors inside the insect gut different from those bound by Cry1A proteins (Lee *et al.*, 2003). Once bound, the protein inserts into the membrane and forms ion-specific pores, which results in disrupted digestion and subsequent death of the

insect (Yu *et al.*, 1997). The VIP3A bacterial protein has a broader spectrum of activity than the Cry proteins (Estruch *et al.*, 1996; Donovan *et al.*, 2001; Selvapandiyan *et al.*, 2001).

Insecticidal efficacy of VIP3A cotton

Field trials of VIP3A cotton have shown that it provides effective control of *H. armigera* in Australia (Llewellyn *et al.*, 2007) and *Spodoptera* in the USA (Cloud *et al.*, 2004). The bacterially produced protein also resulted in high mortality of *H. punctigera* (Liao *et al.*, 2002). No information is available on the efficacy of VIP3A against *Pectinophora gossypiella*, another important lepidopteran pest on cotton in Vietnam.

Bt cottons expressing the *cry1Fa* and *cry1Ac* genes

cry1Fa *and* cry1Ac *transgene events*

The *cry1Fa* and *cry1Ac* genes are chimeric genes, each combining parts of three different *cry* genes isolated from *B. thuringiensis* (Bt). The part of the chimeric *cry1Fa* gene that corresponds to the active core toxin is derived from the native *cry1Fa* gene of *B. t.* variety *aizawai* (Bta). The part of the chimeric *cry1Ac* gene that corresponds to the active core (functional) toxin is derived from the native *cry1Ac* gene of *B. t.* variety *kurstaki* (Btk) strain HD73. This chimeric *cry1Ac* gene differs from the *cry1Ac* gene present in both INGARD® and Bollgard® II cottons. It is a chimeric gene derived from *cry1Ac* and *cry1Ab* genes of *B. t. kurstaki*. The remainder of each of these genes, encoding the carboxy-terminal portion of the proteins which is cleaved off in the insect gut, is derived from parts of the *cry1Ca3* and *cry1Ab1* genes. The carboxyl-terminal portion is not essential for toxicity and its function appears to be in the maintenance of the unusual solubility of the Cry1 proteins (Luthy and Ebersold, 1981). The chimeric genes were developed to improve the level of expression in plants and the solubility of the encoded Bt toxins in the insect gut. The coding sequence of the chimeric genes has been modified further to achieve optimal expression in plants, without affecting the encoded protein sequences.

The chimeric *cry1Fa* and *cry1Ac* genes encode the Cry1Fa and Cry1Ac proteins respectively, which are very similar to native Cry1Fa and Cry1Ac proteins. Within the core toxin, the amino acid sequences of the native and chimeric proteins are 99.3% and 99.6% identical, respectively, and retain the species specificity of toxicity to larvae of lepidopteran insects characteristic of native Cry1Fa and Cry1Ac proteins.

Expression of the chimeric *cry1Fa* gene is controlled by the (4OCS)Dmas 2' promoter, a synthetic promoter derived from the mannose synthase gene (*mas*) promoter and octopine synthase gene (*ocs*) enhancer of *A. tumefaciens* (Ni *et al.*, 1995). The mRNA termination region is provided by the bidirectional polyadenylation signal of *A. tumefaciens* ORF 25 (Barker *et al.*, 1983). Expression of the chimeric *cry1Ac* gene is controlled by the ubiquitin promoter of maize *Zea mays* (Christensen *et al.*, 1992). The mRNA termination region is also provided by the bidirectional polyadenylation signal of *A. tumefaciens* ORF 25 (Barker *et al.*, 1983).

Method of gene transfer in cottons expressing the cry1Fa *and* cry1Ac *genes*
The two chimeric *cry1* genes were each introduced separately into a US commercial cotton variety, GC510, in combination with one copy of the *pat* gene, by *Agrobacterium*-mediated DNA transformation (Zambryski, 1992). The plasmids used contain well-characterized DNA segments required for their replication and selection in bacteria and for transfer from *Agrobacterium* and integration into the cotton genome (Bevan, 1984; Wang *et al.*, 1984). Following co-cultivation, cotton cells were cultured in the presence of glufosinate ammonium to select for those cells containing the inserted gene construct (since the *pat* gene confers tolerance to glufosinate ammonium). Cotton plants containing the insecticidal genes were regenerated from these transgenic cells.

The chimeric *cry1Ac* gene with one *pat* gene were introduced into cotton from plasmid pMYC3006, leading to transformation event 3006-210-23 (Cry1Ac cotton). The chimeric *cry1Fa* gene with one *pat* gene were introduced into cotton cells from plasmid pAGM281, leading to transformation event 281-24-236 (Cry1Fa cotton). The two Bt cotton plants containing the single insecticidal traits were then crossed and repeatedly backcrossed individually to another elite US commercial cotton variety, PSC355 (the 'recurrent parent' in the breeding programme) to generate a third Bt cotton line, referred to as WideStrike™ cotton. Thus, the WideStrike™ cotton line contains two insecticidal genes, chimeric *cry1Fa* and chimeric *cry1Ac*, and two copies of the herbicide tolerance *pat* gene.

Transgene characterization of cottons expressing the cry1Fa *and*
cry1Ac *genes*
Southern blot analysis using probes from each gene and from regulatory sequences (promoters and termination region) demonstrated that the transformation event 3006-210-23 contained one intact copy of the chimeric *cry1Ac* and one intact copy of the herbicide tolerance gene *pat*. Transformation event 281-24-236 contained one intact copy of the chimeric *cry1Fa* and one intact copy of the *pat* gene plus an additional small fragment (Green, 2002; Green *et al.*, 2002a,b). The DNA sequences from each transformation event have also been confirmed by DNA sequence analysis (Song, 2002a,b). The gene constructs in the two transformation events have been shown to be stable over several generations, both by phenotypic (insecticidal and glufosinate ammonium tolerance) and Southern blot analysis, adhering to Mendelian inheritance ratios (Narva *et al.*, 2001a,b). The WideStrike™ cotton contains all of the introduced genetic material of transformation events 281-24-236 and 3006-210-23 (Green, 2002).

Transgene expression of the Cry1Fa and Cry1Ac proteins
Expression levels of Cry1Fa, Cry1Ac and Pat proteins in various plant tissues and in processed cottonseed fractions of these three cotton lines have been determined by enzyme-linked immunosorbent assay (ELISA) (OGTR, 2003b; Tables 4.1 to 4.3). The Cry1Fa protein was detected in Cry1Fa cotton and WideStrike™ cotton in all tissues and processed fractions, except nectar, meal and oil. The Cry1Ac protein in Cry1Ac cotton and WideStrike™ cotton was expressed at lower levels than the Cry1Fa protein in all tissues and fractions,

except pollen. The Pat protein was detected in most of the samples of Cry1Fa cotton and WideStrike™ cotton, but was not detectable in most tissues and fractions of Cry1Ac cotton.

Insecticidal efficacy of cottons expressing Cry1Fa and Cry1Ac proteins
The insecticidal efficacy of the Bt cotton lines (Cry1Fa, Cry1Ac and WideStrike™ cottons) has been assessed in field trials using either artificial or natural infestation with tobacco budworm (*Heliothis virescens*). One trial was conducted using artificial infestation with pink bollworm (*P. gossypiella*). Their performance was compared to non-Bt cotton with and without chemical spray control for the insect pest. Each of these cotton lines was found to perform better than non-Bt cotton that was not sprayed to control lepidopteran insect pests and at least as well as sprayed non-Bt cotton (OGTR, 2003b).

Later trials compared WideStrike™ cotton with sprayed non-Bt cotton, variety PSC355, and found that WideStrike™ cotton could provide effective control of major insect pests of cotton at various locations in the USA (Haile *et al.*, 2004), including tobacco budworm (*H. virescens*), cotton bollworm (*H. zea*) and pink bollworm (*P. gossypiella*). WideStrike™ cotton was also effective against beet armyworm (*Spodoptera exigua*), southern armyworm (*S. eridania*), fall armyworm (*S. frugiperda*), soybean looper (*Pseudoplusia includens*) and cabbage looper (*Trichoplusia ni*). This suggests that the WideStrike™ cotton can provide effective control of a wide range of lepidopteran pests of cotton and is expected to control *H. armigera* (OGTR, 2003b), the main pest of cotton in Vietnam.

4.6. Recommendations

The following is a summary of some of the main recommendations of this chapter:

- To minimize the requirements of risk assessment, transgene locus structure should have: one transgene locus per plant, only one copy of the transgene at the locus, no extra delivered DNA in the transgene locus, no disruption of an existing ORF or expression of other plant genes, no spurious ORFs and minimal rearrangements of genomic DNA flanking the integrated transgene.
- Marker genes should be eliminated from the final transgene or, if this is not possible, then the marker genes that are used should be ones for which risk assessment is easy to conduct.
- Characterization of transgene locus structure should include: number of transgene loci, the number of copies of the transgene at each locus, the location of the transgene and the sequence of each transgene locus, including the genomic DNA flanking the transgene sequences.
- Stable inheritance of the transgene locus should be determined at least into an F_4 or F_5 segregating generation.
- Develop molecular monitoring tools for Bt cotton events, e.g. PCR primers or Southern probes, and use them to check transgene structure after variety production.

- Design and develop a monitoring system for detection and identification of transgenic events in crops and food.
- Measure transgene expression in different tissues over the whole crop season, and in different environments, and use to assess efficacy against Vietnamese target pests and possible exposure of non-target organisms.
- Check whether the measurements of protein expression are accurate and repeatable for different plant tissues and varieties.
- Develop methods to measure transgene product in the environment, e.g. in roots and soil.
- Data on the expression of Bt proteins in several commercialized Bt cottons are summarized.

References

Abremski, K., Hoess, R. and Sternberg, N. (1983) Studies on the properties of P1 site-specific recombination: evidence for topologically unlinked products following recombination. *Cell* 32, 1301–1311.

Adamczyk, J.J. Jr and Meredith, W.R. Jr (2004) Genetic basis for variability of Cry1Ac expression among commercial transgenic *Bacillus thuringiensis* (*Bt*) cotton cultivars in the United States. *Journal of Cotton Science* 8, 17–23.

AgBios (2005) GM Crop Database: MON531/757/1076. Essential Biosafety Edition 2. http://www.agbios.com/dbase.php?action=Submit&evidx=15 (accessed 24 November 2007).

Akin, D.S., Stewart, S.D., Knighten, K.S. and Adamczyk, J.J. Jr (2002) Quantification of toxin levels in cottons expressing one and two insecticidal proteins of *Bacillus thuringiensis*. *Beltwide Cotton Conference Proceedings 2003*, Atlanta, Georgia.

Ammitzboll, H., Mikkelsen, T.N. and Jorgensen, R.B. (2005) Transgene expression and fitness of hybrids between GM oilseed rape and *Brassica rapa*. *Environmental Biosafety Research* 4(1), 3–12.

An, A.Q., McDowell, J.M., Huang, S., McKinney, E.C., Chambliss, S. and Meagher, R.B. (1996) Strong constitutive expression of the *Arabidopsis* ACT2/ACT8 actin subclass in vegetative tissue. *Plant Journal* 10(1), 107–121.

Andow, D. and Hilbeck, A. (2004) Science-based risk assessment for non-target effects of transgenic crops. *BioScience* 54, 637–649.

Assaad, F.F., Tucker, K.L. and Signer, E.R. (1993) Epigenetic repeat-induced gene silencing (RIGS) in *Arabidopsis*. *Plant Molecular Biology* 22, 1067–1085.

Barker, R.F., Idler, K.B., Thompson, D.V. and Kemp, J.D. (1983) Nucleotide sequence of the TDNA region from the *Agrobacterium tumefaciens* octopine Ti plasmid pTi15955. *Plant Molecular Biology* 2, 335–350.

Bayley, C., Morgan, M., Dale, E.C. and Ow, D.W. (1992) Exchange of gene activity in transgenic plants catalyzed by the Cre-lox site-specific recombination system. *Plant Molecular Biology* 18, 353–361.

Bender, J. (2004) DNA methylation and epigenetics. *Annual Review of Plant Biology* 55, 41–68.

Bevan, M. (1984) Binary *Agrobacterium* vectors for plant transformation. *Nucleic Acids Research* 12, 8711–8721.

Breitler, J.C., Labeyrie, A., Meynard, D., Legavre, T. and Guiderdoni, E. (2002) Efficient microprojectile bombardment-mediated transformation of rice using gene cassettes. *Theoretical and Applied Genetics* 104, 709–719.

Britt, A.B. and May, G.D. (2003) Re-engineering plant gene targeting. *Trends in Plant Science* 8(2), 90–95.

Butaye, K.M.J., Cammue, B.P.A., Delauré, S.L. and Bolle, M.F.C. de (2005) Approaches to minimize variation of transgene expression in plants. *Molecular Breeding* 16(1), 79–91.

Cellini, F., Chesson, A., Colquhoun, I., Constable, A., Davies, H.V., Engel, K.H., Gatehouse, A.M.R., Kärenlampi, S., Kok, E.J., Leguay, J.-J., Lehestranta, S., Noteborn, H.P.J.M., Pedersen, J. and Smith, M. (2004) Unintended effects and their detection in genetically modified crops. *Food and Chemical Toxicology* 42, 1089–1125.

Christensen, A.H., Sharrock, R.A. and Quail, P.H. (1992) Maize polyubiquitin genes: structure, thermal perturbation of expression and transcript splicing, and promoter activity following transfer to protoplasts by electroporation. *Plant Molecular Biology* 18, 675–689.

Christou, P. (1997) Rice transformation: bombardment. *Plant Molecular Biology* 35, 197–203.

Cloud, G.L., Minton, B. and Grymes, C. (2004) Field evaluations of VipCotTM for armyworm and looper control. *Proceedings of the 2004 Beltwide Cotton Conference*. National Cotton Council of America, Memphis, Tennessee.

Dale, E.C. and Ow, D.W. (1990) Intra- and intermolecular site-specific recombination in plants cells mediated by bacteriophage P1 recombinase. *Gene* 91, 79–85.

Dale, E.C. and Ow, D.W. (1991) Gene transfer with subsequent removal of the selection gene from the host genome. *Proceedings of the National Academy of Sciences of the USA* 88, 10558–10562.

De Block, M. and Debrouwer, D. (1991) Two T-DNAs co-transformed into *Brassica napus* by a double *Agrobacterium tumefaciens* infection are mainly integrated at the same locus. *Theoretical and Applied Genetics* 82, 257–263.

De Buck, S., Van Montagu, M. and Depicker, A. (2001) Transgene silencing of invertedly repeated transgenes is released upon deletion of one of the transgenes involved. *Plant Molecular Biology* 46(4), 433–445.

De Buck, S., Windels P., De Loose, M. and Depicker, A. (2004) Single-copy T-DNAs integrated at different positions in the *Arabidopsis* genome display uniform and comparable beta-glucuronidase accumulation levels. *Cell Molecular Life Science* 61(19–20), 2632–2645.

Depicker, A., Stachel, S., Dhaese, P., Zambryski, P. and Goodman, H.M. (1982) Nopaline synthase: transcript mapping and DNA sequence. *Journal of Molecular and Applied Genetics* 1, 561–573.

Depicker, A., Herman, L., Jacobs, A., Schell, J. and Van Montagu, M. (1985) Frequencies of simultaneous transformation with different T-DNAs and their relevance to the *Agrobacterium*/plant cell interaction. *Molecular and General Genetics* 201, 477–484.

Dong, H.Z. and Li, W.J. (2007) Variability of endotoxin expression in Bt transgenic cotton. *Journal of Agronomy and Crop Science* 193, 21–29.

Donovan, W.P., Donovan, J.C. and Engleman, J.T. (2001) Gene knockout demonstrates that *vip3A* contributes to the pathogenesis of *Bacillus thuringiensis* toward *Agrotis ipsilon* and *Spodoptera exigua*. *Journal of Invertebrate Pathology* 78, 45–51.

Eamens, A.L., Blanchard, C.L., Dennis, E.S. and Upadhyaya, N.M. (2004) A bidirectional gene trap construct suitable for T-DNA and *Ds*-mediated insertional mutagenesis in rice (*Oryza sativa* L.). *Plant Biotechnology Journal* 2, 367–380.

Estruch, J.J., Warren, G.W., Mullins, M.A., Nye, G.J., Craig, J.A. and Koziel, M.G. (1996) Vip3A, a novel *Bacillus thuringiensis* vegetative insecticidal protein with a wide spectrum of activities against lepidopteran insects. *Proceedings of the National Academy of Sciences of the USA* 93(11), 5389–5394.

European Commission (2002) Decision 2002/623/EC of 24 July 2002 of the European Commission establishing guidance notes supplementing annex II of Directive 2001/18/EC. http://www.biosafety.be/Menu/BiosEur2.html (accessed 24 November 2007).

EU (European Union) (2004) Summary Notification Information Format for the placing on the market (import) of 281-24-236/3006-210-23 cotton in accordance with Directive

2001/18/EC. European Union Joint Research Centre. http://gmoinfo.jrc.it/gmp_browse. aspx (accessed 7 March 2007).

Framond, A.J. de, Back, E.W., Chilton, W.S., Kayes, L. and Chilton, M.-D. (1986) Two unlinked T-DNAs can transform the same tobacco plant cell and segregate in the F1 generation. *Molecular and General Genetics* 202, 125–131.

FSANZ (Food Standards Australia New Zealand) (2004) 1 6-04 4 August 2004 Final Assessment Report Application A509 Food Derived from Insect-protected Cotton Line COT102. Food Standards Australia New Zealand. http://www.foodstandards.gov.au/_srcfiles/A509_GM_ Cotton_FAR_Final.pdf#search=%22A509%22 (accessed 2 January 2007).

Forsbach, A., Shubert, D., Lechtenberg, B., Gils, M. and Schmidt, R. (2003) A comprehensive characterisation of single-copy T-DNA insertions in the *Arabidopsis thaliana* genome. *Plant Molecular Biology* 52, 161–176.

Fu, X., Duc, L.T., Fontana, S., Bong, B.B., Tinjuangjun, P., Sudhakar, D., Twyman, R.M., Christou, P. and Kohli, A. (2000) Linear transgene constructs lacking vector backbone sequences generate low-copy-number transgenic plants with simple integration patterns. *Transgenic Research* 9, 11–19.

Gertz, J.M., Vencill, W.K. and Hill, N.S. (1999) Tolerance of transgenic soybean (Glycine max) to heat stress. In: *1999 Brighton Crop Protection Conference: Weeds*. Proceedings of an International Conference, Brighton, UK, 15–18 November 1999, Volume 3. British Crop Protection Council, Farnham, UK, pp. 835–840.

Green, S.B. (2002) Molecular characterisation of Cry1F (synpro)/Cry1Ac (synpro) stacked transgenic cotton line 281-24-236/3006-48-81. 010075.01. Dow AgroSciences LLC, Indianapolis, Indiana.

Green, S.B., Ernest, A.D. and Bevan, S.A. (2002a) Molecular characterisation of Cry1Ac (synpro) transgenic cotton event 3006-210-23. 010053.01. Dow AgroSciences LLC, Indianapolis, Indiana.

Green, S.B., Ernest, A.D. and Bevan, S.A. (2002b) Molecular characterization of Cry1F(synpro) transgenic cotton event 281-24-236. 010007.01. Dow AgroSciences LLC, Indianapolis, Indiana.

Greenplate, J.T., Mullins, W., Penn, S. and Embry, K. (2001) Cry1Ac levels in candidate commercial Bollgard(R) varieties as influenced by environment, variety, and plant age: 1999 gene equivalency field studies. *Beltwide Cotton Conference Proceedings 2001*. National Cotton Council of America, Memphis, Tennessee, 790 p.

Grossi-de-Sa, M.F., Lucena, W., Souza, M.L., Nepomuceno, A.L., Osir, E.O., Amugune, N., Hoa, Thi Thu Cuc, Hai, Truong Nam, Somers, D.A. and Romano, E. (2006) Transgene expression and locus structure of Bt cotton. In: Hilbeck, A., Andow, D.A. and Fontes, E. M.G. (eds) *Environmental Risk Assessment of Genetically Modified Organisms, Volume 2: Methodologies for Assessing Bt Cotton in Brazil*. CAB International, Wallingford, UK, pp. 93–107.

Gupta, V.V.S.R. and Watson, S. (2004) *Ecological Impacts of GM Cotton on Soil Biodiversity: Below Ground Production of Bt by GM Cotton and Bt Cotton Impacts on Soil Biological Processes*. CSIRO Land and Water, Australia. http://www.deh.gov.au/settlements/ publications/biotechnology/gm-cotton/index.html (accessed 12 September 2006).

Gupta, V.V.S.R., Roberts, G.N., Neate, S.M., McClure, S.G., Crisp, P. and Watson, S.K. (2002) Impact of Bt-cotton on biological processes in Australian soils. In: Akhurst, R.J., Beard, C.E. and Hughes, P.A. (eds) *Biotechnology of Bacillus thuringiensis and its Environmental Impact*. CSIRO Entomology, Canberra, pp. 191–194.

Haile, F.J., Braxton, L.B., Flora, E.A., Haygood, B., Huckaba, R.M., Pellow, J.W., Langston, V.B., Lassiter, R.B., Richardson, J.M. and Richburg, J.S. (2004) Efficacy of WideStrike cotton against non-heliothine Lepidopteran insects. *Proceedings of the 2004 Beltwide Cotton Conference*. National Cotton Council of America, Memphis, Tennessee.

Hansen, G. and Wright, M. (1999) Recent advances in the transformation of plants. *Trends in Plant Science* 4, 226–231.

Hashimoto, W., Momma, K., Katsube, T., Ohkawa, Y., Ishige, T., Kito, M., Utsumi, S. and Murata, K. (1999a) Safety assessment of genetically engineered potatoes with designed soybean glycinin: compositional analyses of the potato tubers and digestibility of the newly expressed protein in transgenic potatoes. *Journal of the Science of Food and Agriculture* 79, 1607–1612.

Hashimoto, W., Momma, K., Yoon, H.-J., Ozawa, S., Ohkawa, Y., Ishige, T., Kito, M., Utsumi, S. and Murata, K. (1999b) Safety assessment of transgenic potatoes with soybean glycinin by feeding studies in rats. *Bioscience Biotechnology and Biochemistry* 63, 1942–1946.

Hauser, T.P., Jørgensen, R.B. and Østergård, H. (1998) Fitness of backcross and F_2 hybrids between weedy *Brassica rapa* and oilseed rape (*B. napus*). *Heredity* 81, 436–443.

Hei, Y., Komari, T. and Kubo, T. (1997) Transformation of rice mediated by *Agrobacterium tumefaciens*. *Plant Molecular Biology* 35, 205–218.

Hernández, M., Pla, M., Esteve, T., Prat, S., Puigdomèmech, P. and Ferrando, A. (2003) A specific real-time quantitative PCR detection system for event MON810 in maize YieldGard® based on the 3′-transgene integration sequence. *Transgenic Research* 12(2), 179–189.

Hoa, T.T.C., Bong, B.B., Huq, E. and Hodges, T.K. (2002) Cre/*lox* site-specific recombination controls the excision of a transgene from the rice genome. *Theoretical and Applied Genetics* 104, 518–525.

Hoa, T.T.C., AlBabili, S., Schaub, P., Potrykus, I. and Beyer, P. (2003) Golden indica japonica rice lines amendable to registration. *Plant Physiology* 133, 161–169.

Hoess, R., Wierzbicki, A. and Abremski, K. (1986) The role of the *lox*P space region in P1 site-specific recombination. *Nucleic Acids Research* 14, 2287–2300.

Hooykaas, P.J. and Schilperoort, R.A. (1992). *Agrobacterium* and plant genetic engineering. *Plant Molecular Biology* 19, 15–38.

Jeong, D.H., An, S., Kang, H.G., Moon, S., Han, J.J., Park, S., Lee, H.S., An, K. and An, G. (2002) T-DNA insertional mutagenesis for activation tagging in rice. *Plant Physiology* 130, 1636–1644.

Jiang, L.J., Duan, L.S., Tian, X.O., Wang, B.M., Zhang, H.F., Zhang, M.C. and Li, Z.H. (2006) NaCl salinity stress decreased *Bacillus thuringiensis* (Bt) protein content of transgenic Bt cotton (*Gossypium hirsutum* L.) seedlings. *Environmental and Experimental Botany* 55(3), 315–320.

Joersbo, M., Donaldson, I., Kreiberg, J., Peterson, S.G., Brunstedt, J. and Okkels, F.T. (1998) Analysis of mannose selection used for transformation of sugar beet. *Molecular Breeding* 4, 111–117.

Kleter, G.A., Peijnenburg, A.A.C.M. and Aarts, H.J.M. (2005) Health considerations regarding horizontal transfer of microbial transgenes present in genetically modified crops. *Journal of Biomedicine and Biotechnology* 4, 326–352.

Kohli, A., Leech, M., Vain, F., Laurie, D. and Christou, P. (1998) Transgene organization in rice engineered through direct DNA transfer supports a two-phase integration mechanism mediated by the establishment of integration hot spots. *Proceedings of the National Academy of Sciences of the USA* 95, 7203–7208.

Komari, T., Halperin, W. and Nester, E.W. (1986) Physical and functional map of supervirulent *Agrobacterium tumefaciens* tumor-inducing plasmid pTiBo542. *Journal of Bacteriology* 166, 88–94.

Kononov, M.E., Bassuner, B. and Gelvin, S.B. (1997) Integration of T-DNA binary vector 'backbone' sequences into the tobacco genome: evidence for multiple complex patterns of integration. *Plant Journal* 11, 945–957.

Kranthi, K.R., Naidu, S.R., Dhawad, C.S., Tatwawadi, A., Mate, K., Patil, E., Bharose, A.A., Behere, G.T., Wadaskar, R.M. and Kranthi, S. (2005) Temporal and intra-plant variability of

Cry1Ac expression in Bt cotton and its influence on the survival of the cotton bollworm, *Helicoverpa armigera* (Hübner) (Noctuidae: Lepidoptera). *Current Science* 89(2), 291–298.

Kuiper, H.A., Kleter, G.A., Noteborn, H.P.J.M. and Kok, E.J. (2001) Assessment of the food safety issues related to genetically modified foods. *Plant Journal* 27(6), 503–528.

Labra, M., Savini, C., Bracale, M., Pelucchi, N., Colombo, L., Bardini, M. and Sala, F. (2001) Genomic changes in transgenic rice (*Oryza sativa* L.) plants produced by infecting calli with *Agrobacterium tumefaciens*. *Plant Cell Reports* 20, 325–330.

Latham, J., Wilson, A.K. and Steinbrecher, R.A. (2006) The mutational consequences of plant transformation. *Journal of Biomedicine and Biotechnology* Article ID 25376, 1–7.

Lee, M.K., Walters, F.S., Hart, H., Palekar, N. and Chen, J.S. (2003) The mode of action of the *Bacillus thuringiensis* vegetative insecticidal protein VIP3A differs from that of Cry1Ab delta-endotoxin. *Applied and Environmental Microbiology* 69(8), 4648–4657.

Liao, C., Heckel, D.G. and Akhurst, R. (2002) Toxicity of *Bacillus thuringiensis* insecticidal proteins for *Helicoverpa armigera* and *Helicoverpa punctigera* (Lepidoptera: Noctuidae), major pests of cotton. *Journal of Invertebrate Pathology* 80(1), 55–63.

Llewellyn, D.J., Mares, C.L. and Fitt, G.P. (2007) Field performance and seasonal changes in the efficacy against *Helicoverpa armigera* (Hübner) of transgenic cotton (VipCot) expressing the insecticidal protein Vip3A. *Agricultural and Forest Entomology* 9(2), 93–101.

Loc, T.N., Tinjuangjun, P., Gatehouse, A.M.R., Christou, P. and Gatehouse, J.A. (2002) Linear transgene constructs lacking vector backbone sequences generate transgenic rice plants which accumulate higher levels of proteins conferring insect resistance. *Molecular Breeding* 9, 231–244.

Lu, H., Zhou, X., Gong, Z. and Upadhyaya, N. (2001) Generation of selectable marker-free transgenic rice using a double right border. *Australian Journal of Plant Physiology* 28, 241–248.

Lucca, P., Xu, Y. and Potrykus, I. (2001) Efficient selection and regeneration of transgenic rice plants with mannose as selective agent. *Molecular Breeding* 7, 43–49.

Luthy, P. and Ebersold, H.R. (1981). *Bacillus thuringiensis* delta-endotoxin: histopathology and molecular mode of action. In: Davidson, E.W. (ed.) *Pathogenesis of Invertebrate Microbial Disease*. Allenheld, Osmun and Co., Totowa, New Jersey, pp. 235–267.

McKnight, T.D., Lillis, M.T. and Simpson, R.B. (1987) Segregation of genes transferred to one plant cell from two separate *Agrobacterium* strains. *Plant Molecular Biology* 8, 439–445.

Makarevitch, I., Svitashev, S.K. and Somers, D.A. (2003) Complete sequence analysis of transgene loci from plants transformed via microprojectile bombardment. *Plant Molecular Biology* 52, 421–432.

Martineau, B., Voelker, T.A. and Sanders, R.A. (1994) On defining T-DNA. *Plant Cell* 6, 1032–1033.

Matzke, M.A. and Matzke, A.J.M. (1995) How and why do plants inactivate homologous (trans) genes? *Plant Physiology* 107, 679–685.

Mercer, K.L., Wyse, D.L. and Shaw, R.G. (2006) Effects of competition on fitness of wild and crop-wild hybrid sunflower from a diversity of wild populations and crop lines. *Evolution* 60(10), 2044–2055.

Mette, M.F., van der Winden, J., Matzke, M.A. and Matzke, A.J.M. (1999) Production of aberrant promoter transcripts contributes to methylation and silencing of unlinked homologous promoters *in trans*. *EMBO Journal* 18, 241–248.

Miles, J.S. and Guest, J.R. (1984) Nucleotide sequence and transcriptional start point of the phosphomannose isomerase gene (*pmi*) of *Escherichia coli*. *Gene* 32, 41–48.

Murray, F., Llewellyn, D., McFadden, H., Last, D., Dennis, E.S. and Peacock, W.J. (1999). Expression of the *Talaromyces flavus* glucose oxidase gene in cotton and tobacco reduces fungal infection, but is also phytotoxic. *Molecular Breeding* 5, 219–232.

Narva, K.A., Palta, A. and Pellow, J.W. (2001a) Product characterisation data for *Bacillus thur-ingiensis* var. *aizawai* Cry1F (synpro) insect control protein as expressed in cotton. GH-C 5304. Dow AgroSciences LLC, San Diego, California.

Narva, K.A., Palta, A. and Pellow, J.W. (2001b) Product characterisation data for *Bacillus thur-ingiensis* var. *kurstaki* Cry1Ac (synpro) insect control protein as expressed in cotton. GH-C 5303. Dow AgroSciences LLC, San Diego, California.

Ni, M., Cui, D., Einstein, J., Narasimhulu, S., Vergara, C.E. and Gelvin, S. (1995). Strength and tissue specificity of chimeric promoters derived from the octopine and mannopine synthase genes. *The Plant Journal* 7, 661–676.

NRC (National Research Council) (2002) *Environmental Effects of Transgenic Plants: The Scope and Adequacy of Regulation*. National Academies Press, Washington, DC.

NRC (National Research Council) (2004) *Biological Confinement of Genetically Engineered Organisms*. National Academies Press, Washington, DC.

Oard, J., Cohn, M.A., Linscoombe, S., Gealy, D.R. and Gravois, K. (2000) Field evaluation of seed production, shattering, and dormancy in hybrid populations of transgenic rice (*Oryza sativa*) and the weed, red rice (*Oryza sativa*). *Plant Science* 157(1), 13–22.

OGTR (2002) Risk Assessment and Risk Management Plan. Commercial release of insecticidal (INGARD event 531) cotton. DIR 022/2002. Office of the Gene Technology Regulator, Australian Government. http://www.ogtr.gov.au/ir/index.htm (accessed 7 March 2007).

OGTR (2003a) Risk Assessment and Risk Management Plan. Field trial of genetically modified cotton (*Gossypium hirsutum*) expressing an insecticidal gene (*vip3A*). DIR 034/2003. Office of the Gene Technology Regulator, Australian Government. http://www.ogtr.gov.au/ir/index.htm (accessed 7 March 2007).

OGTR (2003b) Risk Assessment and Risk Management Plan. Agronomic assessment and seed increase of GM cottons expressing insecticidal genes (*cry1Fa* and *cry1Ac*) from *Bacillus thuringiensis*, DIR 044/2003. Office of the Gene Technology Regulator, Australian Government. http://www.ogtr.gov.au/ir/index.htm (accessed 7 March 2007).

Ohba, T., Yoshioka, Y., Machida, C. and Machida, Y. (1995) DNA rearrangement associated with the integration of T-DNA in tobacco: an example for multiple duplications of DNA around the integration target. *The Plant Journal* 7(1), 157–164.

Olsen, K.M. and Daly, J.C. (2000) Plant-toxin interactions in transgenic Bt cotton and their effect on mortality of *Helicoverpa armigera* (Lepidoptera: Noctuidae). *Journal of Economic Entomology* 93(4), 1293–1299.

Olsen, K.M., Daly, J.C., Finnegan, E.J. and Mahon, R.J. (2005) Changes in Cry1Ac Bt trans-genic cotton in response to two environmental factors: temperature and insect damage. *Journal of Economic Entomology* 98(4), 1382–1390.

Pawlowski, W.P. and Somers, D.A. (1996) Transgenic inheritance in plants genetically engineered using microprojectile bombardment. *Molecular Biotechnology* 6, 17–30.

Pawlowski, W.P., Torbert, K.A., Rines, H.W. and Somers, D.A. (1998) Irregular patterns of transgene silencing in allohexaploid oat. *Plant Molecular Biology* 38, 597–607.

Puchta, H. (2003) Marker-free transgenic plants. *Plant Cell, Tissue and Organ Culture* 74(2), 123–143.

Qu, Q. and Takaiwa, F. (2004) Evaluation of tissue specificity and expression strength of rice seed component gene promoters in transgenic rice. *Plant Biotechnology Journal* 2, 113–125.

Rang, A., Linke, B. and Jansen, B. (2005) Detection of RNA variants transcribed from the trans-gene in Roundup Ready soybean. *European Food Research and Technology* 220, 438–443.

Rochester, I.J. (2006) Effects of genotype, edaphic, environmental conditions, and agronomic practices on Cry1Ac protein expression in transgenic cotton. *Journal of Cotton Science* 10(4), 252–262.

Ruf, S., Karcher, D. and Bock, R. (2007) Determining the transgene containment level provided by chloroplast transformation. *Proceedings of the National Academy of Sciences of the USA* 104(17), 6998–7002.

Russell, S.H., Hoopes, J.L. and Odell, J.T. (1992) Directed excision of a transgene from the plant genome. *Molecular and General Genetics* 234, 49–59.

Sala, F., Arencibia, A., Castiglione, S., Yifan, H., Labra, M., Savini, C., Bracale, M. and Pelucchi, N. (2000) Somaclonal variation in transgenic plants. *Acta Horticulturae* 530, 411–419.

Saxena, D., Stewart, C.N., Altosaar, I., Shu, Q. and Stotzky, G. (2004) Larvicidal Cry proteins from *Bacillus thuringiensis* are released in root exudates of transgenic *B. thuringiensis* corn, potato, and rice but not of *B. thuringiensis* canola, cotton, and tobacco. *Plant Physiology and Biochemistry* 42(5), 383–387.

SBB (2003) Guidelines for molecular characterization of GM plants: Part C Commercial release. Division of Biosafety and Biotechnology (SBB), Scientific Institute of Public Health, Federal Public Health Service, Belgium. http://www.biosafety.be/gmcropff/EN/TP/partC/GuideMGC_PartB_C.htm (accessed 24 November 2007).

Schocher, R.J., Shillito, R.D., Saul, M.W., Paszkowski, J. and Potrykus, I. (1986) Co-transformation of unlinked foreign genes into plants by direct gene transfer. *Bio/Technology* 4, 1093–1096.

Selvapandiyan, A., Arora, N., Rajagopal, R., Jalali, S.K., Venkatesan, T., Singh, S.P. and Bhatnagar, R.K. (2001) Toxicity analysis of N- and C-terminus-deleted vegetative insecticidal protein from *Bacillus thuringiensis*. *Applied and Environmental Microbiology* 67(12), 5855–5858.

Shewmaker, C.K., Sheely, J.A., Daley, M., Colburn, S. and Ke, D.Y. (1999) Seed-specific over-expression of phytoene synthase: increase in carotenoids and other metabolic effects. *The Plant Journal* 20(4), 401–412.

Simpson, R.B., Spielmann, A., Margossian, L. and McKnight, T.D. (1986) A disarmed binary vector from *Agrobacterium tumefaciens* functions in *Agrobacterium rhizogenes*. *Plant Molecular Biology* 6, 403–415.

Snow, A., Pilson, D., Rieseberg, L.H., Paulsen, M.J., Pleskac, N., Reagon, M.R., Wolf, D.E. and Selbo, S.M. (2003) A Bt transgene reduced herbivory and enhances fecundity in wild sunflowers. *Ecological Applications* 13(2), 279–286.

Somers, D.A. and Makarevitch, I. (2004) Transgene integration in plants: poking or patching holes in promiscuous genomes? *Current Opinion in Biotechnology* 15(2), 126–131.

Song, P. (2002a) Cloning and characterisation of DNA sequences in the insert and flanking border regions of Bt Cry1Ac cotton 3006-210-23. GH-C 5522. Dow AgroSciences LLC, Indianapolis, Indiana.

Song, P. (2002b) Cloning and characterisation of DNA sequences in the insert and flanking border regions of Bt Cry1F cotton 281-24-236. GH-C 5529. Dow AgroSciences LLC, Indianapolis, Indiana.

Stegemann, S., Hartmann, S., Ruf, S. and Bock, R. (2003) High-frequency gene transfer from the chloroplast genome to the nucleus. *Proceedings of the National Academy of Sciences of the USA* 100(15), 8828–8833.

Stuurman, J., de Vroomen, M.J., Nijkamp, H.J.J. and Haaren, M.J.J. (1996) Single-site manipulation of tomato chromosomes *in vitro* and *in vivo* using Cre-lox site-specific recombination. *Plant Molecular Biology* 32, 901–913.

Svitashev, S.K. and Somers, D.A. (2001) Genomic interspersions determine the size and complexity of transgene loci in transgenic plants produced by micro-projectile bombardment. *Genome* 44, 691–697.

Svitashev, S.K. and Somers, D.A. (2002) Characterization of transgene loci in plants using FISH: a picture is worth a thousand words. *Plant Cell, Tissue and Organic Culture* 69(3), 205–214.

Svitashev, S.K., Pawlowski, W.P., Makarevitch, I., Plank, D.W. and Somers, D.A. (2002) Complex transgene locus structures implicate multiple mechanisms for transgene rearrangement. *The Plant Journal* 32, 433–445.

UK ACRE (2002) Guidance on best practice for the presentation and use of molecular data in submissions to the Advisory Committee on Releases to the Environment. UK Advisory Committee on Releases into the Environment (ACRE), DEFRA, British Government, London. http://www.defra.gov.uk/environment/acre/molecdata/pdf/acre_mdr_guidance. pdf (accessed 24 November 2007).

US APHIS (2000) Petition for the determination of non-regulated status: Bollgard II cotton event 15985 (*Gossypium hirsutum* L.) producing the Cry2Ab insect control protein derived from *Bacillus thuringiensis* subsp. *kurstaki*. US Animal and Plant Health Inspection Service, Department of Agriculture. http://www.aphis.usda.gov/brs/not_reg.html (accessed 7 March 2007).

US APHIS (2003) Petition for the determination of non-regulated status: Lepidopteran insect protected VIP3A cotton transformation event COT102. US Animal and Plant Health Inspection Service, Department of Agriculture. http://www.aphis.usda.gov/brs/not_reg. html (accessed 7 March 2007).

US APHIS (2004a) Petition for the determination of non-regulated status: *Bt* Cry1F insect-resistant cotton event 281-24-236. US Animal and Plant Health Inspection Service. http://www.aphis. usda.gov/brs/not_reg.html (accessed 7 March 2007).

US APHIS (2004b) Petition for the determination of non-regulated status: *Bt* Cry1Ac insect-resistant cotton event 3006-210-23. US Animal and Plant Health Inspection Service. http://www.aphis.usda.gov/brs/not_reg.html (accessed 7 March 2007).

US EPA (2001) Biopesticides registration action document (BRAD) – *Bacillus thuringiensis* plant-incorporated protectants. 10/16/2001. US Environmental Protection Agency. http:// www.epa.gov/pesticides/biopesticides/pips/bt_brad.htm (accessed 7 March 2007).

Verhalen, L.M., Greenhagen, B.E. and Thacker, R.W. (2003) Lint yield, lint percentage, and fiber quality response in Bollgard, Roundup Ready, and Bollgard/Roundup Ready cotton. *Journal of Cotton Science* 7(2), 23–38.

Wan, P., Zhang, Y.-J., Wu, K.-M. and Huang, M.-S. (2005) Seasonal expression profiles of insecticidal protein and control efficacy against *Helicoverpa armigera* for Bt cotton in the Yangtze River valley of China. *Journal of Economic Entomology* 98(1), 195–201.

Wang, K., Herrera-Estrella, L., Van Montagu, M. and Zambryski, P. (1984) Right 25 bp terminus sequence of the nopaline T-DNA is essential for and determines direction of DNA transfer from *Agrobacterium* to the plant genome. *Cell* 38, 455–462.

Weigel, D., Ahn, J.H., Blazquez, M.A., Borevitz, J.O., Christensen, S.K., Fankhauser, C., Ferrandiz, C., Kardailsky, I., Malancharuvil, E.J., Neff, M.M., Nguyen, J.T., Sato, S., Wang, Z.Y., Xia, Y., Dixon, R.A., Harrison, M.J., Lamb, C.J., Yanofsky, M.F. and Chory, J. (2000) Activation tagging in *Arabidopsis*. *Plant Physiology* 122, 1003–1013.

Wenck, A., Czako, M., Kanevski, I. and Marton, L. (1997) Frequent collinear long transfer of DNA inclusive of the whole binary vector during *Agrobacterium*-mediated transformation. *Plant Molecular Biology* 34, 913–922.

Wilson, K., Long, D., Swinburne, J. and Coupland, G. (1996) A dissociation insertion causes a semidominant mutation that increases expression of TINY, an *Arabidopsis* gene related to APETALA2. *Plant Cell* 8(4), 659–671.

Wilson, A., Latham, J. and Steinbrecher, R. (2004) *Genome Scrambling – Myth or Reality? Transformation-induced Mutations in Transgenic Crop Plants*. Technical Report – October 2004. EcoNexus, Brighton, UK.

Windels, P., Taverniers, I., Depicker, A., Van Bockstaele, E. and De Loose, M. (2001) Characterisation of the Roundup Ready soybean insert. *European Food Research and Technology* 213(2), 107–112.

Wright, W., Dawson, J., Dunder, E., Suittie, J., Reed, J., Kramer, C., Chang, Y., Novitzky, R., Wang, H. and Artim-Moore, L. (2001) Efficient biolistic transformation of maize (*Zea mays* L.) and wheat (*Triticum aestivum* L.) using the phosphomannose isomerase gene, *pmi*, as the selectable marker. *Plant Cell Reports* 20, 429–436.

Ye, F. and Signer, E.R. (1996) RIGS (repeat-induced gene silencing) in *Arabidopsis* is transcriptional and alters chromatin configuration. *Proceedings of the National Academy of Sciences of the USA* 93, 10881–10886.

Ye, X., Al-Babili, S., Klöti, A., Lucca, P., Beyer, P. and Potrykus, I. (2000) Engineering the provitamin A (beta-carotene) biosynthetic pathway into (carotenoid-free) rice endosperm. *Science* 287, 303–305.

Yoder, J.I. and Goldsbrough, A.P. (1994) Transformation systems for generating marker-free transgenic plants. *Bio/Technology* 12, 263–267.

Yu, C.G., Mullins, M.A., Warren, G.W., Koziel, M.G. and Estruch, J.J. (1997) The *Bacillus thuringiensis* vegetative insecticidal protein Vip3A lyses midgut epithelium cells of susceptible insects. *Applied Environmental Microbiology* 63, 532–536.

Zakharov, A., Giersberg, M., Hosein, F., Melzer, M., Müntz, K. and Saalbach, I. (2004) Seed-specific promoters direct gene expression in non-seed tissue. *Journal of Experimental Botany* 55, 1463–1471.

Zambryski, P. (1992) Chronicles from the *Agrobacterium*-plant cell DNA transfer story. *Annual Review of Plant Physiology and Plant Molecular Biology* 43, 465–490.

Zhang, J., Cai, L., Cheng, J., Mao, H., Fan, X., Meng, Z., Chan, K.M., Zhang, H., Qi, J. and Ji, L. (2008) Transgene integration and organization in cotton (*Gossypium hirsutum* L.) genome. *Transgenic Research* (online early) 10.1007/s11248-007-9101-3.

5 Non-target and Biological Diversity Risk Assessment

Angelika Hilbeck, Salvatore Arpaia, A. Nicholas
E. Birch, Yolanda Chen, Eliana M.G. Fontes,
Andreas Lang, Lê Thị Thu Hồng, Gabor L. Lövei,
Barbara Manachini, Nguyễn Thị Thu Cúc, Nguyễn Văn
Huỳnh, Nguyễn Văn Tuất, Phạm Văn Lầm, Pham Van
Toan, Carmen S.S. Pires, Edison R. Sujii, Tràc Khương
Lai, Evelyn Underwood, Ron E. Wheatley, Lewis
J. Wilson, Claudia Zwahlen and David A. Andow

Biological diversity supports and comprises ecological functions that are vital for natural ecosystems and crop production in sustainable agricultural systems. It is comprised of the variation among genotypes, species and ecosystems (CBD, 1992). Changes in biological diversity can have an adverse effect on natural and agricultural ecosystems. Adverse environmental effects are undesirable changes to valued structural or functional characteristics of ecosystems or their components (US EPA, 1998). Adverse effects on crop production could include yield loss, yield instability, reduced crop quality, increased pest management activity, more chemical pesticide use and additional labour to compensate for the loss of ecological services (Andow and Hilbeck, 2004). Failure of biological control can lead to pest outbreaks and yield loss, and reduced pollination services can reduce yield. Moreover, resource-poor farmers may be more vulnerable to such adverse effects, as the chemicals or equipment necessary for remedial action are considerably more costly to them. Adverse effects on non-monetary species and services, such as endangered species or species of cultural significance (religious, ritual, medicinal, traditional purposes), or the food and feed basis of subsistence agriculture, are also important. Different societies may attach different values to these.

In any crop field, many thousands of species occur. Not all have the same importance for ecosystem functioning, nor should all be tested for the potential impacts of transgenic crops. Only a few can be selected and tested in a manner that yields enough information for risk evaluation. Risks to biological diversity should be assessed by examining the effects of transgenic plants on individual species or specific ecological processes. Thus, the selection of species and processes should be done in such a way that risk to all of biological diversity can be considered.

Several risk assessment models have been suggested for selecting species or ecological processes for non-target testing of insecticidal transgenic crops (Cowgill and Atkinson, 2003; Dutton *et al.*, 2003; Andow and Hilbeck, 2004; Poppy and Sutherland, 2004; Garcia-Alonso *et al.*, 2006; Romeis *et al.*, 2006). Some of them are modelled after a pesticide assessment approach (e.g. Dutton *et al.*, 2003) and propose that the selection of appropriate arthropod natural enemies is done according to three criteria: (i) the species' economic and ecological importance for natural pest regulation in the crop; (ii) experimental evidence of exposure of the species to the insecticidal protein; and (iii) information on the specificity of the insecticidal protein. The first criterion is obviously important. The second criterion may be more useful when developing risk hypotheses (see below). The third criterion is useful for selecting test organisms known to be susceptible to the insecticidal protein, but may exclude organisms that are believed not to be susceptible yet suffer negative impacts.

Specificity of the insecticidal protein is often poorly understood because only a small number of species are tested for susceptibility (Lövei *et al.*, submitted). Frequently, toxin specificity is inferred from the known mode of action of the toxin, but such research on modes of action, e.g. of Bt toxins, is almost always conducted on known susceptible species targeted for control by these insecticidal proteins (Hilbeck and Schmidt, 2006). Hilbeck and Schmidt (2006) illustrated several additional uncertainties regarding the specificity of Bt toxins, one of the most studied insecticidal proteins. Many more uncertainties will be associated with less studied novel transgene compounds. Furthermore, no commonly agreed definition of 'specificity' and 'susceptibility' exists. Susceptibility is often defined narrowly as 'can be killed within a short time' (this can be 2 days in the case of Bt toxins) when feeding on a high concentration of toxin (acute toxicity, i.e. one or two short feeding events). A broader, more ecologically encompassing definition of susceptibility would be 'deleterious effect during its life cycle'. This definition is superior because it includes sublethal effects that could accumulate over a longer period of time, such as the entire juvenile stage. Such effects can be ecologically as disruptive as acute toxicity. For instance, a non-target parasitoid population may be able to recover quickly, with little loss of its biological control function, from a short-term acute toxicity that caused a 30% population reduction, but if its development were delayed by the toxin, it may no longer be synchronized with its host and biological control may be lost almost entirely. Moreover, acute toxicity often does not predict sublethal effects reliably (Elmegaard and Jagers op Akkerhuis, 2000; Suter, 2007). Such adverse effects would go unnoticed if decisions for excluding potential non-target species were based on a definition of 'susceptibility' relying on known acute toxicity.

5.1. Non-target Risk Assessment Model

Scientists of the GMO ERA Project have developed a non-target risk assessment model (Andow and Hilbeck, 2004; Birch *et al.*, 2004; Hilbeck *et al.*,

2006) and, in this volume, have applied it to the case study of Bt cotton in Vietnam. This model uses a selection process to identify the species and ecosystem process in the receiving environment (where the transgenic plant is to be grown) that are most associated with a potential risk from the given transgenic plant, and then uses standard risk assessment methods to focus the assessment on specified potential risks that can be assessed experimentally (Fig. 5.1). The procedure begins by identifying the most important ecological functions and values that are associated with possible adverse effects of the transgenic plant in the recipient agricultural ecosystem. These functions and values enable prioritization and selection of relevant and important functional groups (Step 1). In Step 2, taxa or ecological processes within functional groups are prioritized and selected. This is done in several substeps: first, listing the taxa or processes that belong to the functional group, next using a 'Selection Matrix' (to be detailed later) to prioritize them according to their association with the crop and their functional significance in the crop, associated crops and natural areas, and then selecting a small subset for use in risk assessment. Steps 1 and 2 yield a significantly reduced list of taxa or ecosystem processes that are critical for the execution of an important function in the given cropping system and environment and where a possible adverse effect can be expected to inflict significant damage. In Step 3, information on toxin specificity, transgene concentration in specific plant tissues over the growing season, ecological interactions of the selected species or processes and published experimental evidence is used to generate and prioritize plausible risk hypotheses for the selected taxa and processes. This is done in several substeps: (i) identifying potential direct and indirect exposure pathways to the transgenic crop and its transgene products; (ii) identifying potential adverse effects pathways leading from exposure to an adverse effect; (iii) combining the exposure and adverse effects pathways into plausible risk hypotheses; and (iv) prioritizing and selecting plausible risk hypotheses to focus the risk assessment. As a result, critical risk hypotheses are identified for these prioritized species. In Step 4, an analysis plan is developed and experiments are designed that aim at verifying or refuting the selected risk hypotheses. By focusing on the species and processes most likely to be associated with the greatest potential adverse effects and then focusing on the most likely risk hypotheses, this method assesses the worst-case risk scenarios for the crop, transgene and receiving environment. These scenarios are the ones that regulatory bodies and stakeholders will most want to avoid or minimize. Other risks to biological diversity that have not been tested explicitly are likely to be smaller than these worst-case scenarios. The following text explains these steps and associated concepts in detail.

5.2. Step 1: Identifying Relevant Functional Groups of Biological Diversity Associated with Adverse Effects (Fig. 5.1)

A functional group is a group of organisms that carries out the same ecological function. They may belong to the same feeding guild, have the same impact on the ecosystem structure, or have the same pest control function, but belong

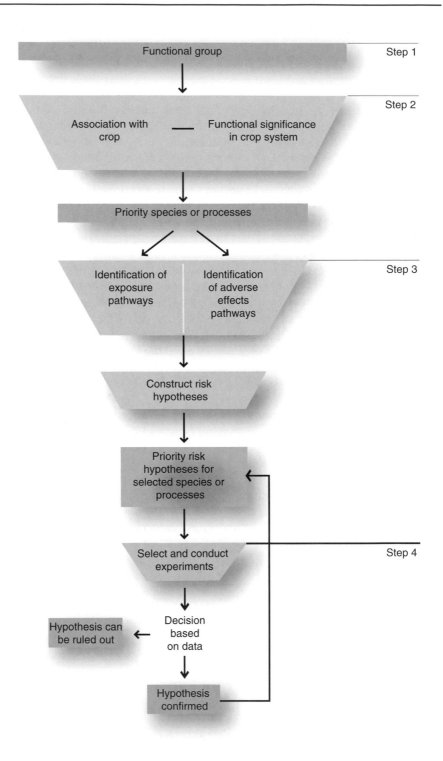

Fig. 5.1. Schematic flow diagram of GMO ERA method for non-target and biodiversity environmental risk assessment of transgenic plants.

to different taxa. The different taxa or processes in the functional group contribute to maintain the ecological function. We can connect ecological functions to possible adverse environmental effects (if affected by the GMO). As the adverse effects may differ for each transgenic crop, the functional groups of concern may also differ. The identification of relevant functional groups that should be considered in environmental risk assessment take into account the components that constitute a 'case', as outlined in the Cartagena Protocol (Annex III, CBD, 2000): (i) the novel trait and intended effect of the transgenic crop, (ii) characteristics of the crop, and (iii) the receiving environment, including changes in farming practices due to use of the transgenic crop plant. A broad perspective at this initial stage is essential in order to avoid overlooking critical functional groups and thereby excluding them entirely from the following risk assessment.

Kinds of adverse effects

Several kinds of potential adverse effects associated with non-target and biological diversity have been identified for transgenic plants. These are: (i) adverse effects on crop production (e.g. Morandin and Winston, 2005); (ii) reduced soil health or quality (DEFRA, 2004); (iii) reduced value of economic activities not related directly to the production of the crop in question, such as honey or silk production (e.g. Li *et al.*, 2002); (iv) reduced cultural value (spiritual, aesthetic, etc.) (e.g. Losey *et al.*, 1999); (v) increased conservation concern (biodiversity loss) (e.g. Firbank, 2003); (vi) reduced environmental quality (e.g. impaired ecosystem services or reduced sustainability); and (vii) increased human disease via environmental change (WHO, 2000). All should be considered, but only some of these are likely to be relevant for any transgenic plant.

Kind of novel trait and intended effect

The choice of functional groups is guided by the intended effect of the transgenic plant, as determined by the kind of transgenic trait. Currently, the majority of commercial transgenic crops worldwide have been developed with only two kinds of traits: resistance against certain insect pests (lepidopteran pests or coleopteran pests) using modified endotoxins from *Bacillus thuringiensis* (Bt toxins), and tolerance to broad-spectrum herbicides (mainly glyphosate and glufosinate ammonium). These traits are sometimes combined in one crop variety, known as a stacked or multiple trait variety.

The intended effect of these traits may entail changes in crop management practices, which can have environmental impacts within the crop ecosystem and in surrounding habitats. Herbicide tolerance will change the herbicide regime, which will alter the weed community in the crop, and which may affect species strongly associated with those weeds (Haughton *et al.*, 2003). Insect-resistant transgenic crops may reduce or shift insecticide use and increase yields (Chapter 1, this volume). This may bring environmental, economic and health benefits. However, whether these benefits are actually achieved depends on how much insecticide is usually used on that crop, the farmers' perception of the efficacy of the transgenic crop and the pest pressure from other pests that are not controlled by the transgenic crop (Pemsl *et al.*, 2005; Yang *et al.*, 2005; Marvier *et al.*, 2007).

Recipient crop biology
The relevant functional groups to be included in risk assessment will vary with crop species and biology and factors limiting crop production. For example, if the crop requires insect pollinators to ensure or enhance yields, then the function of pollinators for crop production is important. For wind pollinated plants, bees that feed on pollen could be evaluated in relation to potential adverse effects on production of honey and other bee products. Both transgenic insect resistance and herbicide tolerance are achieved through the expression of novel transgenic proteins in plant tissues. Non-target organisms, including natural enemies, pollen and nectar feeders and non-target pests come into contact with the transgenic protein-containing plant tissues through either feeding on the crop and/or through indirect contacts via the food web associated with the crop. If crop residues are incorporated into the soil after harvest, in some cases the transgenic material can be present in the soil ecosystem for more than 1 year after crop cultivation (e.g. Baumgarte and Tebbe, 2005). The temporal and spatial scale of possible impacts is therefore defined by the crop plant, the cropping cycle and the management of the field and crop products after harvest.

Receiving environment and intended use
The characteristics of the receiving environment where the transgenic crop will be grown, including management practices, will determine the major functional groups that come into contact with the crop. Different functional groups can be exposed to the crop material if crop residues are transported off the field versus left on the field, or if crop products are stored on-farm, processed immediately, fed to animals, or stored off-farm. Examples of functional groups that can come into contact with the transgenic crop material through these pathways include storage pests, the decomposer community in other fields where crop residues are deposited, or detritivores on animal manure.

Selecting relevant functional groups for Bt cotton in Vietnam

Functional groups that are possibly affected by transgenic Bt cotton, and limit or stimulate crop production significantly, are considered key functional groups for Vietnam. Functional groups that limit cotton production can include invertebrate pests of the crop, crop pathogens and weeds. Functional groups that stimulate cotton production can include pollinators of the crop or plant growth-promoting organisms, such as mycorrhizal fungi, endophytic bacteria, symbiotic nitrogen fixing bacteria or beneficial rhizosphere bacteria. Other key functional groups impact the population dynamics of the above groups, including the function known as biological control. Biological control of pests usually is provided by predators, parasitoids and pathogens of pest species. Because of the value accorded to crop production, four functional groups were selected for Bt cotton in Vietnam: non-target herbivore pests (including insect disease vectors) (Chapter 6, this volume), predators (Chapter 7, this volume), parasitoids (Chapter 8, this volume) and flower visitors (Chapter 9, this volume).

It is also important to consider functional groups associated with ecological functions that support the agricultural ecosystem as a whole, especially in the soil. These include soil processes that help maintain a healthy soil structure and sustain beneficial nutrient cycling (Allan *et al.*, 1995). Soil contains a substantial component of the biodiversity in agroecosystems (Wardle, 2002); the taxonomic range of organisms involved in carrying out soil ecological functions is very wide (Wardle *et al.*, 1999) and taxonomic information is often unavailable (Curtis *et al.*, 2002). Analysis of ecosystem processes in soil, which are facilitated by many organisms, is an alternative to taxon-based analysis (Bengtsson, 2002; DEFRA, 2004). The risk assessment can then be based on an assessment of ecological processes rather than species (Chapter 10, this volume).

Of the remaining kinds of potential adverse effects, the main species of economic concern are honeybees, which will be evaluated under flower visitors (Chapter 9, this volume). There are species with cultural value in Vietnam (e.g. the Bhodi tree, *Ficus religiosa*), but none are associated with cotton or are near cotton fields. The status of rare or endangered species is poorly known in Vietnam, but no endangered species are known to be associated with cotton or nearby habitats. Bt cotton is intended to have a beneficial impact on environmental quality through the reduction of insecticide use (Chapter 1, this volume), although this benefit may not be realized, as mentioned earlier. However, increased cultivation of Bt cotton could also affect the environmental quality of the landscape negatively. An increase in large-scale monoculture cotton cropping could reduce crop and habitat diversity compared to current cotton intercropping systems (Chapter 9, this volume).

5.3. Step 2: Listing and Prioritizing Species or Ecological Processes (Fig. 5.1)

Assigning non-target species or ecological processes to functional groups

The species associated with the crop agricultural ecosystem are classified into the selected functional categories using the information and expertise available. Sources of information include written reports and records of the literature, but also the expert knowledge of researchers, extension personnel and agricultural consultants. It is important to use species information from a range of different production systems, including low pesticide input and organic production, where appropriate. The species composition in production systems with use of insecticides is likely skewed towards the survivors of the pesticide treatments, possibly excluding a number of species with great significance for low chemical input, small-scale mixed cropping systems.

All species that are associated with the ecological function should be listed, though it may not be feasible to list more than 30–50 (e.g. Chapter 6, this volume). Some non-target species may be listed in more than one functional group. This should be expected, because many species have multiple functions in an ecosystem. Species that have multiple ecological functions can be involved in more than one environmental risk. Such species may be key species. For example, a herbivore species may be any of the following: (i) a pest of the crop or of other crops, (ii) a disease vector, (iii) seed disperser, (iv) a decomposer of plant residues, (v) important food for natural enemies, or (vi) a biocontrol organism of weedy plants outside and/or inside the cropping system.

Selecting species or ecological processes using the 'Selection Matrix'

The compiled list of species in each identified functional category is prioritized using a systematic Selection Matrix, which is a tool to identify species/processes within each functional group that have the closest association with the crop and the most significant role in the functioning of the agroecosystem. An effect on these species potentially has the greatest adverse effect on the function, which could lead to adverse consequences. The Selection Matrix will identify these species or processes.

Within each functional group, the species or processes are compared and ranked for their importance based on five criteria that assess their association with the crop and a number of other criteria assessing their functional significance in the crop, associated crops or natural areas. The rankings should take into account published information and expert knowledge within the country. By recording group assessments as individual ranks and requiring group consensus, the process is transparent and the evaluations will be defendable more readily and acceptable by others. The criteria apply during the time the taxon is using the crop and, if the species has more than one generation per year, the generation(s) associated most closely with the crop should be considered. In tropical or subtropical countries, there is often a marked difference in biodiversity between dry season irrigated crop production and rainy season production, and ranking may be done separately for each season. Chapters 6–10 (this volume) illustrate the use of the Selection Matrix for each functional group.

Association with the crop
The association of the non-target species with the relevant crop is determined by considering the association of the taxon or ecosystem process with the crop. Exposure to the specific transgene product or its metabolites in specific crop tissues or in the environment is not yet considered in detail, but will be addressed in Step 3. Because of the large variation in different parts of a country, it is often worthwhile to carry out the ranking separately for each region/agroecological zone/cropping system being considered, and these regions need to be defined before initiating the ranking procedure. The criteria for association with the crop are geographic distribution in the cropping regions, prevalence on the crop, abundance on the crop, phenological (temporal) overlap between the crop and taxon and habitat specialization during the crop cycle. Some of these criteria may not be relevant for some functional groups and additional criteria can be considered for a given functional group (e.g. number of generations per crop season for non-target herbivores).

'Geographic distribution' asks the question: What is the degree of overlap between the species geographic range and the crop production range? The degree of overlap is ranked at the country or region or agroecological zone scale. The geographic boundaries of the area must be specified for the analysis. Species or taxa that are distributed widely are likely to be in greater contact with the transgenic crop. Those species that occur in one of several cropping regions only are not relevant for that function in the other regions; other species may carry out that function in the other regions. For example, a herbivore pest that occurs throughout a cropping region is considered more important than one that is restricted to one part of the region only. Species or biotypes whose range is expanding rapidly, due to relatively recent invasion, may need to be considered in terms of likely geographic distribution, so that the selection process does not become invalid too quickly. Under this heading, the temporal dimension is omitted (this is a separate assessment component later; see below). Here, only the presence or absence of a species is considered.

'Prevalence on the crop' asks the question: What is the proportion of crop habitat in which the species can be found reliably? The species should be given a high rank if the species occurs in all fields of the crop in the country, region or agroecological zone, and a low rank when it occurs in some of the fields only. A predator population that is more prevalent (e.g. occurs in all fields of that crop in the particular region) is likely to be more exposed to the transgenic crop than one that occurs only on a small proportion of the crop in a region. A reduction in a predator that is more prevalent on the crop is also more likely to be related to an adverse effect on biological control in the crop or nearby crops.

'Abundance on the crop' asks the question: How abundant is the species when it does occur on the crop? The average or typical density of each species or taxon is compared. A species or taxon that is more abundant on a crop will mean more individuals exposed to the crop. Higher abundance may also be correlated with greater functional significance. For example, an effect on a predator that is more abundant may be more likely to be related to an adverse effect on biological control in a crop or nearby crops than one that is present in low numbers only. However, this is not always the case.

Some ways to compare abundances within a functional group are by using an estimate of relative biomass or relative abundance among related taxa. Absolute densities are difficult to compare across species because of differences in size, behaviour and ecology that can cause vast differences in average densities. For example, cotton aphids, *Aphis gossypii*, are small and can occur in populations of many hundreds on each plant, while the caterpillar, *Helicoverpa armigera*, is much larger and occurs in smaller densities. Measures of relative abundance are also easily biased by sampling methods with different sampling efficiency for different species. For example, pitfall traps catch high proportions of surface-active, mobile species such as carabids and ants, while under-representing less mobile species such as web-spinning spiders (Duelli *et al.*, 1999; Lang, 2000).

'Phenology' on the crop asks the question: When does the species occur on the crop (phenological or temporal overlap)? The degree of phenological overlap is ranked from both the perspective of the crop (proportion of crop cycle during which the species is present) and of the species (proportion of the species life cycle associated with the crop). Effects on populations that are exposed to the crop throughout their life cycle are often greater than on ones that are exposed only as adults. This is a less important criterion for pollinators because all are present as adults during the flowering period. For parasitoids, the rank is determined by the number of host generations on the crop that can be parasitized successively.

'Habitat specialization' during the crop period asks the question: In which crop systems does the species mostly occur? This criterion ranks the preference of a species for different crop habitats in the receiving environment (i.e. region of release). This can include other crops, field margins, fallow areas and natural areas. A habitat specialist occurs only in the target crop habitat; a habitat generalist occurs in many other habitats. Therefore, a greater proportion of the population of a habitat specialist is likely to be in closer contact with the transgenic crop than that of a habitat generalist. The crop habitat is defined as the crop field and its margins and includes all of the species associated with these, including the crop, any intercrop and weeds. The criterion should be evaluated for the time when the species or taxon is using the crop; for example, the generation associated most closely with the crop. The criterion does not rank habitat preference within the crop, e.g. whether the species lives on the crop plant or on the ground.

Association with crop for soil ecosystem processes

Soil processes in crop ecosystems are driven by the types and amounts of carbon-containing materials entering the soil from plants (Wardle *et al.*, 1999; Wheatley *et al.*, 2001). The constituents of these inputs vary according to the type of plant cultivated, its physiology, structure and the stage of growth. These inputs are primarily from plant residues and from organic compounds released by the roots of growing plants. Soil ecosystem processes therefore can be ranked according to their association with these 'hotspots', which are the main potential exposure pathways (Chapter 10, this volume).

Functional significance

Functional significance is assessed according to the primary function of the functional group to which the species or process is classified. Although species are listed because they likely contribute to the ecological function, this step further distinguishes the listed species into those associated most closely and those associated less closely with the ecological function. In addition, the significance of the species for other functions can also be assessed. For example, in Chapter 6, the species listed in the functional group 'Herbivore pests' were assessed for their function as 'pests in the crop', but also as 'disease vectors', 'food for natural enemies' or 'pests in other (inter)crops'. The selection criteria for functional significance therefore vary for each functional group. A natural enemy species listed in the biological control functional category will be ranked based on what is known about its importance as a biological control agent in the particular cropping system. Estimation of the functional significance of a parasitoid species for biological control would, for example, be determined partly by the significance of its host species as a pest. Within each functional group, species can also be ranked according to their functional importance in nearby agricultural or natural ecosystems. In Chapter 7, the significance of an important predator species is evaluated for this function in nearby crops and natural areas. Functional significance will highlight taxa with multiple important functions. The significance of soil eco-system processes is assessed by their value as indicators of soil health (Arshad and Martin, 2002) and their significance in the crop ecosystem (Chapter 10, this volume).

Prioritization of species or ecosystem processes

Ranking the criteria initiates an ecologically based prioritization process of the listed non-target species or processes. All of the species/processes in the initial list are ranked comparatively for each criterion. The ranks are therefore relative to the other species within the group, and a rank of '1' is high and a rank of '3' is low. It is important to identify gaps in knowledge by designating a rank as '?' (= unknown or insufficient information). Knowledge gaps can be addressed using a precautionary methodology by assigning a 'precautionary rank' as 'high' or '1', or using comparative methods. This way, we can understand better the limitations of the analysis and realize how gaps in knowledge affect the risk assessment process. It will also help to identify key research needs.

The ranks for the criteria for 'Association with Crop' and those for the criteria for 'Functional Significance' are averaged separately to give the overall mean rank for 'Association with Crop' and for 'Functional Significance'. These two means are added together to give an overall summed score for the species. For the herbivore pests in Vietnam, this was done separately for the rainy season cotton crop and for the dry season cotton crop (Chapter 6, this volume). A summed score between 2 and 3 indicates highest priority, a score between 3 and 4 medium priority and between 4 and 6 lower priority.

The highest priority species and processes are taken to the next steps of the methodology (Chapters 6–10, this volume). These are the species and processes that are considered the most important for the particular cropping system, i.e. they are geographically widespread, abundant, present in the agro-ecosystem every year throughout the growing season, associated closely with the crop habitat and have a vital role in the specified ecological function. This approach overcomes the simplistic assumption that species abundance is a direct measure of ecological significance. It is important not to exclude or include species on the basis of only one criterion.

Another criterion for species selection is taxonomic diversity (Chapters 7 and 8, this volume). Similar taxa are not all selected, even if they are all ranked highly, because often they fulfil similar ecological roles. Instead, broad taxonomic representation is important to assess different potential adverse effects of a transgenic plant.

5.4. Step 3: Identify Potential Exposure Pathways and Adverse Effects Pathways and Use These to Formulate and Prioritize Risk Hypotheses (Fig. 5.1)

Identify potential exposure pathways

In this substep, the possible pathways of exposure to the GM crop and its trans-gene products (including their metabolites) are identified for the smaller number of species or processes selected in the previous step. Exposure can occur through many pathways (Hilbeck, 2002; Andow and Hilbeck, 2004; Birch and Wheatley, 2005) and is a function of concentration (how much), persistence (how long), frequency (how often) and movement (where to). Currently, available transgenic plants contain constitutive promoters that induce constant transgene expression in most tissues of the plant, although expression levels vary between plant tissues, individual plants and over time (Chapter 4, this volume). Transgenic plant materials and products can move in the agroecosystem by being transferred between organisms in the food web and enter the soil as plant residues. The harvested transgenic plant products are often transported, which may involve losses during transport and storage, coming into contact with storage pests and, when processed, produce waste products that may be used in another receiving environment (e.g. composts, press cakes, etc.). Crop residues may also be transported and fed to animals. Some of this movement may result in concentration and increased persistence of the transgenic product in the ecosystem.

Exposure via the food chain
Species may be exposed to the transgene product and/or its metabolites by feeding on the transgenic plant (bitrophic exposure: leaves, stem, roots, buds and bracts, flower tissues, fruit, seed, pollen and root exudates), or through another organism, such as a herbivore or detritivore prey or host (tritrophic or multitrophic exposure via feeding on prey or hosts; Hilbeck *et al.*, 1998a,b, 1999; Birch *et al.*, 1999; Harwood *et al.*, 2005). Tri- or multitrophic exposure can also occur via herbivore excretion products, such as honeydew from aphids, thrips or leafhoppers (Bernal *et al.*, 2002; Obrist *et al.*, 2005), earthworm casts (Saxena and Stotzky, 2001; Ahmad *et al.*, 2006), lepidopteran frass (Raps *et al.*, 2001) or arthropod faeces (Wandeler *et al.*, 2002; Howald *et al.*, 2003; Pont and Nentwig, 2005; Torres *et al.*, 2006). Less information is available about the presence of transgene products in phloem and xylem, and nectar, guttation fluids and other plant exudates. Almost all of these saps and fluids have been reported to contain proteins or large polypeptides (French *et al.*, 1993; Peumans *et al.*, 1997; Borisjuk *et al.*, 1999; Imlau *et al.*, 1999; Komarnytsky *et al.*, 2000; Saxena *et al.*, 2002; Buhtz *et al.*, 2004). Aphids feeding on phloem of certain Bt cottons acquire Cry toxin (Zhang *et al.*, 2006). Phloem and/or xylem of Bt rice may express Bt (Bernal *et al.*, 2002), but it is not detectable in Bt 11 and Event 176 maize phloem (Raps *et al.*, 2001). Exposure can change if the organism alters its feeding behaviour. In three well-studied cases,

insect predators and parasitoids avoided Bt-containing prey and hosts (Schuler *et al.*, 1999; Meier and Hilbeck, 2001; Rovenska *et al.*, 2005), which would reduce exposure levels for predators but, at the same time, possibly reducing the degree of biological control on the respective transgenic plants.

Exposure in the agroecosystem
Transgenic plant material and transgene products and metabolites can enter the soil as plant residues during the growing season and after harvest, as sloughed roots and root exudates (Zwahlen *et al.*, 2003; Gupta and Watson, 2004; Saxena *et al.*, 2004). Bt proteins adsorb to clays and humic acids in soil and can persist for considerable amounts of time in soil aggregates, retaining their insecticidal activity. Pollen and anthers can land on other plants in the crop field or along the margins, leading to exposure of species on these plants (Pleasants *et al.*, 2001; Lang *et al.*, 2004), or may coat spiderwebs and thus be ingested by the spider (Ludy and Lang, 2006). The uptake, spread and persistence of transgene products in carabid beetles in a Bt maize agroecosystem was found even if no transgenic crop had been grown there for 1 year (Zwahlen and Andow, 2005).

Exposure on other plants after gene flow
If the transgenic trait is expressed in other plants (crops or weeds) that have received the transgene via gene flow, organisms in habitats other than the transgenic crop may be exposed to transgenic plant material and the transgene protein (Letourneau *et al.*, 2003). Hence, it is important for a comprehensive environmental risk assessment to consider this possibility.

Detection and verification of exposure
Commercially available methods (strip tests and ELISA test kits) for detecting Cry proteins in Bt crops were developed originally for detection of intact Cry proteins in plants. However, the ELISA kits are capable of detecting a large range of different sized Bt proteins. For example, Lutz *et al.* (2005) found that the 65 kDa protein was processed in the digestive tract of cows to 34 and 17 kDa fragments. In the cow faeces, only the 17 kDa fragment could be detected by ELISA. The same ELISA procedures can detect the 91 kDa sized Bt protein expressed in Mon810. It must be noted that ELISA detection of the Bt proteins does not prove any bioactivity of the protein. Only bioassays using known susceptible insects can do that. Proven methods for the detection of Bt metabolites, such as the 34 or 17 kDa fragment, in highly diverse environmental media, such as animal faeces and soils, still need to be developed (Chapter 4, this volume).

Identify potential adverse effects pathways

An adverse effect pathway is a causal chain that starts with an exposed entity and ends with an adverse effect. For species, an adverse effect pathway could begin with a change in a population parameter or a behaviour of the species and end with loss of crop production. For an ecological process (e.g. crop residue decomposition), it could begin with a change in the timing, rate or magnitude of the process and end with a reduction in soil quality.

It is useful to start first by identifying the adverse effects that might result from the identified exposure pathways and then develop ecologically plausible adverse effects pathways that connect the exposure pathway to the adverse effect. Second, pathways that result from indirect effects of transgenic crops are identified ('knock-on' pathways). For example, some pathways may involve changes in cropping practice. The adoption of transgenic crops by farmers may lead to changes in agricultural practices other than those that are directly intended from the transgenic trait. The adoption of a transgenic crop may change crop area, crop rotation or intercropping practices, seasonal timing of crop production and irrigation, tillage and pest management practices. These changes in agricultural practices have a range of environmental effects, some at the landscape level (e.g. land use patterns). Some of the potential effects of these changes are addressed in Chapters 6–10.

The complementary methods of fault- and event-tree analysis are useful techniques for establishing systematically the causal links between the transgenic crop and adverse effect(s) and identifying multiple pathways by which an adverse effect could occur (NRC, 2002; Hayes *et al.*, 2004; Meier and Hilbeck, 2005). Fault-tree analysis identifies possible pathways by tracing backwards from the adverse effect through suspected causal chains to the transgenic crop. It can be helpful for clarifying the events that could cause the adverse effect, in what sequence they must occur and what assumptions must be made for the pathway to be likely to occur. Event-tree analysis uses a forward logic to identify pathways from an initiating exposure event, through a causal chain to adverse effects. It can be helpful for clarifying the different possible consequences of the exposure and what must happen for exposure to end in an adverse effect. They are graphical models of the causal chains of events that lead to the adverse effect. All branches of the causal event chain can be displayed systematically and evaluated. Typically, probabilities are associated with each causal event (NRC, 2002; Hayes *et al.*, 2004). Chapters 6–10 include diagrams of potential risk pathways for each functional group but, due to large knowledge gaps, they mostly do not associate probability estimates to the events.

Knowledge gaps and key links in exposure and adverse effects pathways
It is likely that an analysis of possible exposure and adverse effects pathways for a species will reveal knowledge gaps, which creates uncertainty in the assessment. If exposure cannot be assessed otherwise, the following procedure may be used to address cases of large uncertainty. First, construct a worst-case (maximum adverse effect) scenario consistent with the available information. It is possible that, under this worst-case scenario, the expected exposure is not significant, so it can be concluded that, even given the high uncertainty, exposure is insignificant and, hence, a resulting potential adverse effect from this exposure is unlikely (Fig. 5.2). It is also possible that the worst-case scenario implies that exposure could be significant. In this case, construct a best-case scenario (a minimum adverse effect scenario). It is possible that even under this

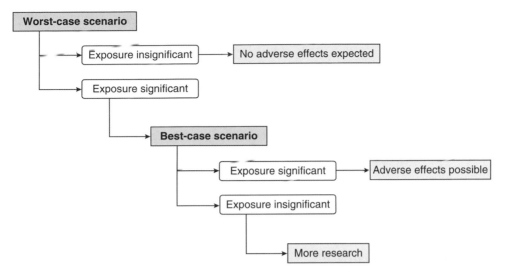

Fig. 5.2. Procedure for addressing uncertainty in analysis of exposure pathways.

best-case scenario, the expected level of exposure is significant, so it could be concluded that, even with the given uncertainty, exposure is significant and an adverse effect cannot be ruled out. Finally, it is possible that the worst-case scenario implies that exposure is significant and the best-case scenario implies exposure is not significant. In this case, additional research would be needed to characterize the most probable pathways. But this uncertainty analysis would allow the identification of the critical key knowledge gaps that must be closed in order to be able to confirm or refute reliably the suspected exposure or adverse effect scenario. This procedure will help to focus research funding to critical issues and possibly avoid costly experiments.

For example, one of the main exposure pathways for egg parasitoids on Bt crops, such as *Trichogramma* spp., is via Cry toxins in host eggs. It is not known if eggs of insect herbivores on Bt crops contain Cry toxins. So, the worst-case scenario assumes that they do contain the toxin, exposure is high and the *Trichogramma* parasitoid developing in the egg might be affected adversely. The best-case scenario assumes that the eggs would not contain the toxins and, therefore, exposure would be nil or low and no adverse effect could materialize. Thus, more research would be needed. However, the most cost-effective approach would be to conduct a simple and inexpensive experiment to determine whether or not the host eggs contain Cry toxins, rather than running a full, and far more costly, experiment with the potentially adversely affected parasitoid. The hosts for some *Trichogramma* spp. are lepidopteran species such as *Spodoptera* spp. A *Spodoptera* sp. could be reared on Bt maize or Bt cotton, eggs collected and tested for the presence of Cry toxins. If they do not contain any Bt toxin (as might be expected), exposure to *Trichogramma* would be non-existent and all adverse-effect scenarios for this important biological control species that involve exposure to Bt would be irrelevant. If the eggs are found to contain Bt toxins, it would be important to carry the assessment of *Trichogramma* spp. further to investigate whether the effects on *Trichogramma* could result in a risk (Fernandes, 2003).

Formulate potential risk hypotheses

The possible exposure pathways and potential adverse effects pathways are combined to create potential risk hypotheses (Box 5.1). This is a statement that defines a specific chain or network of causal linkages, starting with a stressor (e.g. the transgenic crop) and ending in an adverse effect in the receiving environment (US EPA, 1998). For example, a risk hypothesis may be: 'Feeding on aphids raised on Bt cotton will reduce fertility of a particular coccinellid species and reduce its biological control efficacy, leading to increase in pest population(s) and subsequently to greater yield loss and/or increased pest management problems'.

Prioritize the risk hypotheses

Because the potential risks identified in this step are hypothetical (as this step occurs prior to a release of the transgenic crop) and many such risks can be

---

Box 5.1. Risk hypothesis

A risk hypothesis is a statement that defines a specific chain or network of causal linkages, starting with a stressor and ending in an adverse effect in the receiving environment (US EPA, 1998). Risk hypotheses may be based on theory and logic, empirical data, mathematical models, or probability models. They are formulated using a combination of professional judgement and available information on the agroecosystem at risk, characteristics of the transgenic crop and observed or predicted ecological effects that could end in the defined adverse effect.

These risk hypotheses are not equivalent to the traditional statistical null hypothesis. However, predictions generated from risk hypotheses can be tested in a variety of ways allowing for standard statistical approaches.

Null hypothesis testing using frequentist statistics is often not asking the right questions for risk assessment. As stated by Suter (2007): 'In ecological risk assessment we are not interested in testing the hypothesis that a chemical [or transgenic plant] has no toxicity; we want to know what its effects will be at given levels of exposure. Similarly, we are not interested in testing the null hypothesis that two streams [or the transgenic and non-transgenic crop] are identical; we know they are not. We want to know how they differ. Even when researching scientific hypotheses, it is better to compare genuine alternative hypotheses than to test a null hypothesis that nobody believes.'

In risk related problems, Type II errors – the error when the null hypothesis is not rejected when in actuality it should have been rejected – can be more serious than Type I errors, when the null hypothesis is rejected mistakenly (Andow, 2003). When possible, risk assessment experiments should be based on a prospective power analysis, i.e. the power of the experimental design to find a difference of a specified minimum size (Bourguet *et al.*, 2002; Marvier, 2002); otherwise, equivalence tests are suitable for formulating and testing credible null hypotheses (Andow, 2003).

plausible, it is often useful to prioritize them, so that the most likely or most severe ones can be assessed first. Exposure can be judged to be likely or unlikely, or very restricted in timing or scope; for example, if only one life cycle stage is exposed for a short period at a non-critical time, or if only a small proportion of the population is likely to be exposed. The potential adverse effects can be judged for whether the potential effect is likely to be readily reversible or irreversible (this will determine whether risk management measures could mitigate the consequences), whether the effect would occur over an extensive spatial scale or a very local scale and whether the affected people may consider the potential consequences to be major and/or unacceptable, or minor and/or acceptable (OGTR, 2005). The prioritization process should specify the rationale for prioritizing or omitting hypotheses, including acknowledgement of knowledge gaps and uncertainties (US EPA, 1998). A prioritization process can be found in Chapters 6–10.

5.5. Step 4: Develop an Analysis Plan and Suggest Designs for Experiments to Test Risk Hypotheses (Fig. 5.1)

Develop an analysis plan

An analysis plan is the description of how the prioritized risk hypotheses will be analysed to determine the probability of the risks, including data needs, measures, sequencing of experiments and comparisons (US EPA, 1998). Chapters 6–10 outline case-specific analysis plans.

The analysis plan highlights the links within the selected hypothesis pathways that will be analysed experimentally. Risk hypotheses can be analysed by evaluating links that are arguably unlikely, so making the whole risk hypothesis unlikely, or that are easy to test (Andow *et al.*, 2006). In the aphid–coccinellid hypothesis described above, an arguably unlikely link is whether Bt toxin occurs in the phloem of Bt cotton and exposes the coccinellid to this toxin. Some links in risk hypotheses can be tested initially in laboratory or greenhouse conditions, but some aspects of the hypotheses might be tested under field conditions with greater cost-effectiveness, provided that the field trial is isolated and managed so as to restrict unwanted effects on the surrounding environment. After commercialization, monitoring of anticipated adverse effects may be desirable, in order to provide a basis on which to confirm the conclusions in the assessment, detect changes and anticipate ways to address any problems that emerge (e.g. Chapters 6 and 8, this volume). Monitoring may also be a reliable means to identify potential risks arising from landscape-level changes.

Design experiments

Based on the analysis plan, experiments can be designed that falsify or quantify the risk hypothesis. The goal is to determine if a hypothesized risk is an actual risk and, if so, to quantify the risk. If the potential risk cannot be quantified or refuted, the risk hypotheses should be revised based on the new information and alternative experiments should be developed. This iterative process should continue until the potential risk can be quantified or refuted. Laboratory and greenhouse experiments are insufficient to prove that the risk actually occurs in the environment, but they may refute the risk. Appropriate field experiments should be designed to validate the findings at the laboratory level, unless an effect has been refuted strongly at the laboratory or greenhouse level (Birch and Wheatley, 2005; Birch et al., 2007). Examples are provided in Chapters 6–10.

5.6. Conclusion

The non-target model developed by the GMO ERA Project supports hypothesis-driven risk assessment and is in compliance with the provisions put forward in the Cartagena Protocol on Biosafety, the International Plant Protection Convention and EU Directive 2001/18. Our approach starts by considering as many locally relevant species and ecological processes as possible and narrows these down systematically to a smaller number based on ecological criteria. This selection process focuses on those species and ecological processes that are critical for important ecosystem function(s) and could be exposed directly or indirectly to the transgenic plant and its transgene products. The model focuses on ecologically relevant goals by identifying relevant risk hypotheses. This approach is flexible enough to address direct, indirect, immediate, delayed and cumulative effects. Importantly, it also places emphasis on the local ecological interactions and allows a rapid assessment of potential impacts on those, while guiding necessary empirical tests. This methodology should benefit both developing and developed countries alike.

References

Ahmad, A., Wilde, G.E. and Zhu, K.Y. (2006) Evaluation of effects of coleopteran-specific Cry3Bb1 protein on earthworms exposed to soil containing corn roots or biomass. *Environmental Entomology* 35, 976–985.

Allan, D.L., Adriano, D.C., Bezdicek, D.F., Cline, R.G., Coleman, D.C., Doran, J.W., Haberern, J., Harris, R.F., Juo, A.S.R., Mausbach, M.J., Peterson, G.A., Schuman, G.E., Singer, M.J. and Karlen, D.J. (1995) Soil quality: a conceptual definition. *Soil Science Society of America Agronomy News* June 7.

Andow, D.A. (2003) Negative and positive data, statistical power, and confidence intervals. *Environmental Biosafety Research* 30, 625–629.

Andow, D. and Hilbeck, A. (2004) Science-based risk assessment for non-target effects of transgenic crops. *BioScience* 54, 637–649.

Andow, D.A., Birch, A.N.E., Dusi, A., Fontes, E.M.G., Hilbeck, A., Lang, A., Lövei, G.L., Pires, C.S.S., Sujii, E.R., Underwood, E. and Wheatley, R.E. (2006) Principles for risk assessment – an international perspective. In: *9th International Symposium on the Biosafety of Genetically Modified Organisms: Biosafety Research and Environmental Risk Assessment, Jeju Island, Korea*. International Society for Biosafety Research (ISBR), pp. 68–73.

Arshad, M.A. and Martin, S. (2002) Identifying critical limits for soil quality indicators in agro-ecosystems. *Agriculture, Ecosystems and Environment* 88, 153–160.

Baumgarte, S. and Tebbe, C.C. (2005) Field studies on the environmental fate of the Cry1AB Bt toxin produced by transgenic maize (MON810) and its effect on bacterial communities in the maize rhizosphere. *Molecular Ecology* 14, 2539–2551.

Bengtsson, J. (2002) Disturbance and resilience in soil animal communities. *European Journal of Soil Biology* 38, 119–125.

Bernal, C.C., Aguda, R.M. and Cohen, M.B. (2002) Effect of rice lines transformed with *Bacillus thuringiensis* toxin genes on the brown planthopper and its predator *Cyrtorhinus lividipennis*. *Entomologia Experimentalis et Applicata* 102, 21–28.

Birch, A.N.E. and Wheatley, R.E. (2005) GM pest-resistant crops: assessing environmental impacts on non-target organisms. In: Hester, R.E. and Harrison, R.M. (eds) *Sustainable Agriculture*. Issues in Environmental Science and Technology, Volume 21. RSC Publishing, Cambridge, UK, pp. 31–57.

Birch, A.N.E., Geoghegan, I.E., Majerus, M.E.N., McNicol, J.W., Hackett, C.A., Gatehouse, A.M.R. and Gatehouse, J. (1999) Tritrophic interactions involving pest aphids, predatory 2-spot ladybirds and transgenic potatoes expressing snowdrop lectin for aphid resistance. *Molecular Breeding* 5, 75–83.

Birch, A.N.E., Wheatley, R.E., Anyango, B., Arpaia, S., Capalbo, D., Getu Degaga, E., Fontes, E., Kalama, P., Lelmen, E., Lövei, G., Melo, I.S., Muyekho, F., Ngi-Song, A., Ochieno, D., Ogwang, J., Pitelli, R., Schuler, T., Sétamou, M., Sithanantham, S., Smith, J., Van Son, N., Songa, J., Sujii, E., Tan, T.Q., Wan, F.-H. and Hilbeck, A. (2004) Biodiversity and non-target impacts: a case study of Bt maize in Kenya. In: Hilbeck, A. and Andow, D.A. (eds) *Environmental Risk Assessment of Genetically Modified Organisms, Volume 1: A Case Study of Bt Maize in Kenya*. CAB International, Wallingford, UK, pp. 117–186.

Birch, A.N.E., Griffiths, B.S., Caul, S., Thompson, J., Heckmann, L.-H., Krogh, P.H. and Cortet, J. (2007) The role of laboratory, glasshouse and field scale experiments in understanding the interactions between genetically modified crops and soil ecosystems: a review of the ECOGEN project. *Pedobiologia* 15, 171–272.

Borisjuk, N.V., Borisjuk, L.G., Logendra, S., Petersen, F., Gleba, Yu and Raskin, I. (1999) Production of recombinant proteins in plant root exudates. *Nature Biotechnology* 17, 466–469.

Bourguet, D., Chaufaux, J., Micoud, A., Delos, M., Naibo, B., Bombarde, F., Marque, G., Eychenne, N. and Pagliari, C. (2002) *Ostrinia nubilalis* parasitism and the field abundance of non-target insects in transgenic *Bacillus thuringiensis* corn (*Zea mays*). *Environmental Biosafety Research* 1, 49–60.

Buhtz, A., Kolasa, A., Arlt, K., Walz, Ch. and Kehr, J. (2004) Xylem sap protein composition is conserved among different plant species. *Planta* 219, 610–618.

CBD (1992) Convention on Biological Diversity. Secretariat of the Convention on Biological Diversity, Montreal, Canada. www.cbd.int/convention/default.shtml (accessed 24 October 2007).

CBD (2000) Cartagena Protocol on Biosafety to the Convention on Biological Diversity: Text and Annexes. Secretariat of the Convention on Biological Diversity, Montreal, Canada. www.cbd.int/biosafety/protocol.shtml (accessed 24 October 2007).

Cowgill, S.E. and Atkinson, H.J. (2003) A sequential approach to risk assessment of transgenic plants expressing protease inhibitors: effects on non-target herbivorous insects. *Transgenic Research* 12, 439–449.

Curtis, T.P., Sloan, W.T. and Scannell, J.W. (2002) Estimating prokaryotic diversity and its limits. *Proceedings of the National Academy of Sciences of the USA* 99, 10494–10499.

DEFRA (2004) Mechanisms for investigating changes in soil ecology due to GMO releases. Ref EPG 1/5/214. Department of Environment, Food and Rural Affairs, UK Government. www.defra.gov.uk/environment/gm/research/epg-1-5-214.htm (accessed 24 October 2007).

Duelli, P., Obrist, M.K. and Schmatz, D.R. (1999) Biodiversity evaluation in agricultural land-scapes: above-ground insects. *Agriculture, Ecosystems and Environment* 74, 33–64.

Dutton, A., Romeis, J. and Bigler, F. (2003) Assessing the risks of insect resistant transgenic plants on entomophagous arthropods: Bt-maize expressing CryAb as a case study. *BioControl* 48, 611–636.

Elmegaard, N. and Jagers op Akkerhuis, G.A.J.M. (2000) *Safety Factors in Pesticide Risk Assessment: Differences in Species Sensitivity and Acute-Chronic Relations.* NERI Technical Report 325. National Environmental Research Institute, Silkeborg, Denmark.

Fernandes, O.D. (2003) Effect of genetically modified corn (MON810) on *Spodoptera frugiperda* (J.E. Smith, 1797) and on egg parasitoid *Trichogramma* spp. Doutor em Ciências, Área de Concentração: Entomologia. Escola Superior de Agricultura "Luiz de Queiroz", Universidade de São Paulo, Piracicaba, Brazil.

Firbank, L.G. (2003) Introduction. *Philosophical Transactions of the Royal Society of London Series B* 358, 1777–1778.

French, C.J., Elder, M. and Skelton, F. (1993) Recovering and identifying infectious plant viruses in guttation fluid. *Hortscience* 28, 746–747.

Garcia-Alonso, M., Jacobs, E., Raybould, A., Nickson, T.E., Sowig, P., Willekens, H., Van der Kowe, P., Layton, R., Amijee, F., Fuentes, A.M. and Tencalla, F. (2006) A tiered system for assessing the risk of genetically modified plants to non-target organisms. *Environmental Biosafety Research* 5, 57–65.

Gupta, V.V.S.R. and Watson, S. (2004) Ecological impacts of GM cotton on soil biodiversity: Below ground production of Bt by GM cotton and Bt cotton impacts on soil biological processes. CSIRO Land and Water, Australia. www.deh.gov.au/settlements/publications/biotechnology/gm-cotton/index.html (accessed 12 September 2006).

Harwood, J.D., Wallin, W.G. and Obrycki, J.J. (2005) Uptake of Bt endotoxins by non-target herbivores and higher order arthropod predators: molecular evidence from a transgenic corn ecosystem. *Molecular Ecology* 14, 2815–2823.

Haughton, A.J., Champion, G.T., Hawes, C., Heard, M.S., Brooks, D.R., Bohan, D.A., Clark, S.J., Dewar, A.M., Firbank, L.G., Osborne, J.L., Perry, J.N., Rothery, P., Roy, D.B., Scott, R.J., Woiwod, I.P., Birchall, C., Skellern, M.P., Walker, J.H., Baker, P., Browne, E.L., Dewar, A.J.G., Garner, B.H., Haylock, L.A., Horne, S.L., Mason, N.S., Sands, R.J.N. and Walker, M.J. (2003) Invertebrate responses to the management of genetically modified herbicide-tolerant and conventional spring crops. II. Within-field epigeal and aerial arthropods. *Philosophical Transactions of the Royal Society of London Series B* 358, 1863–1877.

Hayes, K.R., Gregg, P.C., Gupta, V.V.S.R., Jessop, R., Lonsdale, M., Sindel, B., Stanley, J. and Williams, C.K. (2004) Identifying hazards in complex ecological systems. Part 3: Hierarchical holographic model for herbicide-tolerant oilseed rape. *Environmental Biosafety Research* 3, 1–20.

Hilbeck, A. (2002) Transgenic host plant resistance and non-target effects. In: Letourneau, D. and Burrows, B. (eds) *Genetically Engineered Organisms: Assessing Environmental and Human Health Effects.* CRC Press LLC, Boca Raton, Florida, pp. 167–185.

Hilbeck, A. and Schmidt, J.E.U. (2006) Another view on Bt proteins – how specific are they and what else might they do? *Biopesticides International* 2, 1–50.

Hilbeck, A., Baumgartner, M., Fried, P.M. and Bigler, F. (1998a) Effects of transgenic Bt corn-fed prey on immature development of *Chrysoperla carnea* (Neuroptera: Chrysopidae). *Environmental Entomology* 27, 480–487.

Hilbeck, A., Moar, W., Pusztai-Carey, M., Filipini, A. and Bigler, F. (1998b) Toxicity of *Bacillus thuringiensis* Cry1Ab toxin to the predator *Chrysoperla carnea* (Neuroptera: Chrysopidae). *Environmental Entomology* 27, 1255–1263.

Hilbeck, A., Moar, W., Pusztai-Carey, M., Filipini, A. and Bigler, F. (1999) Prey-mediated effects of Cry1Ab toxin and protoxin and Cry2A protoxin on the predator *Chrysoperla carnea* (Neuroptera: Chrysopidae). *Entomologia Experimentalis et Applicata* 91, 305–316.

Hilbeck, A., Andow, D.A., Arpaia, S., Birch, A.N.E., Fontes, E.M.G., Lövei, G.L., Sujii, E., Wheatley, R.E. and Underwood, E. (2006) Methodology to support non-target and biodiversity risk assessment. In: Hilbeck, A., Andow, D.A. and Fontes, E.M.G. (eds) *Environmental Risk Assessment of Genetically Modified Organisms, Volume 2: Methodologies for Assessing Bt Cotton in Brazil.* CAB International, Wallingford, UK, pp. 108–132.

Howald, R., Zwahlen, C. and Nentwig, W. (2003) Evaluation of Bt oilseed rape on the non-target herbivore *Athalia rosae. Entomologia Experimentalis et Applicata* 106, 87–93.

Imlau, A., Truernit, E. and Sauer, N. (1999) Cell-to-cell and long-distance trafficking of the green fluorescent protein in the phloem and symplastic unloading of the protein into sink tissues. *The Plant Cell* 11, 309–322.

Komarnytsky, S., Borisjuk, N.V., Borisjuk, L.G., Alam, M.Z. and Raskin, I. (2000) Production of recombinant proteins in tobacco guttation fluid. *Plant Physiology* 124, 927–933.

Lang, A. (2000) The pitfalls of pitfalls: a comparison of pitfall trap catches and absolute density estimates of epigeal invertebrate predators in arable land. *Journal of Pest Science* 73, 99–106.

Lang, A., Ludy, C. and Vojtech, E. (2004). Dispersion and deposition of Bt maize pollen in field margins. *Journal of Plant Diseases and Protection* 111, 417–428.

Letourneau, D.K., Robinson, G.S. and Hagen, J.A. (2003) Bt crops: predicting effects of escaped transgenes on the fitness of wild plants and their herbivores. *Environmental Biosafety Research* 2, 219–246.

Li, W.-D., Ye, G.-Y., Wu, K.-M., Wang, X.-Q. and Guo, Y.Y. (2002) Evaluation of impact of pollen grains of Bt, Bt/CPTI transgenic cotton and Bt corn on the growth and development of the mulberry silkworm *Bombyx mori* Linnaeus (Lepidoptera: Bombyxidae). *Scientia Agricultura Sinica* 35, 1543–1549 (in Chinese with English abstract).

Losey, J.E., Rayor, L.S. and Carter, M.E. (1999) Transgenic pollen harms monarch larvae. *Nature* 399, 214.

Lövei, G.L., Andow, D.A. and Arpaia, S. (submitted) Transgenic insecticidal crops and natural enemies: a detailed update. *Environmental Entomology*, submitted.

Ludy, C. and Lang, A. (2006) Bt maize pollen exposure and impact on the garden spider, *Araneus diadematus. Entomologia Experimentalis et Applicata* 118, 145–156.

Lutz, B., Wiedemann, S., Einspanier, R., Mayer, J. and Albrecht, C. (2005) Degradation of Cry1Ab protein from genetically modified maize in the bovine gastrointestinal tract. *Journal of Agricultural and Food Chemistry* 53, 1453–1456.

Marvier, M. (2002) Improving risk assessment for nontarget safety of transgenic crops. *Ecological Applications* 12, 1119–1124.

Marvier, M., McCreedy, C., Regetz, J. and Kareiva, P. (2007) A meta-analysis of effects of Bt cotton and maize on nontarget invertebrates. *Science* 316, 1475–1477.

Meier, M.S. and Hilbeck, A. (2001) Influence of transgenic *Bacillus thuringiensis* corn-fed prey on prey preference of immature *Chrysoperla carnea* (Neuroptera: Chrysopidae). *Basic and Applied Ecology* 2, 35–44.

Meier, M.S. and Hilbeck, A. (2005) *Faunistische Indikatoren für das Monitoring der Umweltwirkungen gentechnisch veränderter Organismen (GVO).* Naturschutz und Biologische Vielfalt 29. Landwirtschaftsverlag, Münster, Germany (in German).

Morandin, L.A. and Winston, M.L. (2005) Wild bee abundance and seed production in conventional, organic, and genetically modified canola. *Ecological Applications* 15, 871–881.

National Research Council (2002) *Environmental Effects of Transgenic Plants: the Scope and Adequacy of Regulation.* National Academy Press, Washington, DC.

Obrist, L.B., Klein, H., Dutton, A. and Bigler, F. (2005) Effects of Bt maize on *Frankliniella tenuicornis* and exposure of thrips predators to prey-mediated Bt toxin. *Entomologia Experimentalis et Applicata* 115, 409–416.

OGTR (2005) Risk analysis framework. Office of the Gene Technology Regulator, Australia. www.ogtr.gov.au/pubform/riskassessments.htm (accessed 5 May 2007).

Pemsl, D., Waibel, H. and Gutierrez, A.P. (2005) Why do some Bt-cotton farmers in China continue to use high levels of pesticides? *International Journal of Agricultural Sustainability* 3, 44–56.

Peumans, W.J., Smeets, K., van Nerum, K., van Leuven, F. and van Damme, E.J.M. (1997) Lectin and alliinase are the predominant proteins in nectar from leek (*Allium porrum* L.) flowers. *Planta* 201, 298–302.

Pleasants, J.M., Hellmich, R.L., Dively, G.P., Sears, M.K., Stanley-Horn, D.E., Mattila, H.R., Foster, J.E., Clark, P. and Jones, G.D. (2001) Corn pollen deposition on milkweeds in and near cornfields. *Proceedings of the National Academy of Sciences of the USA* 98, 11919–11924.

Pont, B. and Nentwig, W. (2005) Quantification of Bt-protein digestion and excretion by the primary decomposer *Porcellio scaber*, fed with two Bt-corn varieties. *Biocontrol Science and Technology* 15, 341–352.

Poppy, G.M. and Sutherland, J.P. (2004) Can biological control benefit from genetically-modified crops? Tritrophic interactions on insect-resistant transgenic plants. *Physiological Entomology* 29, 257–268.

Raps, A., Kehr, J., Gugerli, P., Moar, W.J., Bigler, F. and Hilbeck, A. (2001) Immunological analysis of phloem sap of *Bacillus thuringiensis* corn and of the non-target herbivore *Rhopalosiphum padi* (Homoptera: Aphididae) for presence of Cry1Ab. *Molecular Ecology* 10, 525–534.

Romeis, J., Meissle, M. and Bigler, F. (2006) Transgenic crops expressing *Bacillus thuringiensis* toxins and biological control. *Nature Biotechnology*, 24(1), 63–71.

Rovenska, G., Zemek, R., Schmidt, J.E.U. and Hilbeck, A. (2005) Effects of transgenic Bt-eggplant expressing Cry3Bb toxin on host plant preference of *Tetranychus urticae* and prey preference of its predator *Phytoseiulus persimilis* (Acari: Tetranychidae, Phytoseiidae). *Biological Control* 33, 293–300.

Saxena, D. and Stotzky, G. (2001) *Bacillus thuringiensis* (Bt) toxin released from root exudates and biomass of Bt corn has no apparent effect on earthworms, nematodes, protozoa, bacteria, and fungi in soil. *Soil Biology and Biochemistry* 33, 1225–1230.

Saxena, D., Flores, S. and Stotzky, G. (2002) Bt toxin is released in root exudates from 12 transgenic corn hybrids representing three transformation events. *Soil Biology and Biochemistry* 34, 133–137.

Saxena, D., Stewart, C.N., Altosaar, I., Shu, Q. and Stotzky, G. (2004) Larvicidal Cry proteins from *Bacillus thuringiensis* are released in root exudates of transgenic B. *thuringiensis* corn, potato, and rice but not of B. *thuringiensis* canola, cotton, and tobacco. *Plant Physiology and Biochemistry* 42, 383–387.

Schuler, T.H., Potting, R.P.J., Denholm, I. and Poppy, G.M. (1999) Parasitoid behaviour and Bt plants. *Nature* 40, 825–826.

Suter, G.W. II. (2007) *Ecological Risk Assessment*, 2nd edn. CRC Press, Boca Raton, Florida.

Torres, J.B., Ruberson, J.R. and Adang, M.J. (2006) Expression of *Bacillus thuringiensis* Cry1Ac protein in cotton plants, acquisition by pests and predators: a tritrophic analysis. *Agricultural and Forest Entomology* 8, 191–202.

US EPA (United States Environmental Protection Agency) (1998) *Guidelines for Ecological Risk Assessment*. EPA/630/R095/002F. United States Environmental Protection Agency, Risk Assessment Forum, Washington, DC.

Wandeler, H., Bahylova, J. and Nentwig, W. (2002) Consumption of two Bt and six non-Bt corn varieties by the woodlouse *Porcelio scaber*. *Basic and Applied Ecology* 3, 357–365.

Wardle, D.A. (2002) *Communities and Ecosystems: Linking the Aboveground and Belowground Components*. Monographs in Population Biology, Vol. 34. Princeton University Press, Princeton and Woodstock, USA.

Wardle, D.A., Nicholson, K.S., Bonner, K.I. and Yeates, G.W. (1999) Effects of agricultural intensification on soil-associated arthropod population dynamics, community structure, diversity and temporal variability over a seven-year period. *Soil Biology and Biochemistry* 31, 1691–1706.

Wheatley, R.E., Ritz, K., Crabb, D. and Caul, S. (2001) Temporal variations in potential nitrification dynamics related to differences in rates and types of carbon inputs. *Soil Biology and Biochemistry* 33, 2135–2144.

WHO (World Health Organization) (2000) *The Release of Genetically Modified Organisms in the Environment: is it a Health Hazard?* Report of a Joint WHO/EURO – ANPA Seminar, Rome-Italy, 7–9 September 2000. World Health Organization, Regional Office for Europe, and European Centre for Environment and Health. www.euro.who.int/foodsafety/Otherissues/20020402_5 (accessed 19 October 2007).

Yang, P., Iles, M., Yan, S. and Joliffe, F. (2005) Farmers knowledge, perceptions and practices in transgenic Bt cotton in small producer systems in Northern China. *Crop Protection* 24, 229–239.

Zhang, G.-F., Wan, F.-H., Lövei, G.L., Liu, W.-X. and Guo, J.-Y. (2006) Transmission of Bt toxin to the predator *Propylaea japonica* (Coleoptera: Coccinellidae) through its aphid prey feeding on transgenic Bt cotton. *Environmental Entomology* 35, 143–150.

Zwahlen, C. and Andow, D.A. (2005) Field evidence for the exposure of ground beetles to Cry1Ab from transgenic corn. *Environmental Biosafety Research* 4, 1 5.

Zwahlen, C., Hilbeck, A., Gugerli, P. and Nentwig, W. (2003) Degradation of the Cry1Ab protein within transgenic *Bacillus thuringiensis* corn tissue in the field. *Molecular Ecology* 12, 765–775.

6 Potential Effect of Transgenic Cotton on Non-target Herbivores in Vietnam

Nguyễn Thị Thu Cúc, Edison R. Sujii, Lewis J. Wilson, Evelyn Underwood, David A. Andow, Mai Văn Hào, Baoping Zhai and Hồ Văn Chiến

About 30 species of herbivorous insects and mites are considered pests on cotton crops in Vietnam. The most important are *Helicoverpa armigera* Hübner, *Thrips palmi* Karny, *Aphis gossypii* Glover, *Amrasca devastans* (Distant), *Pectinophora gossypiella* (Saunders), *Spodoptera exigua* (Hübner) and *Tetranychus urticae* Koch (Chapter 2, this volume). Outbreaks of these and other pests as a result of the introduction of Bt cotton could be a challenge to integrated pest management (IPM) systems. For some pests such as *A. gossypii*, the main vector of cotton blue disease (CBD), increased abundance could be disastrous. The effects of Bt cotton on non-target herbivores could be positive (i.e. fewer pests or more non-pest herbivores) or negative (i.e. more pests or fewer non-pest herbivores) and reduced or more selective insecticide use on Bt cotton could allow greater biological control. In this chapter, we focus on the assessment of the risk of greater crop damage by non-target pest herbivores or disease transmitted by these pests. Here, we focus on the undesirable effects arising from increased abundance of a pest species, but there are additional issues to consider, such as reductions in the abundance of non-target herbivores that are of conservation significance, or have economic or cultural value (see Chapter 9, this volume). A well-known example is the monarch butterfly on Bt maize in the USA (Losey et al., 1999; Jesse and Obrycki, 2000).

The definition of target and non-target herbivores species has important implications in this context and is problematic. Registrations for many existing Bt crops are very specific in the species listed as controlled. For instance, only *H. armigera* and *H. punctigera* are listed as target species for Bollgard II® in Australia (Fitt and Wilson, 2005). Nevertheless, effects on other species will occur, including control, or partial control, of other lepidopteran pests or non-pest lepidopteran species susceptible to the Cry or VIP proteins. The former may be regarded as target species as they would be targeted for control when warranted by their abundance, whereas the latter would be truly non-target species. The lack of specific registrations for these minor pests may, in part, reflect the lack of willingness of the registrants to meet the additional costs associated with registration for species that are less important economically. However, classification of minor lepidopteran pests as targets requires a level of data or knowledge about the control efficacy of the transgenic crop that may not be available in Vietnam. Thus, for regulatory purposes, it may be less ambiguous to consider only the specified controlled species as target species and all other species as non-target species. Included in the non-target group would be any non-lepidopteran insect or mite species that feeds on cotton crops. However, although minor pest lepidopteran species may be regarded as non-targets in the context of this chapter, it may be necessary to consider them as targets in the development of insecticide resistance strategies for Bt cotton crops (Chapter 12, this volume).

Bt cotton has proven to be successful in other parts of the world (Chapter 1, this volume). Nevertheless, careful evaluation of identified potential environmental risks associated with non-target herbivores allows identification of potentially significant adverse effects and the development of strategies to prevent or ameliorate such effects. Potential adverse effects pathways that result in greater crop damage by pests, or greater damage from a crop disease transmitted by pests, are illustrated in Figs. 6.1 and 6.2 and are described below. Pathways 3–5 (below) are discussed further in Chapters 7 and 8 (this volume).

1. Bt cotton may be a better food resource and/or more attractive for non-target pests, leading to increased abundance of pests, such as mirids. This could be because of the lack of feeding damage that the target Lepidoptera would cause; for example, Bt cotton may have higher retention of squares and higher density of flowers (e.g. Whitehouse *et al.*, 2007).
2. Reduced or more selective insecticide use on Bt cotton may allow better survival of pests that formerly were controlled by insecticide treatments against the *Heliothis/Helicoverpa* species that Bt cotton targets; for example, mirids in China (Wu *et al.*, 2002), the USA (Head *et al.*, 2005; Greene *et al.*, 2006) and Australia (Lei *et al.*, 2003; Ward, 2005) and stink bugs, such as *Nezara viridula*, in Australia (Khan and Bauer, 2002) and the USA (Greene *et al.*, 2001, 2006).
3. Reduced abundance of target Lepidoptera on Bt cotton could lead to a reduced abundance of predators or parasitoids that feed on these Lepidoptera (Dively, 2005). This could release a non-target herbivore species from biological control.
4. Feeding on Bt cotton could have a sublethal effect on certain herbivores, particularly non-target Lepidoptera, either due to the effect of the Bt toxin or poorer food quality, or a mix of both effects, and this could affect the predator or parasitoid negatively (e.g. Schuler *et al.*, 2004), resulting in release of another herbivore. For example, *Pseudoplusia includens* (Lepidoptera: Noctuidae) has reduced growth rate and lower body weight on Cry1Ac cotton and this affects its parasitoids (Baur and Boethel, 2003).
5. Herbivores may be unaffected by feeding on Bt cotton but accumulate the Bt toxin in their bodies, which could affect their predators or parasitoids negatively, resulting in release of another herbivore. For example, spider mites and

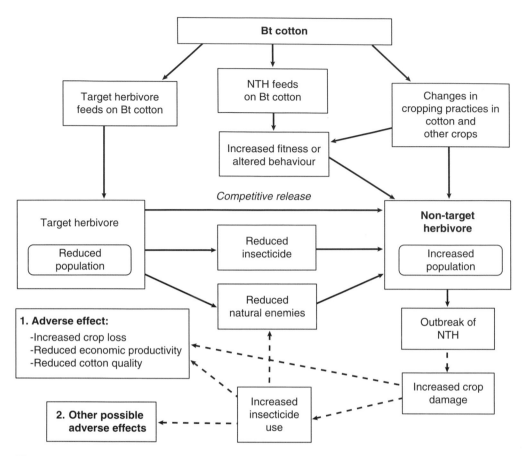

Fig. 6.1. Diagram of possible risk hypotheses for non-target herbivores associated with Bt cotton that could lead to pest outbreaks. Arrows represent causal connections. Additional risk hypotheses are examined in Chapters 7 and 8 (this volume). Large 'non-target herbivore' box includes the possible response (measurement end point) of a non-target herbivore (entity) to Bt cotton (stressor). Arrows leading into the 'non-target herbivore' box may affect the measurement end point in the box. Dashed lines indicate pathways that are not discussed in detail in this chapter. NTH = non-target herbivore.

thrips can accumulate much higher levels of Bt than are present in the plant (e.g. Dutton *et al.*, 2002; Torres and Ruberson, 2008).

For assessment of the potential risks associated with the effects of Bt cotton on non-target herbivores in cotton ecosystems in Vietnam, we used an environmental risk assessment model described in previous volumes of this series (Birch *et al.*, 2004, Sujii *et al.*, 2006). It uses five main steps: (i) list non-target herbivores and identify species for further evaluation; (ii) identify potential exposure pathways and potential adverse effect pathways; (iii) formulate risk hypotheses and design and conduct risk experiments based on the identified exposure pathways and adverse effect pathways to try to reject the risk hypothesis; (iv) reformulate the remaining risk hypotheses; and (v) design experiments

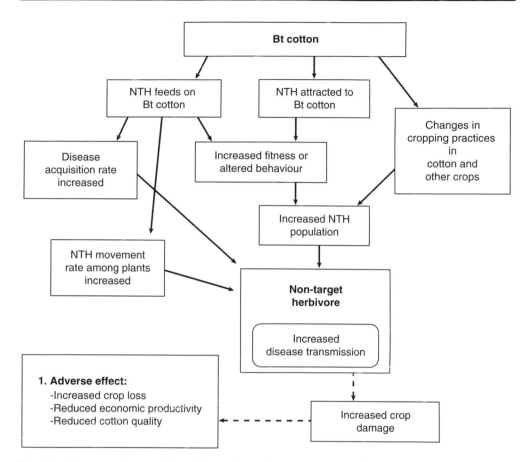

Fig. 6.2. Diagram of a possible risk hypothesis for non-target herbivores associated with Bt cotton that could lead to increased disease transmission. Arrows represent causal connections. Large 'non-target herbivore' box includes the possible response (measurement end point) of a non-target herbivore (entity) to Bt cotton (stressor). Arrows leading into the 'non-target herbivore' box may affect the measurement end point in the box. Dashed lines indicate pathways that are not discussed in detail in this chapter. NTH = non-target herbivore.

to test these hypotheses and characterize risk. This chapter will concentrate on the first three of these steps.

6.1. List Non-target Herbivores and Identify Species for Further Evaluation

Around 30 species of herbivorous insects and mites on cotton in Vietnam are common enough to be considered potential pests. As the target species of Bt cotton in Vietnam was identified as *H. armigera* (Liao *et al.*, 2002; Wan *et al.*,

2005; Llewellyn *et al.*, 2007), this species was removed from the list. For choosing the most relevant non-target herbivores from this list, a Selection Matrix was used to estimate association with crop and to estimate functional significance (Table 6.1). The criteria for association with the crop included geographic distribution, habitat specialization, prevalence (proportion of habitat occupied), abundance on cotton and phenology or synchrony between cotton and insect life cycle (defined in Chapter 5, this volume). The criteria correspond to a preliminary qualitative exposure assessment of the species. Each criterion is ranked in relation to the other species. The ranking was based on literature and information from experts from Vietnam and other cotton producing countries present in the group. In the absence of information for a particular species, some criteria were ranked with ? to indicate a knowledge gap that may deserve further study.

Normally, the ranking should be carried out separately for each cotton growing region of Vietnam. However, the northern cropped area is small compared to other regions and there is less knowledge or experience with the pests or species on cotton in this region, so we focused on the coastal lowlands, the central highlands and the south-eastern region (see Chapter 2, this volume). These areas are similar enough in the range of herbivores present to be considered together.

Pest phenology varies greatly between the rainy and dry seasons, so both were evaluated separately when sufficient information was available, with emphasis on those species that varied the most between seasons. As part of the phenology criteria, we also ranked the number of generations per season, because herbivores with faster life cycles will have more generations exposed to Bt cotton. There are gaps in the knowledge about the herbivore species listed in the matrix. This was addressed through a comparison with the same species in other crops or related species in cotton elsewhere in the world. When certain pieces of information were totally lacking, a precautionary approach assuming the highest rank was used.

All species were ranked for potential functional significance (Table 6.1), using the same procedure as for association with crop. Characteristics included role of species as a vector of disease, significance as a source of food for beneficials, significance as pests on other crops and damage level in the dry and rainy seasons. The issue of intercropping was considered important as herbivores may flow between these crops, both spatially and temporally. We placed emphasis on the major crops in typical crop systems in Vietnam, but also considered minor crops.

The averages for association with crop and for functional significance are then added to give an overall score, to select those species that are both associated most closely with the crop and where an effect associated with the species would have the greatest adverse impact. Thirteen species had a summed score less than four for dry and rainy season, which gave them high or intermediate priority. Five of these are Lepidoptera that will probably have high mortality on any Bt cotton that contains Cry1Ac: *P. gossypiella* (Wan *et al.*, 2004), *Anomis flava* (Cui and Xia, 2000), *Sylepta derogata* (Huang and Liu, 2005),

Table 6.1. Selection Matrix for herbivore pest species associated with cotton in Vietnam, showing scores for six association criteria and four significance criteria. High rank = 1, medium rank = 2, low rank = 3, gap of knowledge = ? Life stage that is a herbivore pest on cotton: all = all feeding life stages, L = larva, N = nymph, A = adult. Association with crop criteria: A1 = geographic distribution; A2 = habitat specialization: degree of association with cotton habitat; A3 = prevalence: proportion of suitable cotton habitat occupied; A4 = abundance on cotton crop; A5 = phenology: degree of overlap of herbivore life history with cotton; A6 = phenology: proportion of cotton growing season when the species is present: E = early, M = middle, L = late; 1 = All, 2 = E–M; 3= M or ML; $_D$ = dry season; $_R$ = rainy season; S = seedling stage, V = vegetative stage, R = reproductive stage. Functional significance criteria: D_D = damage level as pest of cotton in dry season; D_R = damage level as pest of cotton in rainy season; VD = significance as vector of disease; OC = significance as a pest of other crops; FN = significance as a food for natural enemies. M_A = mean of association with crop criteria; M_S = mean of functional significance criteria. Sum is the sum of the two means and rank is the final priority ranking used in species selection. The highest priority species are in bold.

Species/taxon	Order	Feeding guild	Pest stage	A1	A2	A3	A4	A5	A6	A6$_D$	A6$_R$	M$_{AD}$	M$_{AR}$	D$_D$	D$_R$	OC	VD	FN	M$_{SD}$	M$_{SR}$	Sum$_D$	Sum$_R$	Rank
Pectinophora gossypiella Saunders	Lepidoptera: Gelechiidae	Square, boll	L	1	2	2	2	1	R	?	?	1.60	1.60	2	2	3	3	1	2.25	2.25	3.85	3.85	9
Anomis flava F.	Lepidoptera: Noctuidae	Leaf	L	1	2	2	2	1	V	?	?	1.60	1.60	3	3	1	3	1	2.00	2.00	3.60	3.60	6
Earias vitella F.	Lepidoptera: Noctuidae	Boll	L	1	2	2	2	1	R	?	?	1.60	1.60	3	3	1	3	1	2.00	2.00	3.60	3.60	6
Spodoptera litura (F.)	Lepidoptera: Noctuidae	Leaf	L	1	2	2	2	1	V	1	3	1.50	1.83	3	3	1	3	1	2.00	2.00	3.50	3.83	7
Spodoptera exigua (Hübner)	Lepidoptera: Noctuidae	Leaf	L	?	2	2	2	1	V	1	3	1.60	2.00	3	3	1	3	1	2.00	2.00	3.60	4.00	8
Ostrinia furnacalis (Guenée)	Lepidoptera: Pyralidae	Stem	L	1	2	2	2	2	V	?	?	1.80	1.80	3	3	1	3	1	2.00	2.00	3.80	3.80	8
Sylepta derogata F.	Lepidoptera: Pyralidae	Leaf	L	1	2	2	2	1	V	?	?	1.60	1.60	3	3	1	3	1	2.00	2.00	3.60	3.60	6
Tetranychus spp.	Acarina: Tetranychidae	Leaf	All	1	2	2	2	1	V	?	3	1.60	1.83	2	3	1	3	1	1.75	2.00	3.35	3.83	5
Acrida chinensis (Westwood)	Orthoptera: Acrididae	Leaf	All	1	2	2	2	2	?	?	?	1.80	1.80	3	3	2	3	3	2.75	2.75	4.55	4.55	13
Attractomorpha sp.	Orthoptera: Acrididae	Leaf	All	2	2	2	2	2	?	?	?	2.00	2.00	3	3	2	3	3	2.75	2.75	4.75	4.75	14

Continued

Table 6.1. Continued

Species/taxon	Order	Feeding guild	Pest stage	A1	A2	A3	A4	A5	A6	$A6_D$	$A6_R$	M_{AD}	M_{AR}	D_D	D_R	OC	VD	FN	M_{SD}	M_{SR}	Sum_D	Sum_R	Rank
																					Summary		
				Association with crop										Functional significance									
Oxya sp.	Orthoptera: Acrididae	Leaf	All	2	2	2	2	2	?	?	?	2.00	2.00	3	3	2	3	3	2.75	2.75	4.75	4.75	14
Patanga sp.	Orthoptera: Acrididae	Leaf	All	2	2	2	2	2	?	?	?	2.00	2.00	3	3	2	3	3	2.75	2.75	4.75	4.75	14
Trilophidia annulata Thunberg	Orthoptera: Acrididae	Leaf	All	2	2	2	2	2	?	?	?	2.00	2.00	3	3	2	3	3	2.75	2.75	4.75	4.75	14
***Bemisia tabaci* (Gennadius)**	Homoptera: Aleyrodidae	Leaf	All	1	2	2	2	1	S,V	?	?	1.60	1.60	3	3	1	1	1	1.50	1.50	3.10	3.10	2
***Aphis gossypii* Glover**	Homoptera: Aphididae	Leaf	All	1	2	2	2	1	S,V	1	2	1.50	1.67	1	2	1	1	1	1.00	1.25	2.50	2.92	1
***Amrasca devastans* Distant**	Homoptera: Cicadellidae	Leaf	All	1	2	2	2	1	V	2	3	1.67	1.83	2	3	1	1	1	1.25	1.50	2.92	3.33	3
Pseudococcus citriculus Green	Homoptera: Pseudococcidae	Leaf	All	2	2	2	2	1	?	?	?	1.80	1.80	3	3	2	3	1	2.25	2.25	4.05	4.05	11
Riptortus sp.	Heteroptera: Coreidae	Leaf, square, boll	All	2	2	2	2	1	?	?	?	1.80	1.80	3	3	2	3	3	2.75	2.75	4.55	4.55	13
Cletus spp.	Heteroptera: Coreidae	Leaf	All	2	2	2	2	1	?	?	?	1.80	1.80	3	3	2	3	3	2.75	2.75	4.55	4.55	13
Leptocorisa sp.	Heteroptera: Coreidae	Leaf	All	2	2	2	2	1	?	?	?	1.80	1.80	3	3	2	3	3	2.75	2.75	4.55	4.55	13
Lygus sp.	Heteroptera: Miridae	Leaf, square	All	2	2	2	2	1	V,R	?	?	1.80	1.80	3	3	2	2	3	2.50	2.50	4.30	4.30	12

Species	Order: Family	Plant part																					
Piezodorus rubrofasciatus (F.)	Heteroptera: Pentatomidae	Leaf, square, boll	All	2	2	2	1	?	?	?	1.80	1.80	3	3	2	3	3	2.75	2.75	4.55	4.55	13
Nezara viridula L.	Heteroptera: Pentatomidae	Leaf, square, boll	All	2	2	2	1	R	?	?	1.80	1.80	3	3	2	3	3	2.75	2.75	4.55	4.55	13
Dysdercus cingulatus (F.)	Heteroptera: Pyrrhocoridae	boll	All	2	2	2	1	R	?	?	1.80	1.80	3	3	2	3	3	2.75	2.75	4.55	4.55	13
Scirtothrips dorsalis Hood	Thysanoptera: Thripidae	Leaf	All	1	2	2	1	S	1	3	1.50	1.83	1	3	1	1	3	1.50	2.00	3.00	3.83	4
Thrips palmi Karny	Thysanoptera: Thripidae	Leaf	All	1	2	2	1	S	1	3	1.50	1.83	1	3	1	1	3	1.50	2.00	3.00	3.83	4
Liriomyza phaseoli (Tryon)	Diptera: Agromyzidae	Leaf	All	2	2	2	2	V	?	?	2.00	2.00	3	3	1	3	1	2.00	2.00	4.00	4.00	10
Monolepta spp.	Coleoptera: Chrysomelidae	Leaf	A	2	2	2	2	?	?	?	2.00	2.00	3	3	2	3	3	2.75	2.75	4.75	4.75	14
Aulacophora sp.	Coleoptera: Chrysomelidae	Leaf	A	2	2	2	2	?	?	?	2.00	2.00	3	3	2	3	3	2.75	2.75	4.75	4.75	14
Hypomeces sp.	Coleoptera: Chrysomelidae	Leaf	A	2	2	2	2	?	?	?	2.00	2.00	3	3	2	3	3	2.75	2.75	4.75	4.75	14

Ostrinia furnacalis (He *et al.*, 2006) and *Earias vitella* (Kranthi *et al.*, 2004). The other eight species are:

1. *Aphis gossypii* (Homoptera: Aphididae), cotton aphid.
2. *Amrasca devastans* (Homoptera: Cicadellidae), cotton leafhopper or jassid.
3. *Bemisia tabaci* (Homoptera: Aleyrodidae), tobacco whitefly.
4. *Thrips palmi* (Thysanoptera: Thripidae), melon thrips.
5. *Scirtothrips dorsalis* (Thysanoptera: Thripidae), chilli thrips.
6. *Tetranychus* spp. (Acarina: Tetranychidae), spider mites.
7. *Spodoptera exigua* (Lepidoptera: Noctuidae), beet armyworm.
8. *S. litura* (Lepidoptera: Noctuidae), armyworm or clust caterpillar.

All of the selected species are pests in all the cotton growing areas of Vietnam, so the adverse effect would be an increase in the crop damage they cause and associated costs, such as more intensive pest management measures. *A. gossypii* causes crop damage through transmission of CBD. The biology of these species is described in Chapter 2 (this volume). *B. tabaci* was flagged as important due to potential rather than actual disease risk, based on experience from other countries (Akhtar *et al.*, 2004; Briddon, 2007). The accidental introduction of a whitefly-transmitted cotton disease could be a disaster in Vietnam, especially if the use of Bt cotton led to increases in the abundance of this pest. All the species are also significant pests on other crops that are grown in relay, intercrop or mixed cropping systems with cotton, such as beans, maize or vegetables.

Although heteropteran sucking bugs, including *N. viridula* and mirid species, are ranked low for prevalence, abundance and damage potential in cotton in Vietnam, these species and other sucking bugs have emerged as problems on Bt cotton in other countries, and this is a possible risk for Vietnam (Greene *et al.*, 2001, 2006; Khan and Bauer, 2002; Wu *et al.*, 2002; Ward, 2005; Whitehouse *et al.*, 2007).

6.2. Identify Potential Exposure Pathways

Bt proteins are expressed in all growing tissues of Bt cotton throughout the growing season, including flower and fruit tissues, roots and seed endosperm, and concentrations are generally highest in the leaves (see Chapter 4, this volume, for details). Concentrations can vary considerably in different genetic backgrounds, environmental conditions and over the growing season. The eight selected species were evaluated for their likely exposure to Bt proteins, considering exposure via multiple trophic levels.

The potential for bitrophic exposure due to feeding on Bt cotton plant tissues and parts was evaluated for each species by considering the herbivore feeding habit, expression of the transgene product in the parts it feeds on and any documented detection of the toxins in the herbivore (Table 6.2).

- *A. gossypii* lives on the undersides of cotton leaves and penetrates through cotton tissues directly to feed on phloem sieve tubes. On the way to locating phloem cells, aphids may also penetrate some mesophyll cells; however,

Table 6.2. Potential exposure assessment for Bt cotton. Assessment is based primarily on the literature on Cry1Ac cotton. ? indicates uncertainty. As several of the other Bt cottons use a similar promoter, their expression is expected to be similar, although this should be confirmed.

Selected species		*Aphis gossypii*	*Thrips palmi*	*Scirtothrips dorsalis*	*Tetranychus urticae*	*Amrasca devastans*	*Bemisia tabaci*	*Spodoptera exigua*	*Spodoptera litura*
BITROPHIC EXPOSURE	Which life cycle stages occur in cotton?	All	All	All	All	All	All	Larvae	Larvae
	Growth stage of cotton when present (early, mid, reproductive).	All season	Early	Early	All	All	All	?	?
	Does the species feed on plant tissues or products containing the transgene product?	Yes?	Yes	Yes	Yes	Yes	Yes?	Yes	Yes
	Is bitrophic exposure possible?	Yes?	Yes	Yes	Yes	Yes	Yes?	Yes	Yes
	Have the transgene product or metabolites been detected in the species after feeding?	Yes?	Yes	Yes	Yes	Yes	Yes?	Yes	Yes
	Does bitrophic exposure occur?	Yes?	Yes	Yes	Yes	Yes	Yes?	Yes	Yes
TRITROPHIC EXPOSURE	Does the species feed on animal products or excretions that contain the transgene product?	No	No	No	No	No	No	Yes	Yes
	Is tritrophic exposure through animal products possible?	No	No	No	No	No	No	Yes	Yes
	Does tritrophic exposure through animal products occur?	No	No	No	No	No	No	?	?

Continued

Table 6.2. *Continued*

Selected species		Aphis gossypii	Thrips palmi	Scirtothrips dorsalis	Tetranychus urticae	Amrasca devastans	Bemisia tabaci	Spodoptera exigua	Spodoptera litura
OTHER TROPHIC EXPOSURE	Is the species cannibalistic?	No	Yes?	Yes?	No	No	No	Yes	Yes
	Does it eat other species in its own feeding guild that are exposed?	No	? Mite eggs	?	No	No	No	?	?
	Is other trophic exposure or exposure through cannibalism possible?	No	?	?	No	No	No	Yes	Yes
EXPOSURE VIA GENE FLOW RECIPIENTS	Can gene flow occur?	Only on a small scale in restricted areas (see Chapter 11, this volume)							
	If yes, could the species feed on plants or eat prey on plants that have received the transgene through gene flow?	Yes?	Yes?	Yes?	Yes?	Yes?	Yes?	Yes?	Yes?
	Is exposure via gene flow recipients possible?	? Only on a small scale	? Only on a small scale	? Only on a small scale	? Only on a small scale	? Only on a small scale	? Only on a small scale	? Only on a small scale	? Only on a small scale
EXPOSURE BY CHANGED BEHAVIOUR	Is the transgenic plant less attractive?	?	?	?	?	?	?	?	?
	Does it have a lower food value than the non-transgenic cultivar?	?	?	?	?	?	?	?	?
	Does the species avoid food containing the transgene product or metabolites?	?	?	?	?	?	?	?	?
	Are behaviours likely to increase or decrease exposure?	?	?	?	?	?	?	?	?
SUMMARY	IS EXPOSURE POSSIBLE?	Yes	Yes	Yes	Yes	Yes	Yes	Yes	Yes
	DOES EXPOSURE OCCUR?	Yes?	Yes	Yes	Yes	Yes	Yes?	Yes	Yes

once the stylet is established in the phloem cells, this is where the vast majority of feeding occurs (Leclant and Deguine, 1994). It is not known whether the Cry toxins are present in significant concentrations in Bt cotton phloem fluids. Zhang *et al.* (2006a) found trace quantities of Cry1Ac in cotton aphids fed on Chinese Bt cotton (6.0 ng/g body weight of aphids, which was about 4% of the concentration in the Bt cotton leaf tissue).

- Whiteflies damage leaf mesophyll cells as they penetrate leaf tissues, but feed mainly from phloem (Freeman *et al.*, 2001).
- Spider mites live on the undersides of cotton leaves and use piercing mouthparts to pierce the epidermis and suck out the contents of the underlying mesophyll cells (Reddall *et al.*, 2004). Torres and Ruberson (2008) found that *T. urticae* fed on Bt cotton contained 16.8 times the leaf concentration of Cry1Ac. Similar results have been found with *T. urticae* feeding on Bt maize (Dutton *et al.*, 2002; Obrist *et al.*, 2006a,b).
- Thrips attack seedling cotton cotyledons, young leaves and stems, and young and mature thrips reside along leaf veins, lacerate the epidermis and suck mesophyll cell contents. Thrips also feed on flower tissues, pollen and nectar (Kirk, 1997; Parajulee *et al.*, 2006). Torres and Ruberson (2008) found that *Frankliniella occidentalis* immatures and adults contained two-thirds of the leaf concentration of Cry1Ac after feeding on Bt cotton. On Bt maize, *F. tenuicornis* was shown to acquire high concentrations of Cry1Ab toxin in laboratory studies (Obrist *et al.*, 2005), though only low levels were found in field studies (Obrist *et al.*, 2006b).
- Both immature and adult leafhoppers suck on mesophyll leaf tissue and phloem (Backus *et al.*, 2005) and are therefore likely to be exposed similarly to thrips and mites.
- *Spodoptera* species chew leaf tissue and have been shown to take up and accumulate significant quantities of Cry toxins from cotton (Torres *et al.*, 2006; Zhang *et al.*, 2006b; Torres and Ruberson, 2008).

Therefore, all species are likely to be exposed to Bt proteins by feeding on Bt cotton; whitefly and *A. gossypii* are likely to be exposed to trace amounts only in the cotton mesophyll, but will be exposed if cotton phloem contains Bt proteins.

The possibility of other exposure pathways through feeding on excretions and frass of other herbivores and cannibalism or predation (higher trophic interactions) was also considered. The *Spodoptera* species are known to be cannibalistic and the thrips species can be cannibalistic at high densities and when alternative food is not available (Kirk, 1997), which could increase exposure for survivors. *T. palmi* may feed on mite eggs (Kirk, 1997) and *Spodoptera* eat lepidopteran faeces (frass), so these species additionally could be exposed through these tritrophic pathways. The frass of the *Spodoptera* species has been shown to contain Cry toxins (Raps *et al.*, 2001). It is not known if mites pass on ingested Bt to their eggs, although it is probably unlikely. None of the other selected herbivores are predacious or feed on products or excretions of other herbivores, except perhaps incidentally, and, if this occurs, it is not an important component of the diet. However, all the herbivore species that ingest Bt are likely to also excrete the Cry toxin and/or its metabolites. For example, the Homoptera produce copious amounts of honeydew or similar liquid excretions that can coat plant leaves and Bernal *et al.* (2002) found some evidence of the presence of a Cry toxin in the honeydew of the brown plant hopper feeding on Bt rice.

Exposure on cotton relatives that might receive a Bt gene via gene flow is not considered to be important, as gene flow is likely to be restricted to a small scale in Vietnam (Chapter 11, this volume). Possible changes in exposure due to changes in herbivore behaviour on Bt cotton, related to the oviposition preference, were also considered (Table 6.2). The assessment of herbivore exposure to Bt cotton revealed a lack of basic information about feeding and oviposition preferences. In the absence of this information, we adopted a conservative, precautionary approach and considered that the preferences might exacerbate potential problems, thereby providing, in some cases, some justification for additional studies to clarify these possible risks.

Potential exposure pathways have been evaluated (Table 6.2) and all eight herbivore species are possibly exposed to Bt proteins. All species, except perhaps the aphids and whitefly, are also likely to accumulate significant amounts of Bt during most, or part, of the growing season, depending on what tissues they preferentially feed on and at what cotton growth stage they are abundant.

6.3. Identify Potential Adverse Effect Pathways and Damage Potential

For the eight selected species, the following section identifies the possible pathways leading to potential adverse effects (Figs. 6.1 and 6.2), such as increased fitness and/or increased attraction to Bt cotton, release from insecticide control and possible effects via predators or parasitoids. Available research evidence from other countries for the first three pathways is evaluated here; evidence for effects on natural enemies is considered in Chapters 7 and 8.

Aphis gossypii Glover (Homoptera: Aphididae), Cotton Aphid

The importance of *A. gossypii* as a cotton pest is increasing throughout the cotton producing regions of the world and, in Vietnam, the main adverse effect mediated by *A. gossypii* is as a vector of CBD, which causes yield loss and reduced fibre quality (Chapter 2, this volume). Aphids can also cause economic damage during the boll-opening phase, because of honeydew contamination of open bolls, resulting in sticky lint. Cotton aphids are also pests of many other crops. Although they are typically the targets of insecticide applications in Vietnam, it is still possible that aphid populations would be released later in the season by a reduction in insecticide applications applied to control lepidopteran pests.

Liu *et al.* (2005) found that *A. gossypii* fed on Cry1A + CpTI cotton had significantly shorter reproductive duration in the first generation, but equivalent survival rates and fecundity in the second and third generations, compared to those fed on non-transgenic cotton. However, they did not find any significant differences between aphids on Bt cotton expressing only Cry1Ac compared to the non-Bt cotton, indicating that the transitory effect on the first aphid generation on Cry1A + CpTI cotton could have been related to feeding inhibition from the cowpea trypsin inhibitor, which was compensated for in the subsequent generations. Aphids have not been a significant problem in Bt cotton crops in northern and eastern China (Sun *et al.*, 2003; Wu and Guo, 2003, 2005), but have been reported as a problem in irrigated Bt cotton compared to non-Bt cotton in Xinjiang (Sun *et al.*, 2002). Abundances of aphids on Bt cotton were very variable over three seasons in China (Men *et al.*, 2004). Aphids were the most abundant pests found on unsprayed Cry1Ac cotton in field experiments in Australia (Whitehouse *et al.*, 2005). Lower aphid densities were found on VIP3A cotton in Australia compared to common non-Bt varieties (Whitehouse *et al.*, 2007). No information is available on the effect of any Bt crop on transmissibility of insect-vectored diseases.

Adverse effects associated with *A. gossypii* in Vietnam are:

- Reduction in Bt cotton yield and fibre quality caused by CBD. Transmission by cotton aphids within the first 10 weeks of growth causes severe damage, even when aphid population density is low.
- Reduction in Bt cotton yield and fibre quality through cotton aphid feeding and honeydew contamination during boll maturation. This damage is usually not significant.
- Possible adverse effect pathways involving *A. gossypii* that lead to reduced cotton yield or fibre quality are:
  - Increased fitness of cotton aphids on Bt cotton.
  - Increased attractiveness of Bt cotton for cotton aphids.
  - Increased transmission and prevalence of CBD.
  - Release from insecticides used to control lepidopteran pests on conventional cotton.
  - Release from natural enemies feeding on cotton aphids on Bt cotton.

Amrasca devastans (Distant) (Homoptera: Cicadellidae), Cotton Leafhopper

The adverse effect associated with *A. devastans* on cotton in Vietnam is yield loss caused by their feeding on cotton leaves. They cause damage at low densities early in the season, particularly in the dry season when populations develop rapidly from 10–15 days after sowing onwards, but, at high densities, can also cause damage later in the season (Chapter 2, this volume). A key factor limiting leafhopper damage in Vietnam is leaf hairiness, which is a feature of the most commonly used varieties. It is likely that Bt cotton will be in a similar hairy leaf background and so should maintain a similar level of leafhopper resistance. However, if smooth-leaf Bt cotton varieties were introduced, they would be more susceptible to leafhopper damage. Sharma and Pampapathy (2006) found no evidence of increased susceptibility or resistance of Indian Bt varieties compared to a range of conventional varieties, but found that the degree of hairiness had a big influence.

Adverse effects associated with A. *devastans* in Vietnam are:

- Reduction in Bt cotton yield caused by injury to leaves throughout the season.
- Possible adverse effect pathways involving A. *devastans* that lead to reduced cotton yield.
- Increased fitness of cotton leafhopper on Bt cotton.
- Bt cotton could have lower resistance and/or higher attractiveness to cotton leafhopper.
- Cotton leafhopper populations could be released by reduced pyrethroid insecticide use.
- Release from natural enemies feeding on cotton leafhopper on Bt cotton.

Bemisia tabaci (Gennadius) (Homoptera: Aleyrodidae), Tobacco Whitefly

B. *tabaci* is not yet a major pest on cotton in Vietnam, but it is an important insect pest of cotton worldwide because of its role as a vector for Gemini viruses causing several significant cotton diseases, such as cotton leaf curl virus (CLCuV), its potential to reduce cotton yield and reduce cotton quality by damaging plants and contaminating cotton lint with honeydew and its high potential to develop resistance to insecticides, especially the B-Biotype. This pest is also an effective vector of over 60 viruses on other crops, many of which cause economic damage, particularly to vegetables and ornamental crops.

Tobacco whitefly is an important pest of Bt cotton under irrigation in southwestern USA (Naranjo, 2005). Whitehouse et al. (2007) found that VIP3A cotton had higher whitefly populations, possibly due to the lusher plant canopy and the hairier leaves of the VIP3A variety (Coker 312) compared to non-transgenic Australian cotton varieties. Sharma and Pampapathy (2006) found no evidence of increased susceptibility or resistance of Indian Bt varieties compared to a range of conventional varieties. Guo et al. (2004) found significant differences between the quality of B. *tabaci* as food for larvae of the lacewing (*Chrysopa sinica*) and the coccinellid *Propylaea japonica* when the whiteflies were raised on Cry1Ac cotton (NuCotn 33B), on Bt cotton with the fused Cry1Ab/Ac transgene (GK12) and on conventional parent varieties, but the reasons are unclear and it is not known if the whitefly contained Cry toxins.

Adverse effects associated with B. *tabaci* in Vietnam are:

- Direct damage to cotton currently is not significant, but tobacco whiteflies are significant pests of other crops grown near cotton.
- Future possibility of significant damage to cotton if a tobacco whitefly-transmitted disease becomes prevalent in Vietnam.

Possible adverse effect pathways involving B. *tabaci* that could lead to adverse effects:

- Increased fitness of tobacco whitefly on Bt cotton.
- Bt cotton could have higher attractiveness to tobacco whiteflies.
- Tobacco whitefly populations could be released by reduced pyrethroid insecticide use.

- Release from natural enemies feeding on tobacco whitefly on Bt cotton.
- Increased transmission and prevalence of tobacco whitefly-transmitted disease (although no such diseases are reported presently from Vietnam).

Thrips palmi Karny and *Scirtothrips dorsalis* Hood (Thysanoptera: Thripidae)

During the rainy season in Vietnam, thrips appear only at the end of the growing season and do not affect cotton yield. However, during the dry season, thrips appear early and cause injury to cotton through most of the growing season. They cause most injury to seedling cotton, attacking the cotyledons or young leaves and, if sufficiently abundant, they will damage the apical meristem. Damage to the apical meristem may lead to delays and asynchrony in crop maturation, which can spread out the period of harvest and place portions of the yield at increased risk, depending on the weather. During the boll maturation period, very high numbers of thrips can cause loss of squares and distorted leaves, though no data are available on whether this reduces yield (Chapter 2, this volume).

No information is available on the possible effects of any Bt crop or the specific Bt toxins in Bt cotton on the thrips species of concern. No differences in the population densities of other thrips species were observed in Bt (Cry1Ac and VIP3A) cotton in Australia (Whitehouse *et al.*, 2005, 2007).

Thrips are often controlled by insecticides that target other pests, including imidacloprid seed treatments targeting aphids and insecticides applied against *Helicoverpa* sp. or *Heliothis* spp., so they could be released by decreased spraying. On the other hand, there is evidence that thrips abundance and damage are increased by the application of some insecticides (Etienne *et al.*, 1990), because these insecticides kill their predators but not the thrips (e.g. Shibao *et al.*, 2004), and decreased spraying could allow better biological control. The possibility that these species would be released by decreased insecticide use is therefore uncertain.

Potential adverse effects associated with *T. palmi* and *S. dorsalis* in Vietnam are:

- Possible yield reductions in dry season Bt cotton, particularly from early season injury.
- Reductions of *T. palmi* on Bt cotton could reduce their predation of spider mites, leading to increased spider mite problems.
- An increase in thrips populations in Bt cotton could spill over on to other crops, which could lead to increased thrips damage and prevalence of thrips-transmitted diseases on vegetables and ornamental crops.

Possible adverse effect pathways involving thrips that could lead to reduced cotton yield in Vietnam are:

- Increased fitness of thrips on Bt cotton.
- Bt cotton could have lower resistance or higher attractiveness to thrips.
- Thrips populations could be released by reduced insecticide use.
- Release from natural enemies feeding on thrips on Bt cotton.

Tetranychus urticae Koch and other *Tetranychus* spp. (Acarina: Tetranychidae)

The two-spotted spider mite, *T. urticae*, is a widespread pest of cotton and several species of *Tetranychus* spider mites occur in Vietnam. Spider mites are typically secondary pests in cotton systems, induced by spraying for other primary pests such as *Helicoverpa* or *Heliothis* spp., because the insecticides used often do not affect mites but kill their main predators, which include predatory bugs (e.g. *Geocoris* spp., *Orius* spp. and *Deraeocoris* spp.), coccinellids and thrips (Wilson *et al*., 1998). In addition, some insecticides, such as the pyrethroids, may stimulate the spider mites to disperse (Penman and Chapman, 1988). In Vietnam, mites are only a problem in dry season cotton after applications of insecticides to control thrips and bollworm, but they are important pests of other crops such as cassava, chilli and fruit (Chapter 2, this volume), so any adverse effects may occur primarily in these crops if they are relayed or planted close to Bt cotton.

Research from China found no increase in spider mite populations in unsprayed Bt cotton containing Cry1A + CpTI or Cry1Ac compared with unsprayed conventional cotton (Ma *et al*., 2006) and sprayed Bt versus sprayed non-Bt cotton (Men *et al*., 2004). Deng *et al*. (2003), however, found that cumulative and peak numbers of spider mites were higher on Cry1A + CpTI cotton both on fields that used pesticide control and on fields with no insecticide use, compared to conventional cotton managed with IPM techniques, even though the Bt fields had higher natural enemy populations. In general, it is unlikely that spider mite populations would be released by reduced insecticide use, as typically they are induced pests.

It is unlikely that feeding on Bt cotton has an effect on spider mites (Reddall *et al*., 2004; Torres and Ruberson, 2008); however, as spider mites have been shown to accumulate significant quantities of Cry toxins in an active form (Obrist *et al*., 2006a), they could be part of adverse effect pathways involving effects on their predators, including predatory mites and coccinellids.

Adverse effects associated with *Tetranychus* spp. in Vietnam are:

• Damage in dry season cotton after insecticide applications.
• Damage to other nearby crops such as cassava, chilli and fruit.

Possible adverse effect pathways involving *Tetranychus* spp. that could lead to reduced cotton yield are:

• Increased fitness of spider mites on Bt cotton.
• Increased attractiveness of Bt cotton to spider mites.
• Release from natural enemies feeding on spider mites on Bt cotton.

Spodoptera exigua (Hübner) and *S. litura* (Fabricius) (Lepidoptera: Noctuidae)

The larvae of these moths can be pests of cotton in Vietnam and *S. exigua* (beet armyworm) is more of a problem than *S. litura*. *S. exigua* is more abundant in the dry season than the rainy season; *S. litura* is more abundant in the rainy season. Both species cause damage to many other kinds of crop plants, including maize, tobacco, beans, groundnut, flowers and vegetables.

S. litura is affected sublethally by Cry1Ac in Chinese Bt cotton (Zhang *et al.*, 2006b), though unaffected by the Chinese Cry1Ab/Ac fused transgene (Guo *et al.*, 2003; Zhang *et al.*, 2006b). In the USA, *S. exigua* is affected sublethally by Cry1Ac cottons and suffers high mortality from the combination of Cry1Ac and Cry2Ab (Adamczyk *et al.*, 2001; Stewart *et al.*, 2001; Chitkowski *et al.*, 2003), VIP3A (Cloud *et al.*, 2004), or the combination of Cry1Ac and Cry1F (Haile *et al.*, 2004). The response of *S. exigua* to the Chinese Bt cottons may be similar to that of *S. litura*, although this remains to be confirmed. Similarly, Cry1Ab in Bt maize prolongs *S. littoralis* development, reduces survival and results in smaller adults (Dutton *et al.*, 2002). *S. frugiperda* and *S. eridania*, however, show much lower sensitivity to Bt cottons (Adamczyk *et al.*, 2001; Chitkowski *et al.*, 2003). Thus, it is likely that several of the Bt cottons will cause sublethal effects on *S. exigua* and *S. litura* in Vietnam and will not directly cause these species to be more abundant (see also Chapter 12, this volume). It is possible, however, that the reaction of the two Vietnamese *Spodoptera* species may differ. If the Chinese Cry1Ab/Ac fused transgene is used in Vietnamese cotton, *Spodoptera* species may be unaffected. In this case, their populations in dry season cotton could increase due to reduced use of insecticides that control both *H. armigera* and *Spodoptera* species, including pyrethroids, lufenuron, chlorfluazuron, tebufenozide, chlorfenapyr, abamectin or spinosad (see Chapter 2, this volume).

Adverse effects associated with *Spodoptera* spp. in Vietnam are:

* Damage to seedlings and leaves early in the season, reducing cotton yield.
* Damage to squares and flowers from midseason, reducing cotton yield.
* Damage to other crops near cotton.

Possible adverse effect pathways associated with *Spodoptera* that could lead to reduced cotton yield are:

* Increased fitness of *Spodoptera* spp. on Bt cotton.
* Higher attractiveness of Bt cotton to *Spodoptera* spp.
* Possible release of population from reduction in insecticide use against *H. armigera*.
* Release from natural enemies feeding on *Spodoptera* spp. on Bt cotton.

6.4. Formulate Risk Hypotheses

A risk hypothesis is a chain of causal links, starting from Bt cotton and ending with the characteristic of the selected herbivore that is connected closely to an undesirable adverse effect (US EPA, 1998). Exposure is a necessary prerequisite for any subsequent effects associated directly with the transgene product. The absence of these products renders additional studies on the direct effects of the transgene products unnecessary. However, it is still possible that Bt cotton has effects independent of or interacting with the transgene products. Considering the pathways in Figs. 6.1 and 6.2, and the potential exposure pathways and adverse effects pathways, four generic risk hypotheses can be formulated to describe the ways Bt cotton might result in an adverse effect via effects on non-target herbivores.

Risk hypothesis 1. Increased attractiveness and/or food quality of Bt cotton results in higher herbivore populations and greater damage to cotton

The risk hypothesis requires that Bt cotton has increased nutritional value and/or higher abundance of squares and flowers, providing better food resources for non-target herbivores. This could be because of the lack of feeding damage from the controlled target Lepidoptera and/or because of differences between the transgenic plant and the non-Bt cotton varieties in phenology, morphology, or chemical composition, due to its varietal background or the transformation and subsequent breeding process, as well as expression of the Cry toxins (see Chapter 4, this volume). This hypothesis includes both direct effects of the transgene products, all interactions of the transgene in the Bt cotton plant and other differences between the Bt cotton variety and the non-Bt varieties used as control. It can be subdivided into two subhypotheses that differentiate between increased population fitness (survival and vigour) and increased population size (reproduction and immigration) related to increased attractiveness or improved food quality of Bt cotton. If both are unlikely, then one may conclude that attractiveness and food quality in Bt cotton are unlikely to cause greater damage from non-target pests. The subhypotheses are unlikely for any particular herbivore if the Bt cotton suppresses its population.

Risk hypothesis 1.1
Improved food quality of Bt cotton results in higher immature survival, faster development, reduced pre-reproductive period, larger size or longer lifespan of *A. devastans, A. gossypii, B. tabaci, T. palmi, S. dorsalis* or *Tetranychus* species, which increases their absolute fitness. This may result in a higher population of herbivores, leading to the adverse effects identified for each species.

Risk hypothesis 1.2
Increased attractiveness of Bt cotton increases immigration and/or oviposition of *A. devastans, A. gossypii, B. tabaci, T. palmi, S. dorsalis* or *Tetranychus* species on Bt cotton. This may result in higher herbivore pest populations, leading to the adverse effects identified for each species.

Risk hypothesis 2. Effects of Bt cotton on *Aphis gossypii* results in higher transmission of vectored plant pathogens (Fig. 6.2)

This risk hypothesis requires that *A. gossypii* associated with Bt cotton transmits CBD in greater numbers and/or more rapidly and thereby causes greater damage. Pathogen transmission is higher if there are more vectors, a larger reservoir, or faster transmission. If Bt cotton results in higher populations of *A. gossypii* (Hypotheses 1.1, 1.2 or 3.1), the following hypothesis may apply.

If diseases transmitted by *B. tabaci* become established in Vietnam, a similar hypothesis involving *B. tabaci* could be developed.

Risk hypothesis 2.1
Higher populations of *A. gossypii* on Bt cotton in the first 10 weeks of growth can mean that more vectors carry the CBD pathogen, increasing the probability that each cotton plant becomes infected.

Risk hypothesis 3. Release from insecticides (Fig. 6.1)

Bt cotton in Vietnam is expected to reduce the need to spray insecticides to control *H. armigera*, particularly on dry season cotton from midseason onwards (see Chapter 1, this volume). The possible reduction of two to five pyrethroid or abamectin insecticide sprays in dry season Bt cotton from mid to late season could result in release of herbivore populations that were controlled previously by these insecticide sprays. The main insecticides used to control heliothine pests on cotton in Vietnam which also control some non-target pest species are pyrethroids (cypermethrin, deltamethrin, cyfluthrin and beta-cyfluthrin, lambda-cyhalothrin) and abamectin (see Table 2.2 in Chapter 2, this volume).

Risk hypothesis 3.1
The reduction of two to five pyrethroid or abamectin insecticide sprays in Bt cotton in dry season cotton results in higher populations of *A. devastans*, *T. palmi*, *S. dorsalis*, or *B. tabaci*, and the increased populations cause significant damage to cotton.
      Under this hypothesis, the Bt crop itself has no effect on the non-target herbivore, but the herbivore population increases because they are no longer killed by the insecticide. This risk hypothesis requires that a non-target herbivore is controlled originally by insecticides against the target pest and, when the pesticides and target pest are removed from the cotton, it can increase its population and cause damage to the cotton. It is assumed that increased beneficial populations are insufficient to control these pests.

Risk hypothesis 4. Indirect effects via natural enemies

Risk hypothesis 4.1
S. litura and *S. exigua* are affected sublethally by Bt cotton and have a longer development time, lower larval survival and body weight and decreased pupation rate. This altered fitness may affect negatively the predators or parasitoids feeding on *Spodoptera* spp., and the decreased predator/parasitoid population would lead to the release of a different pest from biological control.

Risk hypothesis 4.2
Tetranychus species, thrips or leafhoppers are unaffected by Bt cotton but accumulate significant amounts of Cry toxins in their bodies. Predators

consuming large quantities of these pests may be affected negatively by the Cry toxins, leading to the release of a different pest from biological control.

Risk hypothesis 4.3
Predators or parasitoids may be affected by the lack of Lepidoptera on Bt cotton which are killed by the Cry toxins, including *H. armigera*, *P. gossypiella*, *S. derogata*, *E. vitella*, *A. flava* and *O. nubilalis*. As a consequence, different pests that were controlled by these predators and parasitoids are released.

6.5. Prioritization of Risk Hypotheses (Table 6.3)

Risk hypothesis 1

Assessment of Risk Hypothesis 1 could focus on *A. gossypii*, *A. devastans* and *B. tabaci*. Published work on other thrips species in Australia and *T. urticae* in China indicate that Bt cotton containing Cry toxins is unlikely to result in population increases of these species, and *Spodoptera* spp. are likely to be affected sublethally to some degree by Bt cotton. However, it is possible that these species may react differently in Vietnam, so potential outbreaks should be monitored after commercialization.

A number of other factors can influence herbivore populations, potentially complicating the interpretation of experimental results. Aphid populations generally are affected strongly by seasonal variation. Published evidence indicates that the quality and attractiveness of Bt cotton for leafhoppers and whitefly may be influenced more by varietal characteristics independent of the Bt transgene, such as degree of hairiness of the leaves.

Risk hypothesis 2

Because *A. gossypii* is the only significant disease vector on cotton in Vietnam, assessment of Risk Hypothesis 2 should focus on this species only. However, if other cotton disease vectors and their associated diseases are identified in Vietnam, then the hypothesis should be broadened to include these as well (e.g. *B. tabaci* and CLCuV). Vietnamese Bt cotton varieties are likely to be susceptible to CBD, as there is no known resistance in any of the *G. hirsutum* genotypes currently available in Vietnam (Chapter 2, this volume). It is important that Bt cotton is not more susceptible than the current non-Bt varieties, as this could increase the risk from the disease and necessitate extra insecticide applications to control aphids and prevent disease transmission, increasing costs, risk of resistance development in the aphids, secondary pest outbreaks and other negative impacts on health and the environment. The hypothesis should be given high priority if greenhouse tests of Hypotheses 1 or 3 for *A. gossypii* indicate that aphid populations can increase in Bt cotton, which will increase the risk of disease transmission, and/or if the Bt cotton varieties show high susceptibility in greenhouse conditions.

Table 6.3. Prioritization of risk hypotheses. Rank values: 1 = highest rank, 2 = intermediate rank, 3 = lowest rank. Product is the multiplication product of the preceding rank numbers within potential consequence. The final product is the product of relative likelihood and consequence product. Ranks were determined by expert evaluation. Potential reversibility of the consequences varies from (1) irreversible to (3) readily reversed. Spatial scale of consequence varies from (1) extensive to (3) local. Potential magnitude varies from severe (1) to minor (3). Final product of the priority risk hypotheses is in bold.

High priority herbivore pest	Risk hypothesis	Relative likelihood	Relative potential consequence			Consequence product	Final product
			Potential reversibility	Potential spatial scale	Potential magnitude		
Aphis gossypii	**RH1**	1	2	2.5	3	15	**15**
Amrasca devastans	**RH1**	1.5	2	2.5	1.5	7.5	**11.25**
Bemisia tabaci	**RH1**	1	2	2.5	3	15	**15**
Thrips palmi	RH1	2.5	2	2.5	2	10	25
Scirtothrips dorsalis	RH1	2.5	2	2.5	2.5	12.5	31.25
Tetranychus urticae	RH1	2.5	2	2.5	2.5	12.5	31.25
Spodoptera exigua & S. litura	RH1	3	2	2.5	2	15	45
Aphis gossypii	**RH2**	1.5	1.5	2	1	3	**4.5**
Aphis gossypii	RH3	1.5	2.5	2	3	15	22.5
Amrasca devastans	**RH3**	2	2.5	2	1.5	7.5	**15**
Bemisia tabaci	RH3	1.5	2.5	2	3	15	22.5
Thrips palmi	**RH3**	2	2.5	2	2	10	**20**
Scirtothrips dorsalis	RH3	2	2.5	2	2.5	12.5	25
Tetranychus urticae	RH3	2.5	2.5	2	3	15	37.5
Spodoptera exigua & S. litura	RH3	2.5	2.5	2	2	10	25

Risk hypothesis 3

To investigate Risk Hypothesis 3, it is necessary to evaluate the control effect on the non-target herbivore species for the insecticides used to control the heliothine pests in cotton. Release of the herbivore population can occur only if the insecticides are highly efficacious against the herbivore and if any increases in beneficial populations as a result of changed insecticide use are insufficient to control increases in the herbivore population. A preliminary evaluation indicates that the hypothesis is of highest priority for *A. devastans* and *T. palmi* because of their potential to cause damage in mid to late season. Release of *A. gossypii* populations is less likely to occur as aphids are the main targets of insecticide applications anyway and, if aphid populations were released late in the season, damage is likely to be insignificant. The evidence for release of whitefly and thrips is not clear or contradictory. *B. tabaci* currently is not considered a significant pest on cotton, but this could change if whitefly-vectored diseases are introduced, in which case Hypothesis 3 for whitefly should be given more attention. Although *Spodoptera* spp. are controlled by the same insecticides as are used for *H. armigera*, it is not necessarily expected that they will be released by reduced insecticide use, as they are likely to be affected sublethally by many of the Bt cottons. Spider mites are unlikely to be released as they are usually well controlled by natural enemies, unless these are disrupted by insecticide use.

Risk hypothesis 4

These risk hypotheses are discussed in Chapters 7 and 8 (this volume). In this chapter, we suggest establishing the presence of Bt in the herbivore species when feeding on Bt cotton as an indication of the potential for effects on higher trophic levels (predators or parasitoids).

6.6. Analysis Plan and Suggested Experimental Protocols for Non-target Herbivores

In what follows, we provide an analysis plan and experimental protocols for evaluating the risk hypotheses, starting with relatively inexpensive screening methods and proceeding as necessary to more complicated methods to quantify risk. Screening methods should be initiated in laboratory or greenhouse conditions, but some aspects of the risk hypotheses might be tested with greater cost-effectiveness in field conditions.

Some aspects of the two subhypotheses (1.1, 1.2) related to Risk Hypothesis 1 can be tested initially in the laboratory or glasshouse (Experiments 1 and 2), but a number of factors, such as associated changes in crop management, can also affect the population dynamics of non-target herbivores and field tests are needed for conclusive inferences about the magnitude of risk (Experiment 7).

Some initial assessments of Risk Hypothesis 2 can be done with laboratory and greenhouse experiments. Disease susceptibility can be tested initially in the greenhouse (Experiment 4). The protocols of Experiments 1 and 2 can be conducted with infective or infected aphids to test whether disease transmission might be altered (also, Experiments 5 and 6). However, the impact of any increase in aphid populations on disease frequency can be tested conclusively only with Bt cotton in the field, where the many interacting factors that may influence early colonization and survival of aphids are in play. Factors such as associated changes in crop management can affect the dynamics of vectored diseases and field observations would be needed for conclusive inferences about the magnitude of risk (Experiment 7).

Evaluation of Risk Hypothesis 3 requires data collected from the field and should be quantified later in the risk assessment process during field trials and before commercialization.

In order to provide data for analysis of Risk Hypothesis 4 for natural enemy species, tests of Bt exposure and accumulation can be started with laboratory or greenhouse experiments before the crop is grown in the field (Experiment 3), but should be confirmed with season-long field tests (Experiment 7).

6.7. Greenhouse or Laboratory Research

Risk hypothesis 1. Increased attractiveness and/or food quality of Bt cotton results in higher herbivore populations and greater damage to cotton

Two sets of experiments are proposed below to test this hypothesis for the eight selected species, keeping in mind that the highest priority might be given to *A. gossypii*, *A. devastans* and *B. tabaci*. Greenhouse experiments can be used to evaluate the effects of differences in morphology, chemical composition and nutritional value between Bt cotton and the non-Bt control. If Bt and non-Bt cotton plants are infested with *H. armigera*, the experiments can test the effect of lack of feeding damage from *H. armigera* on attractiveness; however, this may be tested more efficiently in the field. The greenhouse experiments will not test for effects of differences in crop phenology or other differences in crop management between Bt and non-Bt cotton (for example, farmers may decide to invest more resources and time in fertilization and pest surveillance of Bt cotton). These aspects can be investigated only in large-scale field experiments.

Experiment 1
The absolute fitness of non-target herbivores feeding on Bt cotton is increased (Risk Hypothesis 1.1).

Aim: To test if insects feeding on Bt cotton plants have similar growth, fecundity, survival and lifespan to those on conventional plants (priority species to test are *A. gossypii*, *A. devastans* and *B. tabaci*).

High dose toxicity studies could be conducted in the laboratory if purified Bt protein and suitable rearing methods are available. While these tests, when conducted properly, can be useful screening methods for direct toxicity (or, in this case, direct physiological benefits), they cannot evaluate other possible differences in Bt cotton. The methods proposed below allow evaluation of these other possible differences, but they lack some of the advantages of high dose toxicity tests as screening methods. It is likely that sample size and equivalence thresholds (Andow, 2003) will need to be considered carefully when using these methods as screening methods.

PROPOSED GREENHOUSE/LABORATORY PROTOCOLS FOR EACH SPECIES.   *A. gossypii.* Laboratory feeding trials can be used to compare development, fecundity and lifespan on conventional and transgenic cotton (of the same genetic background) using the methods of Liu *et al.* (2005). Collect aphids (20 clones) from untransformed cotton fields. Maintain clones separately on untransformed cotton under optimal laboratory conditions. Grow Bt and conventional cotton plants singly in the greenhouse, place each plant in a cage to isolate it from other herbivores and natural enemies and inoculate each one with five to ten adult apterous aphids at the 4–6 leaf stage. Remove all of the adults when more than eight offspring are produced within 12h and retain eight to ten first instars to initiate a cohort. Thereafter, record the development and reproduction of aphids on each plant daily (or every other day). During the reproductive period of resulting adults, count the number of neonates daily and remove them daily, until the adult aphids die. Transfer ten of these neonates to a new cotton plant to collect data on the second generation. This procedure can be replicated for the third generation. In each generation, use four or five cohorts as replications on each of the transgenic and untransformed cotton plants. The presence of Bt proteins should be measured in the insects throughout the experiment using qualitative or quantitative ELISA. Record the developmental stages and number of neonates daily. Monitor temperature. Estimate population parameters: mortality, lifespan, developmental time of immature stages, growth rate and fecundity.

   A. devastans. The method uses a hard plastic base on to which are placed wet paper towels, which keep plant tissue alive. The plastic sheet and paper size can be varied to culture just one, or more than one, leafhopper. A leaf disc is placed on the wet paper. A nymph or adult can be placed on the leaf. A small plastic chamber is then placed over the leafhopper to confine it. It can be cylindrical with a fine mesh top to allow air circulation but prevent leafhopper escape. The chamber must be pressed firmly against the leaf disc without damaging the leaf epidermis. The chamber can be removed to inspect the leafhopper and the leaf discs easily can be replaced daily. Follow the same basic experimental design and protocol as for aphids. Because leafhoppers lay eggs into leaf tissue, it is not practical to count eggs. Therefore, it is necessary to retain removed leaf discs or tissues and count hatching nymphs as a measure of fecundity. Perform the experiments on at least 20 leafhoppers and repeat experiments five times over at least two generations. Measure the same parameters as for aphids.

   B. tabaci. Use the methods of Salas and Mendoza (1995). Collect the pupae of *B. tabaci* from untransformed cotton fields and place the pupae in a humidity chamber until the adults emerge. Transfer each adult to the leaf of a young, healthy untransformed cotton plant (free of disease and insects) inside a rearing and reproduction cage to isolate it from other herbivores and natural enemies. Follow the basic methods for aphids and mites. Rear *B. tabaci* on plants in growth chambers at 26 ± 1°C (LD 14:10h, 70% RH), starting with 10–15 adults/plant. Select 50 first instars to study the duration of that stadium and observe 25 of these to determine the duration of the 'crawling' period of the first instar. To determine the number and duration of each instar, count and remove the molted exuviae. Study adult lifespan and fecundity with 30 virgin males and 30 virgin females; place these insects in male/female pairs inside glass vials using cotton leaflets as food. Each day, score the number of eggs laid. Measure the same parameters as for aphids.

Thrips (*T. tabaci* and *S. dorsalis*). Follow the same basic design and protocol as for leafhoppers.

T. urticae. The fitness of mites raised on Bt and conventional cotton can be compared using methods adapted from Wilson (1994) and Skirvin and Williams (1999). Establish plants of Bt cotton and conventional cotton of the same genetic background in a greenhouse. Ensure there are sufficient plants for the experiments and for maintaining mite cultures. Collect spider mites (*T. urticae*) from conventional cotton fields and establish separate cultures, under identical conditions, on Bt and conventional cotton plants. This will allow for any conditioning effects of host plant type. When the potted plants have reached about 8–10 true leaves, and have leaves large enough to work with (at least 5–6cm wide), collect one leaf from the fourth mainstem node leaf below the terminal for each of ten Bt and ten conventional plants. Place each leaf on agar in a separate 13.5cm Petri dish with a ventilated lid. Place five mated adult female *T. urticae* on each leaf and allow them to deposit eggs. After 24h, remove the adults and place the Petri dishes in an incubator at 25°C, 70% RH and a 16:8 L:D photoperiod. Examine the leaves twice daily and, as soon as the eggs hatch, remove all the larvae, except one. Monitor the stage of the mite twice each day until it becomes an adult and, if it is a female, give her a mate and monitor daily oviposition until death. Remove eggs every 3 days and replace the leaves. Repeat the experiment five times. Measure the same parameters as for aphids.

S. litura and *S. exigua*. Collect egg rafts of each species from cotton plants in the field. Use these to establish a culture of each species on conventional cotton. Establish plants of Bt and conventional cotton in individual pots in a glasshouse. Cage mated females on plants for 24h. Remove females and check each plant for the presence of egg rafts. Manually thin each egg raft down to ten eggs and allow the eggs to hatch. Inspect each plant twice daily and record the developmental stage and survival of the larvae. When larvae mature into adults, ensure each female is mated, then follow daily fecundity until the moth dies. Measure the same parameters as for aphids.

Experiment 2

Non-target herbivores prefer (avoid) feeding or ovipositing on Bt cotton (Risk Hypothesis 1.2).

Aim: To establish if non-target herbivores show a preference for Bt cotton plants for feeding or oviposition compared with conventional cotton, as this may increase populations and cotton damage (priority species to test are *A. gossypii*, *A. devastans* and *B. tabaci*).

PROPOSED PROTOCOLS FOR EACH SPECIES. *A. gossypii. Choice experiments.* Set up a culture of aphids on conventional plants. Let the culture develop to high density so that alates (winged forms) develop. Grow Bt and conventional plants with the same parental background in pots. When the plants have about eight to ten nodes, set up the experiments. The design will have a central culture plant. Around this will be two sets of five Bt and five conventional cotton plants arranged equidistantly around the culture plant and not physically contacting the source plant. Monitor the number of alates and apterae on each plant at 24 h, 72 h and weekly for 3 weeks (or until aphids are too abundant to count). This will show preference for selection by the alates for feeding and reproduction. The experiment should be replicated ten times. If there is preference for any treatment, the cause should be investigated; for example, nutritional quality, morphological differences (leaf hardness, thickness), or secondary products (volatiles and non-volatiles).

A. gossypii. No-choice experiments. The same experimental set-up and procedure as above should be used, except only one cotton line will be provided at a time, i.e. either Bt cotton only or non-transgenic cotton only.

A. devastans, B. tabaci, T. tabaci and *S. dorsalis.* Use similar protocol as for aphids. Generate plants infested heavily with thrips, whitefly or leafhoppers. Place these in the centre of the choice area. Stop watering this source plant to encourage the adults to move off. Monitor the numbers of adults and nymphs or larvae on the plants over a similar time period as for aphids. Shorter time periods can be used for those species that move frequently.

T. urticae. Choice experiments. Use the technique published by Wilson (1994). Take leaf discs (4 cm diameter) from each plant type. Cut the discs in half; assemble a new complete disc consisting of half from each plant type. Place reconstructed disc, with the lower leaf surface upwards, in a Petri dish containing wet cotton wool. Ensure the discs are joined closely so mite movement between each half is not impeded. Place ten adult female mites on each half, after 24 and 48 h record the number of mites, feeding damage and eggs on each side. This will show if there is any preference for transgenic or conventional cotton for feeding and oviposition.

T. urticae. No-choice experiments. The same experimental set-up and procedure as above should be used, except only one cotton line will be provided at a time, i.e. either Bt cotton only or non-transgenic cotton only.

S. litura and *S. exigua.* Collect egg rafts of each species from cotton plants in the field. Use these to establish a culture of each species on conventional cotton. Establish plants of Bt and conventional cotton in individual pots in a glasshouse. Place plants in a 6 × 6 grid of Bt and conventional plants, alternating in both directions. Release freshly eclosed and mated females into the glasshouse and allow them to oviposit for 48 h. After this time, record the number of eggs on each plant type.

Experiment 3

The non-target herbivores contain Bt proteins in their bodies during some/any developmental stage (Risk Hypotheses 1 and 4).

Aim: To test if insects feeding on Bt cotton contain Bt proteins in their bodies, to: (i) confirm that there is a likely pathway of Bt through these pests that potentially could affect higher trophic levels; and (ii) test whether herbivores that show any observed fitness effects of Bt cotton (Experiment 1) are exposed to the Bt proteins. If Bt proteins are found in non-target herbivore species, then it may be necessary to consider the potential effects on the predators or parasitoids that feed on these species (see Chapters 7 and 8, this volume).

PROPOSED GREENHOUSE/LABORATORY PROTOCOLS. Grow plants in pots using seed from conventional and transgenic varieties of the same genetic background (ten replications). Rear insects and mites in the laboratory or greenhouse on conventional cotton, or a different host. Transfer insects or mites to cotton plants when they have six to eight true leaves. Collect samples of leaves and insects/mites of a range of developmental stages and test for the presence of Bt proteins using qualitative or quantitative ELISA. Bt expression within the plants should be assessed on at least three occasions during the experiment. It is important to compare the levels of expression in these plants with that known from field-grown plants elsewhere (see Tables 4.1, 4.2 and 4.3 in Chapter 4, this volume). If expression is low, it may be necessary to grow plants for longer or under different conditions to achieve comparable expression to field-grown plants. The timing of testing of the herbivores after infestation will vary according to the development time of the herbivore. An appropriate time for collection for each species after infestation may be:

1. *A. gossypii*, after 2 weeks when plants are well infested and before extensive damage.
2. *A. devastans*, after 4 weeks when plants are well infested and before extensive damage.
3. *B. tabaci*, after 2 weeks when plants are well infested and before extensive damage.
4. Thrips (both species), after 2–3 weeks when plants are well infested and before extensive damage.
5. Mites, after 2 weeks when plants are well infested and before extensive damage.
6. *Spodoptera* spp., after one complete life cycle (about 28 days).

The number of insects or mites tested in a single ELISA test should be large enough to allow for detection of Bt proteins, even at low levels and accounts for their size. Care is needed when removing aphids for detection tests, as carelessness can lead to contamination and spurious results. For aphids and whitefly, the honeydew could be collected on an acetate sheet and evaluated for Bt content. This experiment is simple to conduct in the field, where insects can be sampled from natural populations on Bt cotton.

Risk hypothesis 2. Bt cotton results in greater disease transmission

Experiment 4
Susceptibility of Bt cotton to CBD.
　　Aim: To determine if Bt cotton is more susceptible than non-Bt cotton to CBD vectored by *A. gossypii*.

PROPOSED GREENHOUSE/LABORATORY PROTOCOL. Inoculate equal-aged, disease-free, young Bt cotton and its isoline with CBD in a greenhouse. Use the same infective winged adult aphids in a crossover design (each aphid inoculates four plants, alternating Bt and non-Bt cotton; the initial plant is random), controlling the amount of time the aphid is allowed to feed on the plant. Measure the time until the plant shows disease symptoms (Lampert *et al.*, 1993).

If Experiments 1 or 2 indicate that *A. gossypii* populations might be greater on Bt cotton, or Experiment 4 indicates that the Bt varieties are more susceptible, then the dynamics of CBD transmission may also be important for assessing risk. The protocols of Experiments 1 and 2 can be repeated using infected and infective aphids. Increased transmission can occur if the vector acquires the pathogen more readily or transmits it more quickly. This could occur if expression of Bt proteins alters plant physiology to affect aphid behaviour (and more complicated possibilities for faster transmission also exist, involving interactions between plant and vector physiology). CBD is transmitted in a persistent circulative manner by *A. gossypii* (see Chapter 2, this volume) and, typically, the aphid requires a few hours to acquire the virus, followed by at least a 12 h latent period within the aphid body but, after this, transmission requires less than 1 h on the plant (Hull, 2002). Greenhouse experiments can be used to measure aphid pathogen acquisition and transmission on Bt cotton compared to non-Bt cotton.

Experiment 5
Pathogen acquisition by vector from infective Bt cotton (Risk Hypothesis 2).
   Aim: To determine if *A. gossypii* acquires the CBD pathogen more readily from Bt cotton.

PROPOSED GREENHOUSE/LABORATORY PROTOCOL. Establish at least 10–20 CBD diseased Bt and non-Bt plants (the plants from Experiment 4 can be used) in a greenhouse. Place ten uninfected winged adult aphids on each plant long enough for them to acquire the pathogen. Collect each aphid from each plant and test it to determine if it has acquired the pathogen (Gray *et al.*, 2002) by placing it individually on to a young susceptible potted cotton plant. Monitor the plants for the development of CBD symptoms.

Experiment 6
Pathogen transmission by vector from Bt cotton (Risk Hypothesis 2).
   Aim: To determine if *A. gossypii* transmits CBD more readily from Bt cotton, either by moving more frequently in association with Bt cotton or preferring to settle on Bt cotton.

PROPOSED GREENHOUSE/LABORATORY PROTOCOLS

1. Using non-diseased Bt and non-Bt cotton plants in a greenhouse, place an infective winged adult aphid on each and allow it to settle and begin feeding. Measure the time from the initiation of feeding until the aphid leaves the plant (Zehnder *et al.*, 2001).

2. Using non-diseased Bt and non-Bt cotton plants in a greenhouse, release infective winged adult aphids equidistant from each and count the number of aphids that settle on each plant (Edwards *et al.*, 2003).

6.8. Field Research

Risk hypothesis: interactions in the field may affect herbivore populations in ways not possible to observe in the laboratory or greenhouse

Field research is required to evaluate Risk Hypothesis 3, release from insecticides, and also Risk Hypothesis 1, the interaction between changes in food availability, due to reduced herbivory from the target pests and the suitability of plants as hosts for non-target species. Field research is also required to consider the multitrophic and intraguild interactions that may occur and which cannot be tested effectively in the laboratory. Field research may also be needed to quantify possible effects associated with Risk Hypotheses 1 and 2 if these hypotheses have not been falsified in the laboratory. All of these issues can be evaluated in the same experiment. The field experiment may also provide information relevant to studies of the effects of Bt cotton on predators, parasitoids, flower visitors and soil processes (Chapters 7, 8, 9 and 10, this volume). To support this research, we suggest establishing if Cry toxins are present in the herbivore species when feeding through the season on Bt cotton in the field (e.g. Torres *et al.*, 2006).

Experiment 7
Evaluation of target herbivore population on Bt cotton in the field (Risk Hypotheses 1, 3 and 4).
    Aim: To determine if Bt cotton increases the abundance of non-target herbivore pest species, due to: (i) increased food quantity or quality from reduced competition with the target species (Risk Hypothesis 1); (ii) release from insecticides applied against the target species (Risk Hypothesis 3); and (iii) release from biological control by natural enemies (Risk Hypothesis 4). This will require experimental comparisons under sprayed and unsprayed conditions.

PROPOSED PROTOCOL.   The experiment will have a randomized block design with four replications. CBD is potentially a problem in the unsprayed treatments, as affected plants could be quite different to healthy plants in their attractiveness to insects, which could bias results. If CBD-resistant Bt and conventional varieties with similar genetic backgrounds are available, these should be used; however, this is unlikely. Instead, the experiment must prevent the confounding effects of CBD yet allow for the typical spray practices that might be required if Bt cotton was grown in a CBD affected area. Experiments would need to be done in areas where CBD is not a problem, but spray treatments or seed treatments for aphids, similar to areas where CBD is a problem, should be imposed. Standard agronomic practices for the respective areas should be used. Experimental designs will be similar at different sites.

Plots should be separated as far as possible to avoid the contamination of pesticides in the treatment without pesticide use. This could entail use of buffer cotton to reduce drift effects. Plots should be at least 20 m × 20 m in size. Ideally, the experiments should be carried out in at least three regions which cover a range of the typical cropping systems used by local growers, including the northern region, if possible. The experiments should also be carried out in both the rainy season and the dry season.

The six treatments are the combinations of either Bt or non-Bt variety with three pest control strategies:

Crop variety:

1. Local variety with Bt genes (or an adapted, imported variety if a local variety is unavailable).
2. Local variety conventional (or an adapted imported variety if a local variety is unavailable).

Note: the Bt and conventional varieties should have the same parental background.

Pest control strategy:

1. None (not sprayed).
2. Sprayed by hand for aphids as required according to local thresholds or according to 'typical' timings if aphids are not abundant in the area.
3. Sprayed by hand for other pests as required according to local thresholds.

Note: the Bt and conventional cotton should be sprayed at the same times for aphids in pest control Treatment 2, but individually according to pest pressure in Treatment 3.

Natural infestation by pests and beneficials should be allowed to occur and the treatments compared in terms of plant growth and development, abundance of pests, beneficials and other insects or mites and yield.

SAMPLING OF PEST POPULATIONS AND DAMAGE

1. Monitor pest abundance in each plot twice per week with visual samples of at least 2 m of row per plot. Use these numbers to decide when control should be made on the sprayed treatments using local pest thresholds, as described above. For example, the threshold for *A. devastans* is one nymph per leaf (Chapter 2, this volume). If thresholds are not available, approximate normal commercial practice or agree between local experts and farmers on some thresholds to use. This may require additional specific sampling for mites, aphids, thrips and whitefly due to their more patchy distributions (see descriptions below). When assessing *Spodoptera* spp. eggs and larvae, aphids and whitefly, it is also important to record the level of parasitism and to collect and incubate samples in the laboratory to evaluate the survival of the parasitoids so this can be compared between treatments.
2. Monitor the abundance of all insects and mites on the canopy with weekly 'Dvac' or 'Beat sheet' sampling. Separate to species level where possible, especially for readily identified predators, parasites and pests.

3. Monitor plant growth and development weekly. Growth should be assessed by measuring plant height and the number of mainstem nodes produced from the cotyledons to the first unfurled leaves in the plant terminal for five plants consecutively in a row in each plot. Though it is intended that the research be done in a CBD-free area, it will also be valuable to score the plants for the presence of the disease. Once fruiting begins, the numbers of squares, flowers, green bolls, damaged bolls (and cause) and open bolls should be recorded for 1 m in each plot. The plants used to record nodes and height can be from the same 1 m section or different random sections. Once bolls begin to open, 2 m sections of row should be marked in each plot and progressively hand harvested each week until all bolls are open. This will provide information on both the total yield (by accumulating the hand harvests for each 2 m section over time) and duration from crop sowing to maturity, which defines the length of time that non-target herbivores are exposed to Bt cotton. Collect leaf samples from each plot once per month and dry. Collect separately from node 4 and node 10 below terminal. Carry out nitrogen analysis to indicate nutritional content.

4. Monitor the expression of Bt proteins by the plants. Make collections of squares, young leaves, middle-aged leaves and old leaves from each plot, including conventional plots, and analyse for the presence of Bt proteins using a quantitative ELISA.

PEST SPECIFIC SAMPLING REQUIREMENTS. *Aphids.* Cotton aphid densities will be quantified weekly from seedling emergence. Cotton plant terminal ratings consist of visually estimating the number of aphids on the terminal and separately on the first fully expanded leaf (usually the third or fourth leaf below the terminal) for 20 randomly selected plants in each plot. If aphid numbers per leaf or terminal are > 200 aphids, make a rough estimate to the nearest hundred. For each plant sampled, also score the overall damage level to the plant as:

None – no aphids observed.

Light – aphids are present on an occasional leaf.

Medium – aphids are present on numerous leaves and some leaves are curling on edges. The upper surface of some leaves may have a light deposit of honeydew. Chlorotic leaves are present in the terminal of some plants.

Heavy – aphids are present on most leaves, and leaves are crinkling and curling. A deposit of honeydew is easily visible and the leaves feel sticky. Clumps of stunted plants with chlorotic terminals are fairly common.

Thrips. Record thrips densities weekly from seedling emergence onwards. The numbers of larvae and adult thrips on each plant should be recorded by collecting samples of plants from each plot, placing them in sealed plastic bags and washing the plants using the method described by Sadras and Wilson (1998). Add water to each bag and shake vigorously for 1 min, after which the thrips are separated by pouring the water through a very fine mesh sieve. Repeat the wash for the sample to ensure all thrips have been dislodged and then wash the collected thrips from the sieve, using 100% ethanol, into a small, sealable, labelled jar for storage and later counting. The samples should start as 20 whole plants per plot until plants have four leaves; thereafter, the top of the plants down to the fourth mainstem node below the terminal should be sampled so that the amount of tissue to be processed remains manageable. Subsamples of the thrips should be taken after counting and identified to provide an estimate of the numbers of each species present. Evaluate the thrips injury to the leaves using a six-degree rating scale based on the extent of the symptoms appearing on the leaves:

1. No damage.
2. Few rough brown blisters, less than 5% of the leaf covered.
3. Blisters covering not more than one-third of the leaf.
4. Blisters covering not more than half of the leaf.
5. Blisters covering not more than three-quarters of the leaf.
6. Blisters covering more than three-quarters of the leaf.

Whitefly. Whitefly densities will be recorded weekly from seedling emergence to flowering stage. An estimate of the numbers of adults will be obtained using the leaf-roll technique developed for Arizona (Naranjo and Flint, 1994, 1995). This entails taking hold of the petiole of the fifth mainstem node leaf below the terminal, carefully rolling the leaf over and quickly counting the number of adult whitefly present (whitefly adults will disperse if disturbed). From the same plant, collect a mainstem node leaf from the upper, middle and lower canopy. In the laboratory, score the number of nymphal whitefly on each leaf and take a subsample for later species and biotype identification using molecular techniques. Twenty plants per plot should be sampled, recording data from each separately.

Mites. Visually assess the number of adult female mites on the third, sixth and ninth mainstem node below the terminal for 20 randomly selected plants per plot. Also make an estimate for each leaf of the proportion of the leaf area damaged by mites.

SAMPLING AND TESTING FOR BT CONTENT THROUGHOUT THE SEASON. Collect samples of lower leaves, terminal leaves, squares, flower tissues, pollen and boll tissues at intervals during the growing season and test for the presence of Bt proteins using qualitative or quantitative ELISA (see Chapter 4, this volume for more detail). Collect samples of insects/mites of a range of developmental stages at intervals during the growing season, as in Experiment 3.

6.9. Suggestions for Further Risk Evaluation

1. Collate, synthesize and make available existing knowledge on cotton pests. Cotton has been cultivated for a long time in Vietnam but information on economically important arthropods and their associated beneficial species present in different cotton agroecosystems is fragmented and scattered across a range of publications and documents. Research deals mostly with the central coastal lowlands region of Vietnam. Lack of a document combining all current knowledge hindered the assessment of some species and created information gaps. Synthesis of existing knowledge on cotton pests is the first step in planning for non-target impact assessments.

2. Species identification. For some pests, the species present have not been identified fully and this needs to be done, e.g. non-target sucking insects such as *Lygus* sp., *Dysdercus* sp., *Leptocorisa* sp.

3. Long-term research and monitoring. As the effects of Bt cotton in the system will depend on the proportion of the crop sown to this technology, and because some effects may be subtle and take some time to emerge, it is essential that monitoring continues after Bt cotton has been released commercially. The cotton non-target pest fauna should be monitored at several locations in each cotton growing region. This will provide a basis on which to assess changes and also to plan for ways of addressing any problems that emerge.

4. Understanding the mechanisms of observed effects by Bt cotton (if found) on non-target herbivores. Less is known about the potential non-target effects of VIP3A than the Cry proteins. Some possible changes, such as reduced insecticide spraying, could allow some non-target herbivore populations to increase in pest status and this may need additional research to understand why certain species, and not others, become pests. Other effects are unknown and may require additional study, such as the direct effects of VIP3A on herbivore fitness and interactions with other stacked Cry proteins, or changed food quality.

References

Adamczyk, J.J.J., Adams, L.C. and Hardee, D.D. (2001) Field efficacy and seasonal expression profiles for terminal leaves of single and double *Bacillus thuringiensis* toxin cotton genotypes. *Journal of Economic Entomology* 94(6), 1589–1593.

Akhtar, K.P., Hussain, M., Khan, A.I., Haq, M.A. and Iqbal, M.M. (2004) Influence of plant age, whitefly population and cultivar resistance on infection of cotton plants by cotton leaf curl virus (CLCuV) in Pakistan. *Field Crops Research* 86(1), 15–21.

Andow, D.A. (2003) Negative and positive data, statistical power, and confidence intervals. *Environmental Biosafety Research* 2, 75–80.

Backus, E., Serrano, M.S. and Ranger, C.M. (2005) Mechanisms of hopperburn: an overview of insect taxonomy, behavior, and physiology. *Annual Review of Entomology* 50, 125–151.

Baur, M.E. and Boethel, D.J. (2003) Effect of Bt-cotton expressing Cry1A(c) on the survival and fecundity of two hymenopteran parasitoids (Braconidae, Encyrtidae) in the laboratory. *Biological Control* 26(3), 325–332.

Bernal, C.C., Aguda, R.M. and Cohen, M.B. (2002) Effect of rice lines transformed with *Bacillus thuringiensis* toxin genes on the brown planthopper and its predator *Cyrtorhinus lividipennis*. *Entomologia Experimentalis et Applicata* 102, 21–28.

Birch, A.N.E., Wheatley, R.E., Anyango, B., Arpaia, S., Capalbo, D., Getu Degaga, E., Fontes, E., Kalama, P., Lelmen, E., Lövei, G., Melo, I.S., Muyekho, F., Ngi-Song, A., Ochieno, D., Ogwang, J., Pitelli, R., Schuler, T., Sétamou, M., Sithanantham, S., Smith, J., Van Son, N., Songa, J., Sujii, E., Tan, T.Q., Wan, F.-H. and Hilbeck, A. (2004) Biodiversity and nontarget impacts: a case study of Bt maize in Kenya. In: Hilbeck, A. and Andow, D.A. (eds) *Environmental Risk Assessment of Genetically Modified Organisms, Volume 1: A Case Study of Bt Maize in Kenya*. CAB International, Wallingford, UK, pp. 117–186.

Briddon, R.W. (2007) The leaf curl epidemics: the situation with cotton leaf curl disease. In: ISAAA (ed.) *Regional Consultation on Biotech Cotton for Risk Assessment and Opportunities for Small Scale Cotton Growers*. National Institute for Biotechnology and Genetic Engineering, Faisalabad, Pakistan, 6–8 March. Common Fund for Commodities,

Report CFC/ICAC 34FT, pp. 53–61. http://www.icac.org/projects/CommonFund/cfc_icac_34/ (accessed 1 October 2007).

Chitkowski, R.L., Turnipseed, S.G., Sullivan, M.J. and Bridges, W.C. Jr (2003) Field and laboratory evaluations of transgenic cottons expressing one or two *Bacillus thuringiensis* var. *kurstaki* Berliner proteins for management of noctuid (Lepidoptera) pests. *Journal of Economic Entomology* 96(3), 755–762.

Cloud, G.L., Minton, B. and Grymes, C. (2004) Field evaluations of VipCotTM for armyworm and looper control. *Proceedings of the 2004 Beltwide Cotton Conference*. National Cotton Council, Arizona.

Cui, J.-J. and Xia, J.-Y. (2000) Effects of Bt (*Bacillus thuringiensis*) transgenic cotton on the dynamics of pest population and their enemies. *Acta Phytophylacica Sinica* 27(2), 141–145 (in Chinese with English abstract and tables).

Deng, S.-D., Xu, J., Zhang, Q.-W., Zhou, S.-W. and Xu, G.-J. (2003) Effects of transgenic Bt cotton on population dynamics of the non-target pests and natural enemies of pests. *Acta Entomologica Sinica* 46, 1–5 (in Chinese with English abstract).

Dively, G.P. (2005) Impact of transgenic VIP3A × Cry1Ab lepidopteran-resistant field corn on the nontarget arthropod community. *Environmental Entomology* 34, 1267–1291.

Dutton, A., Klein, H., Romeis, J. and Bigler, F. (2002) Uptake of Bt-toxin by herbivores feeding on transgenic maize and consequences for the predator *Crysoperla carnea*. *Ecological Entomology* 27, 441–447.

Edwards, O.R., Ridsdill-Smith, T.J. and Berlandier, F.A. (2003) Aphids do not avoid resistance in Australian lupin (*Lupinus angustifolius*, L-luteus) varieties. *Bulletin of Entomological Research* 93, 403–411.

Etienne J., Guyot, J. and Van Waetermeulen, X. (1990) Effect of insecticides, predation, and precipitation on populations of *Thrips palmi* on aubergine (eggplant) in Guadeloupe. *Florida Entomologist* 73, 339–342.

Fitt, G.P. and Wilson, L.J. (2005) Integration of Bt Cotton in IPM systems: an Australian perspective. In: Hoddle, M. (ed.) *Second International Symposium on Biological Control of Arthropods, Volume I*. USDA Forest Service Publication FHTET-2005-08, Davos, Switzerland, 12–16 September 2005, pp. 382–389.

Freeman, T.P., Buckner, J.S., Nelson, D.R., Chu, C.C. and Henneberry, T.J. (2001) Stylet penetration by *Bemisia argentifolii* (Homoptera: Aleyrodidae) into host leaf tissue. *Annals of the Entomological Society of America* 94(5), 761–768.

Gray, S.M., Smith, D.M., Barbierri, L. and Burd, J. (2002) Virus transmission phenotype is correlated with host adaptation among genetically diverse populations of the aphid *Schizaphis graminum*. *Phytopathology* 92, 970–975.

Greene, J.K., Turnipseed, S.G., Sullivan, M.J. and May, O.L. (2001) Treatment thresholds for stink bugs in cotton. *Journal of Economic Entomology* 94, 403–409.

Greene, J.K., Bundy, C.S., Roberts, P.M. and Leonard, B.R. (2006) Identification and management of common boll-feeding bugs in cotton. Report EB158 Clemson Extension. Clemson University, South Carolina. http://www.clemson.edu/psapublishing/Pages/Entom/EB158.pdf (accessed 8 October 2007).

Guo, J.-Y., Dong, L. and Wan, F.-H. (2003) Influence of Bt transgenic cotton on larval survival of common cutworm *Spodoptera litura*. *Chinese Journal of Biological Control* 19, 145–148 (in Chinese with English abstract).

Guo, J.-Y., Wan, F.-H. and Dong, L. (2004) Survival and development of immature *Chrysopa sinica* and *Propylea japonica* feeding on *Bemisia tabaci* propagated on transgenic Bt cotton. *Chinese Journal of Biological Control* 20, 164–169 (in Chinese with English abstract).

Haile, F.J., Braxton, L.B., Flora, E.A., Haygood, B., Huckaba, R.M., Pellow, J.W., Langston, V.B., Lassiter, R.B., Richardson, J.M. and Richburg, J.S. (2004) Efficacy of WideStrike cotton against non-heliothine Lepidopteran insects. *Proceedings of the 2004 Beltwide Cotton Conference*. National Cotton Council, Arizona.

He, K., Wang, Z., Bai, S., Zheng, L., Wang, Y. and Cui, H. (2006) Efficacy of transgenic Bt cotton for resistance to the Asian corn borer (Lepidoptera: Crambidae). *Crop Protection* 25(2), 167–173.

Head, G., Moar, W., Eubanks, M., Freeman, B., Ruberson, J., Hagerty, A. and Turnipseed, S. (2005) A multiyear, large-scale comparison of arthropod populations on commercially managed Bt and non-Bt cotton fields. *Environmental Entomology* 34, 1257–1266.

Huang, D.-L. and Liu, H.-Q. (2005) Resistance of three transgenic cotton varieties to *Sylepta derogata. Jiangsu Journal of Agricultural Sciences* 21(2), 98–101 (in Chinese with English abstract and tables).

Hull, R. (2002) *Matthews' Plant Virology*, 4th edn. Academic Press, London.

Jesse, L.C.H. and Obrycki, J.J. (2000) Field deposition of Bt transgenic corn pollen: lethal effects on the monarch butterfly. *Oecologia* 125, 241–248.

Khan, M. and Bauer, R. (2002) Damage assessment, monitoring and action thresholds of stinkbug pests in cotton. *Proceedings of the Australian Cotton Growers Research Conference, Brisbane, August 2002*. Australian Cotton Growers Research Association, Narrabri, New South Wales, Australia, pp. 395–400.

Kirk, W.D.J. (1997) Feeding. In: Lewis, T. (ed.) *Thrips as Crop Pests*. CAB International, Wallingford, UK, pp. 119–174.

Kranthi, S., Kranthi, K.R., Siddhabhatti, P.M. and Dhepe, V.R. (2004) Baseline toxicity of Cry1Ac toxin against spotted bollworm, *Earias vitella* (Fab) using a diet-based bioassay. *Current Science* 87(11), 1593–1597.

Lampert, E.P., Toms, P.M. and Gooding, G.V. (1993) Influence of host plant variety on the acquisition and transmission of tobacco vein mottle virus by *Myzus nicotianae* Blackman to burley tobacco. *Journal of Agricultural Entomology* 10, 45–49.

Leclant, F. and Deguine, J.P. (1994) Aphids. In: Matthews, G.A. and Tunstall, J.P. (eds) *Insect Pests of Cotton*. CAB International, Wallingford, UK, pp. 285–324.

Lei, T., Khan, M. and Wilson, L.J. (2003) Boll damage by sucking pests: an emerging threat, but what do we know about it. In: Swanepoel, A. (ed.) *World Cotton Research Conference III: Cotton for the New Millenium*. Agricultural Research Council – Institute for Industrial Crops, Cape Town, South Africa, pp. 1337–1344.

Liao, C., Heckel, D.G. and Akhurst, R. (2002) Toxicity of *Bacillus thuringiensis* insecticidal proteins for *Helicoverpa armigera* and *Helicoverpa punctigera* (Lepidoptera: Noctuidae), major pests of cotton. *Journal of Invertebrate Pathology* 80(1), 55–63.

Liu, X.D., Zhai, B.P., Zhang, X.X. and Zong, J.M. (2005) Impact of transgenic cotton plants on a non-target pest, *Aphis gossypii* Glover. *Ecological Entomology* 30, 307–315.

Llewellyn, D.J., Mares, C.L. and Fitt, G.P. (2007) Field performance and seasonal changes in the efficacy against *Helicoverpa armigera* (Hübner) of transgenic cotton (VipCot) expressing the insecticidal protein VIP3A. *Agricultural and Forest Entomology* 9(2), 93–101.

Losey, J.E., Rayor, L.S and Carter, M.E. (1999) Transgenic pollen harms monarch larvae. *Nature* 399, 214.

Ma, X.-M., Liu, X.-X., Zhang, Q.-W., Li, J.-J. and Ren, A.-A. (2006) Impact of transgenic *Bacillus thuringiensis* cotton on a non-target pest *Tetranychus* spp. in Northern China. *Insect Science* 13, 279–286.

Men, X., Ge, F., Edwards, C.A. and Yardim, E.N. (2004) Influence of pesticide applications on pest and predators associated with transgenic Bt cotton and nontransgenic cotton plants. *Phytoparasitica* 32, 246–254.

Naranjo, S.E. (2005) Long-term assessment of the effects of transgenic Bt cotton on the function of the natural enemy community. *Ecological Entomology* 34(5), 1211–1223.

Naranjo, S.E. and Flint, H.M. (1994) Spatial distribution of preimaginal *Bemisia tabaci* (Homoptera: Aleyrodidae) in cotton and development of fixed precision sequential sampling plans. *Environmental Entomology* 23(2), 254–266.

Naranjo, S.E. and Flint, H.M. (1995) Spatial distribution of adult *Bemisia tabaci* (Homoptera: Aleyrodidae) in cotton and development and validation of fixed-precision sampling plans for estimating population density. *Environmental Entomology* 24, 261–270.

Obrist, L.B., Klein, H., Dutton, A. and Bigler, F. (2005) Effects of Bt maize on *Frankliniella tenuicornis* and exposure to prey-mediated Bt toxin. *Entomologia Experimentalis et Applicata* 115, 409–416.

Obrist, L.B., Dutton, A., Romeis, J. and Bigler, F. (2006a) Biological activity of Cry1Ab toxin expressed by Bt maize following ingestion by herbivorous arthropods and exposure of the predator *Chrysoperla carnea*. *BioControl* 51(1), 31–48.

Obrist, L.B., Dutton, A., Albajes, R. and Bigler, F. (2006b) Exposure of arthropod predators to Cry1Ab toxin in Bt maize fields. *Ecological Entomology* 31, 143–154.

Parajulee, M.N., Shrestha, R.B. and Leser, J.F. (2006) Influence of tillage, planting date, and Bt cultivar on seasonal abundance and within-plant distribution patterns of thrips and cotton fleahoppers in cotton. *International Journal of Pest Management* 52(3), 249–260.

Penman, D.R. and Chapman, R.B. (1988) Pesticide-induced mite outbreaks: pyrethroids and spider mites. *Experimental and Applied Acarology* 4, 265–276.

Raps, A., Kehr, J., Gugerli, P., Moar, W.J., Bigler, F. and Hilbeck, A. (2001) Immunological analysis of phloem sap of *Bacillus thuringiensis* corn and of the non-target herbivore *Rhopalosiphum padi* (Homoptera: Aphididae) for presence of Cry1Ab. *Molecular Ecology* 10, 525–534.

Reddall, A.A., Sadras, V.O., Wilson, L.J. and Gregg, P.C. (2004) Physiological responses of cotton to two-spotted spider mite damage. *Crop Science* 44, 835–846.

Sadras, V.O. and Wilson, L.J. (1998) Recovery of cotton crops after early season damage by thrips (Thysanoptera). *Crop Science* 38, 399–409.

Salas, J. and Mendoza, O. (1995). Biology of the sweetpotato whitefly (Homoptera: Aleyrodidae) on tomato. *Florida Entomologist* 78(1), 154.

Schuler, T.H., Denholm, I., Clark, S.J., Neal, S.C. and Poppy, G.M. (2004) Effects of Bt plants on the development and survival of the parasitoid *Cotesia plutellae* (Hymenoptera: Braconidae) in susceptible and Bt-resistant larvae of the diamondback moth, *Plutella xylostella* (Lepidoptera: Plutellidae). *Journal of Insect Physiology* 50, 435–443.

Sharma, H.C. and Pampapathy, G. (2006) Influence of transgenic cotton on the relative abundance and damage by target and non-target insect pests under different protection regimes in India. *Crop Protection* 25, 800–813.

Shibao, M., Ehara, S., Hosomi, A. and Tanaka, H. (2004) Seasonal fluctuation in population density of phytoseiid mites and the yellow tea thrips, *Scirtothrips dorsalis* Hood (Thysanoptera: Thripidae) on grape, and predation of the thrips by *Euseius sojaensis* (Ehara) (Acari: Phytoseiidae). *Applied Entomology and Zoology* 39, 727–730.

Skirvin, D.J. and Williams, M.D. (1999) Differential effects of plant species on a mite pest (*Tetranychus urticae*) and its predator (*Phytoseiulus persimilis*): implications for biological control. *Experimental and Applied Acarology* 23, 497–512.

Stewart, S.D., Adamczyk, J.J., Knighten, K.S. and Davis, F.M. (2001) Impact of Bt cotton expressing one or two insecticidal proteins of *Bacillus thuringiensis* Berliner on growth and survival of noctuid (Lepidoptera) larvae. *Journal of Economic Entomology* 94, 752–760.

Sujii, E.R., Lovei, G.L., Setamou, M., Silvie, P., Fernandes, M.G., Dubois, G.S.J. and Almeida, R.P. (2006) Non-target and biodiversity impacts on non-target herbivorous pests. In: Hilbeck, A., Andow, D.A. and Fontes, E.M.G. (eds) *Environmental Risk Assessment of Genetically Modified Organisms, Volume 2: Methodologies for Assessing Bt Cotton in Brazil*. CAB International, Wallingford, UK, pp. 133–154.

Sun, C.-G., Xu, J., Zhang, Q.-W., Feng, H.-B., Wang, F. and Song, R. (2002) Effect of transgenic Bt cotton on population of cotton pests and their natural enemies in Xinjiang. *Chinese Journal of Biological Control* 18(3), 106–110 (in Chinese with English abstract).

Sun, C.-G., Zhang, Q.-W., Xu, J., Wang, Y.-X. and Lui, J.-L. (2003) Effects of transgenic *Bt* cotton and transgenic *Bt* + *CpTI* cotton on population dynamics of main cotton pests and their natural enemies. *Acta Entomologica Sinica* 46, 705–712 (in Chinese with English abstract).

Torres, J.B. and Ruberson, J.R. (2008) Interactions of *Bacillus thuringiensis* Cry1Ac toxin in genetically engineered cotton with predatory heteropterans. *Transgenic Research* online early.

Torres, J.B., Ruberson, J.R. and Adang, M.J. (2006) Expression of *Bacillus thuringiensis* Cry1Ac protein in cotton plants, acquisition by pests and predators: a tritrophic analysis. *Agricultural and Forest Entomology* 8, 191–202.

US EPA (1998) *Guidelines for Ecological Risk Assessment*. United States Environmental Protection Agency, Risk Assessment Forum, Washington, DC, EPA/630/R095/002F.

Wan, P., Wu, K.-M., Huang, M. and Wu, J. (2004) Seasonal pattern of infestation by pink bollworm *Pectinophora gossypiella* (Saunders) in field plots of Bt transgenic cotton in the Yangtze River valley of China. *Crop Protection* 23(5), 463–467.

Wan, P., Zhang, Y.-J., Wu, K. and Huang, M.-S. (2005) Seasonal expression profiles of insecticidal protein and control efficacy against *Helicoverpa armigera* for Bt cotton in the Yangtze River valley of China. *Journal of Economic Entomology* 98(1), 195–201.

Ward, A.L. (2005) Development of a treatment threshold for sucking insects in determinate Bollgard II transgenic cotton grown in winter production areas. *Australian Journal of Entomology* 44, 310–315.

Whitehouse, M.E.A., Wilson, L.J. and Fitt, G.P. (2005) A comparison of arthropod communities in transgenic Bt and conventional cotton in Australia. *Environmental Entomology* 35, 1224–1241.

Whitehouse, M.E.A., Wilson, L.J. and Constable, G.A. (2007) Target and non-target effects on the invertebrate community of Vip cotton, a new insecticidal transgenic. *Australian Journal of Agricultural Research* 58, 273–285.

Wilson, L.J. (1994) Plant-quality effect on the life history parameters of the two-spotted spider mite (Acari: Tetranychidae) on cotton. *Journal of Economic Entomology* 87, 1665–1673.

Wilson, L.J., Bauer, L.R. and Lally, D.A. (1998) Effect of early season insecticide use on predators and outbreaks of spider mites (Acari: Tetranychidae) in cotton. *Bulletin of Entomological Research* 88, 477–488.

Wu, K. and Guo, Y. (2003) Influences of *Bacillus thuringiensis* Berliner cotton planting on population dynamics of the cotton aphid, *Aphis gosspii* Glover, in Northern China. *Environmental Entomology* 32, 312–318.

Wu, K. and Guo, Y. (2005) The evolution of cotton pest management practices in China. *Annual Review of Entomology* 50, 31–52.

Wu, K., Li, W., Feng, H. and Guo, Y. (2002) Seasonal abundance of the mirids, *Lygus lucorum* and *Adelphocoris* spp. (Hemiptera: Miridae) on Bt cotton in northern China. *Crop Protection* 21, 997–1002.

Zehnder, G.W., Nichols, A.J., Edwards, O.R. and Ridsdill-Smith, T.J. (2001) Electronically monitored cowpea aphid feeding behavior on resistant and susceptible lupins. *Entomologia Experimentalis et Applicata* 98, 259–269.

Zhang, G.-F., Wan, F.-H., Lovei, G.L., Liu, W.-X. and Guo, J.-Y. (2006a) Transmission of Bt toxin to the predator *Propylaea japonica* (Coleoptera: Coccinellidae) through its aphid prey feeding on transgenic Bt cotton. *Environmental Entomology* 35(1), 143–150.

Zhang, G.-F., Wan, F.-H., Liu, W.H. and Guo, J.-Y. (2006b) Early instar response to plant-delivered Bt-toxin in a herbivore (*Spodoptera litura*) and a predator (*Propylaea japonica*). *Crop Protection* 25(6), 527–533.

7 Invertebrate Predators in Bt Cotton in Vietnam: Techniques for Prioritizing Species and Developing Risk Hypotheses for Risk Assessment

Phạm Văn Lầm, Lã Phạm Lân, Angelika Hilbeck, Nguyễn Văn Tuất and Andreas Lang

The biodiversity of an agroecosystem is important for the environment and for the farmer because it influences ecosystem functions that are vital for sustainable crop production and the environment surrounding the agroecosystem. Biodiversity is an integral part of agricultural sustainability. For sustainable agriculture, any new technology/practice must operate to enhance environmental health, stable production and economic efficiency. To promote sustainable agriculture, it is necessary to assess the potential environmental risks and benefits of genetically modified organisms on ecosystem functions, such as biological control of pests, before introducing them on a large scale.

Any management measure applied in agroecosystems may increase or decrease some pest populations. For example, several related changes in the rice ecosystems of South-east Asia contributed to an increased frequency of outbreaks of brown planthopper, *Nilaparvata lugens* (Stål), namely widespread planting of high yielding rice varieties, expansion of irrigation schemes and increased use of nitrogen fertilizers and synthetic insecticides (Kenmore, 1979; Heinreichs and Mochida, 1984; Heong and Schoenly, 1998). Invertebrate predators play a key role in limiting and controlling this and other pests (Hagen *et al.*, 1971; van den Bosch *et al.*, 1971; DeBach, 1975) and have further many unknown, direct and indirect effects within the food web (Polis and Holt, 1992; Sih *et al.*, 1998). As a consequence, assessing the potential effects of transgenic Bt cotton on arthropod predators in Vietnamese agroecosystems may not be straightforward.

Potentially, Bt cotton cultivation in Vietnam may cause a decrease in predator populations and a subsequent increase in secondary pests and concomitant crop losses. Natural enemies can be exposed to Bt toxin through direct and indirect exposure (e.g. Harwood *et al.*, 2005; Zwahlen and Andow, 2005). Predators, such as some coccinellid beetles, may feed directly on Bt plant material such as pollen (Behura and Parida, 1979; Nguyen *et al.*, 1980; Omkar and Singh, 2005). Indirect tritrophic exposure can occur via feeding on prey and is possibly the main route of exposure to Bt cotton (Torres *et al.*, 2006; Zhang *et al.*, 2006a, Torres and Ruberson, 2008). In the laboratory, non-target natural enemies can show adverse effects from transgenic plants and/or products (Lövei and Arpaia, 2005); however, laboratory studies on the effect of Bt cotton on predators (excluding parasitoids) are limited (e.g. Ponsard *et al.*, 2002; Guo *et al.*, 2004; Zhang *et al.*, 2006b; Torres and Ruberson, 2008). Indications for possible adverse effects have been reported by Ponsard *et al.* (2002), who examined the effect of Bt cotton and of lepidopteran prey that had ingested Bt cotton on the survivorship of four important heteropteran predators of cotton pests: longevity significantly decreased for *Orius tristicolor* and *Geocoris punctipes*, whereas no effect was found for *Nabis* sp. and *Zelus renardii*.

Compared to laboratory studies, field and semi-field studies with Bt cotton have been conducted more frequently (e.g. Sisterson *et al.*, 2004; Head *et al.*, 2005; Naranjo 2005a,b; Torres and Ruberson, 2005, 2006, 2007; Whitehouse *et al.*, 2005, 2007; Sharma *et al.*, 2007). In general, the majority of populations of non-target organisms studied showed no reductions due to Bt cotton cultivation; nevertheless, in some cases, densities of certain natural enemies were lower in Bt cotton. The latter may be a direct effect of Bt cotton or mediated via reduced prey populations, the actual significance of the predator reduction for biological control remaining unclear (Naranjo, 2005b). This general picture seems to be supported by field studies from China (e.g. Wang and Xia, 1997; Liu *et al.*, 2002, 2004; Wan *et al.*, 2002; Men *et al.*, 2004). However, several of these field studies were restricted in spatial scale, study duration or replication and the data were highly variable; thus, the probability of detecting an effect appeared to be limited – compare the power analyses in Bourguet *et al.* (2002), Perry *et al.* (2003), Lang (2004) and Meissle and Lang (2005). Recently, Marvier *et al.* (2007) published a meta-analysis of peer-reviewed field studies of the effects of Bt cotton (and Bt maize and Bt potato) on the abundance of non-target invertebrates. For Bt Cry1Ac cotton, they found that when comparing the Bt cotton with a no-insecticide control, the Bt cotton had significantly reduced overall abundance of non-target invertebrate groups; whereas compared with the insecticide sprayed control, the Bt cotton had significantly higher non-target invertebrate abundance. In particular, Coleoptera and Hemiptera, orders that include predatory species, appear to be slightly but significantly less common in Cry1Ac cotton than in non-transgenic, insecticide-free cotton.

Transgenic insect-resistant crops may have large benefits and can constitute a powerful tool for pest management (e.g. Wilson *et al.*, 1992; Cui and Xia, 2000; Reed *et al.*, 2001). However, compatibility with integrated pest management and biological control needs to be assessed before commercial release, taking into account the regional biotic, abiotic and agronomic conditions. For instance, although Bt cotton may be protected effectively against the target pest, the increase of secondary pests may require additional insecticide sprayings (Wu *et al.*, 2002; Men *et al.*, 2004; Wang *et al.*, 2006). So far, comprehensive studies and assessments for Bt cotton have not been carried out in Vietnam and the South-east Asian region.

There are hundreds of predator species in an agricultural ecosystem and it is impossible to evaluate the risk to every species. Moreover, in many regions of the world, such as tropical or subtropical areas, the predator species, their species interactions and ecological functions are often not known and/or the effects of changes in agricultural management practices are not understood or cannot be predicted. Here, we begin an assessment of the effects of Bt cotton on predators in Vietnam. We proceed by selecting the most relevant species for risk assessment in a scientifically objective and transparent way, conduct a preliminary exposure and adverse effects characterization, identify key risk hypotheses and provide experimental protocols that allow one to confirm or refute these hypotheses.

7.1. Numbers of Arthropod Natural Enemies in Cotton Ecosystems in Vietnam

Predators are beneficials reducing and/or controlling pest populations in fields; very few pests exist that have no known predators. In Vietnam, 69 species of arthropod predators have been recorded on cotton, belonging to the orders Coleoptera (30 species or genera – henceforth refered to as taxon or taxa), Heteroptera (17 taxa), Diptera (8 taxa), Hymenoptera (4 taxa), Thysanoptera (1 taxon), Neuroptera (1 taxon), Dermaptera (1 taxon), Araneae (4 taxa), and Acari (1 taxon) (Pham, 1993, 1996a,b, 2002; Pham *et al.*, 1993; Ha, 1995; Nguyen, 1996). The parasitoids are dealt with in Chapter 8, this volume.

7.2. Prioritizing Predator Species for Environmental Risk Assessment

A Selection Matrix (Birch *et al.*, 2004; Faria *et al.*, 2006; Hilbeck *et al.*, 2006; Chapter 5, this volume) was used to identify the species at the greatest potential risk by screening the species that exhibit the highest association with cotton ('Association with Crop') and are likely to have the highest functional significance ('Functional Significance'). Of the 69 predators, ten taxa had too little information to be ranked reliably (i.e. less than half of the criteria could be ranked). The uncertainty associated with these species or genera is discussed further below. The remaining taxa ($n = 59$) were ranked according to six criteria related to association with cotton and four criteria related to their functional significance (Table 7.1). Ranks were given to each species based on published information, the expert knowledge of the authors of this chapter and consultation with other experts in Vietnam (see list of co-authors of this book). If no information was available for a certain criterion, a precautionary value of '1?' was assigned. This meant that if information was lacking for any species, the referring criterion received the highest rank, i.e. a value of 1, automatically (also indicating a high priority for further studies). The means for the 'Association with Crop' criteria and for the 'Functional Significance' criteria were added together to give an overall score between 2 and 6 (Table 7.1; 2–3 = high priority, 3–4 intermediate priority, 4–6 = low priority).

Table 7.1. Selection Matrix for predator species associated with cotton in Vietnam, showing scores for six association criteria and four functional significance criteria. High rank = 1, medium rank = 2, low rank = 3, gap of knowledge = 1? Predatory life stage: all = all feeding stages, L = larvae, A = adult. Association with Crop criteria: A1 = geographic distribution; A2 = habitat specialization: degree of association with cotton habitat; A3 = prevalence: proportion of suitable habitat occupied; A4 = abundance on cotton crop; A5 = phenology: degree of overlap of predator life history with cotton; A6 = phenology: proportion of cotton growing season when species is present (E = early; M = middle; L = late; 1 = all; 2 = E–M; 3 = M or ML). Functional Significance criteria: BC = significance as biological control agent in cotton; FN = significance as a food for other natural enemies; OC = significance in association with other crops; NT = significance in natural areas. MA = mean of Association with Crop criteria; MS = mean of Functional Significance criteria. Sum is the sum of the two means and rank is the final priority ranking used in species selection. The highest priority species are in bold.

Species or taxon	Order and family	Life stage	Main prey	A1	A2	A3	A4	A5	A6	MA	BC	FN	OC	NT	MS	Sum	Rank
Oxyopes javanus Thorell	Araneae: Oxyopidae	All	A. gossypii, A. devastans, Lep larvae	1	3	1	2	1	1	1.50	2	3	1	2	2.00	3.50	6
Pardosa pseudo-annulata (Boes. et Str.)	Araneae: Lycosidae	All	A. gossypii, A. devastans, Lep larvae	1	3	1	2	1	1	1.50	2	3	1	2	2.00	3.50	6
Tetragnatha javana (Thor.)	Araneae: Tetragnathidae	All	A. gossypii, A. devastans, Lep larvae	2	3	1	2	1	1	1.67	2	3	1	2	2.00	3.67	8
Tetragnatha virescens (Okuma)	Araneae: Tetragnathidae	All	A. gossypii, A. devastans, Lep larvae	3	3	1	3	1	1	2.00	2	3	1	2	2.00	4.00	11
Ectomocoris biguttulus Stal	Heteroptera: Reduviidae	All	A. gossypii, A. devastans, Lep larvae	3	3	3	3	2	3	2.83	3	2	2	2	2.25	5.08	22
Ectrychotes crudelis F.	Heteroptera: Reduviidae	All	A. gossypii, A. devastans, Lep larvae	3	3	3	3	2	3	2.83	3	2	2	2	2.25	5.08	22
Oncocephalus sp.	Heteroptera: Reduviidae	All	A. gossypii, A. devastans, Lep larvae	3	3	3	3	2	3	2.83	3	2	2	2	2.25	5.08	22

Continued

Table 7.1. *Continued*

Species or taxon	Order and family	Life stage	Main prey	Association with crop							Functional significance					Summary	
				A1	A2	A3	A4	A5	A6	MA	BC	FN	OC	NT	MS	Sum	Rank
Polididus armatissimus Stal	Heteroptera: Reduviidae	All	*A. gossypii, A. devastans,* Lep larvae	3	3	3	3	2	3	2.83	3	2	2	2	2.25	5.08	22
Rhinocoris fuscipes F.	Heteroptera: Reduviidae	All	*A. gossypii, A. devastans,* Lep larvae	3	3	2	3	2	3	2.67	3	2	2	2	2.25	4.92	20
Rhinocoris sp.	Heteroptera: Reduviidae	All	*A. gossypii, A. devastans,* Lep larvae	3	3	1?	3	2	3	2.50	3	2	2	2	2.25	4.75	18
Sycanus croceovittatus Dohrn	Heteroptera: Reduviidae	All	*A. gossypii, A. devastans,* Lep larvae	3	3	2	2	2	3	2.50	3	2	2	2	2.25	4.75	18
Sycanus versicolor Dohrn	Heteroptera: Reduviidae	All	*A. gossypii, A. devastans,* Lep larvae	3	3	2	3	2	2	2.50	3	2	2	2	2.25	4.75	18
Andrallus spinidens F.	Heteroptera: Pentatomidae	All	Lep larvae	2	3	1	2	2	2	2.00	1	2	1	1	1.25	3.25	4
Eocanthecona furcellata Wolff	Heteroptera: Pentatomidae	All	Lep larvae	2	3	1	2	2	2	2.00	1	2	1	1	1.25	3.25	4
Orius spp.	Heteroptera: Anthocoridae	All	*T. palmi, A. gossypii, A. devastans, T. urticae,* Lep larvae	3	1?	2	2	1	1	1.67	2	2	2	2	2.00	3.67	8
Geocoris spp.	Heteroptera: Lygaeidae	All	*T. palmi, A. gossypii, A. devastans,* Lep eggs and larvae	3	3	2	2	1	1	2.00	3	2	2	2	2.25	4.25	14

Species	Order: Family	Stage	Prey							Mean					Mean		Total
Nabis capsiformis Genmar	Heteroptera: Nabidae	All	Lep eggs and larvae	3	3	2	2	2	3	2.50	3	2	2	2	2.25	4.75	18
Antilochus conquebertii (F.)	Heteroptera: Pyrrhocoridae	All	*Dysdercus* sp.	2	2	1	1	1	2	1.50	2	2	2	2	2.00	3.50	6
Scolothrips sexmaculatus (Pergande)	Thysanoptera: Thripidae	All	*T. palmi, T. urticae*	3	3	2	1	2	2	2.17	2	1	2	2	1.75	3.92	10
Chrysopa spp.	Neuroptera: Chrysopidae	L	*A. gossypii, A. devastans,* Lep eggs and larvae	2	3	2	2	2	3	2.33	1	2	2	1	1.50	3.83	9
Asarcina aegrota F.	Diptera: Syrphidae	L	*A. gossypii*	3	1?	1?	2	3	3	2.17	3	1	3	2	2.25	4.42	15
Syrphus corollae F.	Diptera: Syrphidae	L	*A. gossypii*	1?	1?	1?	2	3	3	1.83	3	1	3	2	2.25	4.08	12
***Episyrphus balteatus* (De Geer)**	Diptera: Syrphidae	L	*A. gossypii*	2	3	1	2	1	2	1.83	2	1	2	2	1.75	3.58	7
Helophilus bengalensis Wied.	Diptera: Syrphidae	L	*A. gossypii*	3	1?	3	2	3	3	2.50	3	2	2	2	2.25	4.75	18
***Ischiodon scutellaris* F.**	Diptera: Syrphidae	L	*A. gossypii*	2	3	1	2	1	2	1.83	2	2	1	1	1.50	3.33	5
Lathyrophthalmus quinquelineatus F.	Diptera: Syrphidae	L	*A. gossypii*	3	3	3	2	3	3	2.83	3	2	2	2	2.25	5.08	22
Megaspis sp.	Diptera: Syrphidae	L	*A. gossypii*	3	1?	1?	2	3	1	1.83	3	2	2	2	2.25	4.08	12
Sphaerophoria scripta L.	Diptera: Syrphidae	L	*A. gossypii*	3	1?	1?	2	3	1	1.83	3	1?	2	2	2.00	3.83	9
Icaria marginata Saussure	Hymenoptera: Vespidae	A	Lep larvae	3	1?	3	3	3	3	2.67	3	1?	2	2	2.00	4.67	17
Polistes stigma F.	Hymenoptera: Vespidae	A	Lep larvae	3	1?	3	3	3	3	2.67	3	1?	2	2	2.00	4.67	17

Continued

Table 7.1. Continued

Species or taxon	Order and family	Life stage	Main prey	A1	A2	A3	A4	A5	A6	MA	BC	FN	OC	NT	MS	Sum	Rank
Vespa cincta F.	Hymenoptera: Vespidae	A	Lep larvae	3	1?	3	3	3	3	2.67	3	1?	2	2	2.00	4.67	17
Formiconus braminus Fer.-Senec.	Coleoptera: Anthicidae	All	*A. gossypii*	3	3	2	3	2	2	2.50	3	2	2	2	2.25	4.75	18
Paederus fuscipes Curtis	Coleoptera: Staphylinidae	All	*A. gossypii, A. devastans,* Lep larvae	1	3	1	2	2	1	1.67	2	2	1	1	1.50	3.17	3
Paederus tamulus Erich.	Coleoptera: Staphylinidae	All	Lep larvae	1	3	1	2	2	1	1.67	2	2	1	1	1.50	3.17	3
Callide splendidula (F.)	Coleoptera: Carabidae	All	Lep larvae	1?	1?	3	3	2	3	2.17	3	2	1?	2	2.00	4.17	13
Chlaenius bioculatus Chaudoir	Coleoptera: Carabidae	All	Lep larvae	3	3	3	3	2	3	2.83	3	2	2	2	2.25	5.08	22
Chlaenius hamifer Chaudoir	Coleoptera: Carabidae	All	Lep larvae	3	3	3	3	2	3	2.83	3	2	2	2	2.25	5.08	22
Chlaenius xanthopleurus Chaudoir	Coleoptera: Carabidae	All	Lep larvae	1	3	1	3	2	2	2.00	2	2	1	1	1.50	3.50	6
Desera sp.	Coleoptera: Carabidae	All	Lep larvae	3	3	3	3	2	3	2.83	3	2	3	2	2.50	5.33	24
Pheropsophus jessoensis Moraw.	Coleoptera: Carabidae	All	Lep larvae	3	3	3	3	2	3	2.83	3	2	3	2	2.50	5.33	24
Scarites terricola Bon.	Coleoptera: Carabidae	All	Lep larvae	3	3	2	3	2	3	2.67	3	2	3	2	2.50	5.17	23
Clivina sp.	Coleoptera: Carabidae	All	Lep larvae	2	3	2	3	2	2	2.33	2	2	1	1	1.50	3.83	9

Species	Order: Family	Stage	Prey													
Drypta japonica Bates	Coleoptera: Carabidae	All	Lep larvae	22	5.08	2.25	2	2	2	3	2.83	3	2	3	3	3
Eucolliuris fuscipennis (Chaudoir)	Coleoptera: Carabidae	All	*A. gossypii, A. devastans,* Lep larvae	8	3.67	1.50	1	1	2	2	2.17	2	2	2	2	3
Ophionea indica (Thumb.)	Coleoptera: Carabidae	All	Lep larvae	5	3.33	1.50	1	1	2	2	1.83	2	2	2	1	3
Ophionea ishii Habu	Coleoptera: Carabidae	All	Lep larvae	8	3.67	1.50	1	1	2	2	2.17	2	2	2	2	3
Cicindela angulata F.	Coleoptera: Cicindelidae	All	Lep larvae	17	4.67	2.50	2	3	2	3	2.17	3	2	3	3	1?
Cicindela specularis Chaudoir	Coleoptera: Cicindelidae	All	Lep larvae	21	5.00	2.50	2	3	2	3	2.50	3	2	3	3	1?
Neocollyris sp.	Coleoptera: Cicindelidae	All	Lep larvae	21	5.00	2.50	2	3	2	3	2.50	3	2	3	3	1?
Anisolemnia sp.	Coleoptera: Coccinellidae	All	*A. gossypii,* Lep larvae	16	4.50	2.00	2	1?	2	3	2.50	3	2	3	3	1?
Coccinella transversalis F.	Coleoptera: Coccinellidae	All	*A. gossypii,* Lep larvae	12	4.08	1.75	2	1	2	2	2.33	2	2	3	3	3
Lemnia biplagiata (Swartz)	Coleoptera: Coccinellidae	All	*A. gossypii,* Lep larvae	22	5.08	2.25	2	2	2	3	2.83	3	2	3	3	3
Harmonia octomaculata (F.)	Coleoptera: Coccinellidae	All	*A. gossypii,* Lep larvae	4	3.25	1.25	1	1	2	1	2.00	1	2	2	2	3
Harmonia sedecim-notata (F.)	Coleoptera: Coccinellidae	All	*A. gossypii,* Lep larvae	19	4.83	2.00	2	2	2	2	2.83	3	2	3	3	3
Menochilus sexmaculatus (F.)	Coleoptera: Coccinellidae	All	*A. gossypii,* Lep larvae	1	2.75	1.25	1	1	2	1	1.50	1	2	1	1	3
Micraspis discolor (F.)	Coleoptera: Coccinellidae	All	*A. gossypii,* Lep larvae	1	2.75	1.25	1	1	2	1	1.50	1	2	1	1	3
Micraspis vincta (Gor.)	Coleoptera: Coccinellidae	All	*A. gossypii,* Lep larvae	2	3.00	1.50	1	1	2	1	1.50	1	2	1	1	1?
Scymnus sp.	Coleoptera: Coccinellidae	All	*A. gossypii,* Lep larvae	9	3.83	2.00	2	2	2	2	1.83	2	2	2	1?	3
Propylea japonica (Thunb.)	Coleoptera: Coccinellidae	All	Lep larvae	8	3.67	1.50	1	1	2	2	2.17	2	2	2	3	3

Three taxa received a high priority, 26 taxa received an intermediate priority and 30 taxa received a low priority. From the highest and intermediate taxa, we prioritized taxa initially with a score less than 3.50, resulting in a top priority list which included 14 taxa: four coccinellids (three of which had the highest priority), two staphylinids, two pentatomid bugs, two carabids, two spiders and two syrphid flies (Table 7.2). These were considered of primary importance for further evaluation in risk assessment.

We used another procedure to account for the taxa with high uncertainty. These ten taxa were compared against the top priority list. Authors and associated experts made an 'educated guess' concerning the information gaps and judged whether any of these were likely to receive a rank similar to these top taxa. The reduviid bug taxa *Sirthenea* sp. and *Pristhesancus* sp. (Heteroptera: Reduviidae), the carabid *Odacantha metallica* Fairmaire (Coleoptera: Carabidae) and the coccinellid *Synoncha grandis* Thunberg (Coleoptera: Coccinellidae) were compared against the other ranked confamilial species. The better-known confamilial species were of relatively low priority, so we concluded that these uncertain species were more likely to be of lower priority and they were not retained. Although there were no confamilial species for the other six taxa, we concluded that they were not high priority based on knowledge of the cotton agroecosystem in Vietnam. These taxa were the predatory mite *Amblyseius* (= *Euseius*) *tularensis* Congdon (Acarina: Phytoseiidae), the mantid *Empusa unicornis* (L.) (Mantodea: Empusidae), *Oecanthus* sp. (Orthoptera), the earwig *Labidura* sp. (Dermaptera: Labiduridae), the mirid bug *Campylomma* sp. (Heteroptera: Miridae) and the digging wasp *Sphex luteipennis* Fab. (Hymenoptera: Sphecidae). The mite fauna of cotton is not well known (see Chapters 2 and 6, this volume) and *A. tularensis* may be an important predator of phytophagous mites.

Generally, all 14 top priority species of Table 7.2 should be carried to the next step of the assessment of potential exposure pathways (see below and outline in Chapter 5, this volume). For the purpose of the present chapter, however, we selected seven species as case examples and proceeded with this subset only. We noted that focusing solely on the priority scores (see Table 7.2) in selecting these seven species would have resulted in the choice of four ladybird beetles (Coccinellidae). This would have over-represented this one family and would have resulted in missing the diversity of the agroecosystem and associated ecological services. Thus, we supplemented these priority scores by using additional criteria given in Box 7.1. These criteria were applied to account for biological control relevance, taxonomic diversity of selected predators, ease of species identification and range of different feeding habits, as well as different habitat niches of predators. More generally, these additional criteria of Box 7.1 could always be applied in circumstances where limited resources would restrict the study to a limited subset of species only.

The following group of seven species fulfilled the criteria of Box 7.1 and were selected to be analysed further: the coccinellid beetle *Menochilus sexmaculatus* (F.), the predatory bug *Eocanthecona furcellata* Wolff., the wolf spider *Pardosa pseudoannulata* (Boes. et Str.), the lynx spider *Oxyopes javanus* Thorell, the staphylinid beetle *Paederus fuscipes* Curtis, the syrphid

Table 7.2. The top priority list of one-fifth of the taxa, with scores less than 3.50.

Rank	Taxon	Order and family	Main prey on cotton	Ecological characteristics
1	*Menochilus sexmaculatus*[2]	Coleoptera: Coccinellidae	*A. gossypii*, Lep. larvae	Plant canopy-dwelling; adults and larvae predatory; abundant in bean crops
1	*Micraspis discolor*	Coleoptera: Coccinellidae	*A. gossypii*, Lep. larvae	Plant canopy-dwelling; adults and larvae predatory
2	*Micraspis vincta*	Coleoptera: Coccinellidae	*A. gossypii*, Lep. larvae	Plant canopy-dwelling; adults and larvae predatory
3	*Paederus fuscipes*	Coleoptera: Staphylinidae	*A. gossypii*, *A. devastans*, Lepidopteran eggs and larvae	Surface-dwelling; adults and larvae predatory
3	*Paederus tamulus*	Coleoptera: Staphylinidae	Lepidopteran eggs and larvae	Surface-dwelling; adults and larvae predatory
4	*Harmonia octomaculata*	Coleoptera: Coccinellidae	*A. gossypii*, Lep. larvae	Plant canopy-dwelling; adults and larvae predatory
4	*Andrallus spinidens*	Heteroptera: Pentatomidae	Lep. larvae	Plant canopy-dwelling; adults and nymphs predatory
4	*Eocanthecona furcellata*	Heteroptera: Pentatomidae	Lep. larvae	Plant canopy-dwelling; adults and nymphs predatory
5	*Ophionea indica*	Coleoptera: Carabidae	Lep. larvae	Surface-dwelling; adults and larvae predatory
5	*Ischiodon scutellaris*	Diptera: Syrphidae	*A. gossypii*	Plant canopy-dwelling; predatory larvae
6	*Chlaenius xanthopleurus*	Coleoptera: Carabidae	Lep. larvae	Surface-dwelling; adults and larvae predatory
7	*Oxyopes javanus*	Araneae: Oxyopidae	Various	Plant canopy-dwelling; actively stalk hunts various prey of mass equal or smaller than spider
7	*Pardosa pseudoannulata*	Araneae: Lycosidae	*A. devastans*, *A. gossypii*, various	Surface-dwelling; actively chase various prey of mass equal or smaller than spider
8	*Episyrphus balteatus*	Diptera: Syrphidae	*A. gossypii*	Plant canopy-dwelling; predatory larvae

[2]Now known more commonly as *Cheilomenes sexmaculata* (F.) http://www.itis.gov/servlet/SingleRpt/SingleRpt?search_topic = TSN&search_value=692620.

Box 7.1. Five selection criteria for choosing the predator pool for the risk assessment of Bt cotton.

No.	Selection criteria for the predator pool
1	A high priority score (< 3.50), i.e. high association with Bt cotton and high environmental significance of the species.
2	Taxonomic diversity; to represent predator biodiversity because a high biodiversity of predators may benefit biological control and other ecological services of predators.
3	Ease of species identification under field conditions. This is an important advantage for field experiments and surveys.
4	Diversity in feeding habit: include generalists and specialists, to get a holistic and representative picture of the system to be assessed, in particular with regard to the complex food web. Specialist predators are often important biological control agents if they are specialized on an important specific prey (group), e.g. ladybird beetles on aphids. In contrast, generalist predators have multiple, often unknown links in the food web; thus, it is necessary to incorporate them to account for unidentified and unexpected effects.
5	Diversity in habitat structure preference: plant-living and epigaeic (soil-dwelling) predators. Soil-dwelling predators also prey on soil detritivores and a change of predation pressure on detritivores may result in altered decomposition processes and nutrient cycling essential for crop plants. Plant-living predators are more likely to prey on plant-living herbivores, i.e. the major pests; therefore, any change in their population densities would have consequences for biological control.

---

fly *Ischiodon scutellaris* F. and the carabid beetle *Ophionea indica* (Thunb.). From the list of 14 species (Table 7.2), we selected only one species for each taxonomic family. We selected *O. indica* for its higher rank for association with the crop than the other carabid; *I. scutellaris* for its higher rank for functional significance than the other syrphid (it is also slightly more common, though it ranked the same for association with crop); in Vietnam, *P. fuscipes* seems to have a wider distribution and is more abundant than *P. tamulus* (Pham, 2000); *M. sexmaculatus* is much better studied than *Micraspis discolor* and therefore more information is available on the species (e.g. Pham, 2004; Omkar and Singh, 2005; Omkar *et al.*, 2005a,b); *E. furcellata* is chosen over *Andrallus spinidens* because it is more abundant and is already used as a biological control agent (e.g. Yasuda and Wakamura, 1992; Pham *et al.*, 1994a,b; Vu *et al.*, 1994; DeClercq, 2000). The final choice was balanced between ground predators (*P. fuscipes*, *P. pseudoannulata*, *O. indica*) and more plant-living predators (*M. sexmaculatus*, *E. furcellata*, *O. javanus*, *I. scutellaris*), as well as between aphid predators (*M. sexmaculatus*, *I. scutellaris*), predators of lepidopteran larvae (*E. furcellata*) and more generalistic species (*P. pseudoannulata*, *O. javanus*, *P. fuscipes*, *O. indica*). In addition, all of these species may be important biological control agents of the major pests of cotton in Vietnam and are widely distributed and abundant, as shown by their high rank in Table 7.1. As mentioned before, if time and resources permit, all 14 species of Table 7.2 should be considered in a risk assessment.

7.3. Assessment of Potential Exposure Pathways

The Selection Matrix evaluates possible exposure based on the spatio-temporal overlap of predators with cotton. In Table 7.3, we evaluate the seven species' exposure to transgene products and metabolites due to various trophic links and behaviours, describing six potential direct and indirect exposure pathways (as outlined in Chapter 5, this volume). Bitrophic exposure through pollen feeding is possible for the coccinellid and the staphylinid beetle. Coccinellid larvae and adults consume pollen (e.g. Behura and Parida, 1979; Lundgren *et al.*, 2004, 2005). The main exposure pathway of Bt toxin to the spiders is through prey, although a bitrophic exposure of spiders by feeding on pollen or nectar cannot be ruled out (e.g. Vogelei and Greissl, 1989; Ludy, 2004). The two species do not spin webs so do not ingest pollen plant materials caught on web material (Ludy and Lang, 2006).

All seven species had some uncertainty, i.e. information gaps, associated with demonstrated and proven uptake of transgenic products by non-targets. The key knowledge gaps were bitrophic exposure of *E. furcellata* or *I. scutellaris* and tritrophic exposure of *M. sexmaculatus* or *I. scutellaris* via aphid prey. Few records of plant feeding of *E. furcellata* past the first instar exist (De Clercq, 2000), and none on cotton. The authors are not aware of any evidence of plant feeding of the species on cotton in Vietnam. However, because this behaviour is typical of the family, we considered the possibility of bitrophic exposure through plant fluids (Coll and Guershon, 2002). Bitrophic exposure on Bt cotton has been proved for other predatory Heteroptera (Torres *et al.*, 2006; Torres and Ruberson, 2008). Hoverflies are known to rely on nectar as the energy source for adults (Ambrosino *et al.*, 2006), but it is not known whether Bt cotton floral or extra-floral nectar contains a significant amount of Bt (see Chapters 4 and 9, this volume). According to Zhang *et al.* (2006a), aphids fed on one kind of Bt cotton may transmit the Bt protein in low concentrations to a coccinellid predator, though field studies have not detected Bt proteins in aphids on another kind of Bt cotton (Torres *et al.*, 2006). Equivalent studies with *M. sexmaculatus* and *I. scutellaris* do not exist, to our knowledge. *M. sexmaculatus* can also consume prey other than aphids (Agarwala *et al.*, 2001), which may contribute to tritrophic transmission of Bt toxin to this predator. In conclusion, all seven species listed in Table 7.3 are likely to be exposed to transgene products and/or metabolites.

Case example species for developing risk hypotheses

In the remainder of the chapter, we focus on three species as case examples; however, for a full risk assessment, all species selected after the assessment of potential exposure pathways should be considered. The three species chosen are: *M. sexmaculatus*, *P. pseudoannulata* and *E. furcellata*. As before (Box 7.1), these species were selected to cover a range of characteristics: mobile (adult *Menochilus* and *Eocanthecona*) versus comparably less mobile (adult *Pardosa*, larvae of *Menochilus*), fairly specialist (*Eocanthecona*)

Table 7.3. Exposure table for the seven selected predator species. Assessment is based primarily on the literature on Cry1Ac cotton, and '?' indicates uncertainty. As most of the other Bt cottons use a similar promoter, their expression is expected to be similar, although this should be confirmed.

Selected species		O. javana	P. pseudoannulata	M. sexmaculatus	E. furcellata	P. fuscipes	I. scutellaris	O. indica
BITROPHIC EXPOSURE	1. List the plant tissues or secretions on which it feeds	None	None	Pollen	Not described?	Pollen	Nectar (and pollen?)	Pollen
	2. Which tissues/secretions fed upon express transgene product?	–	–	Pollen	Not described?	Pollen	Pollen; nectar?	Pollen
	3. Is this feeding important for the predator?	–	–	No	?	No	Nectar – yes; pollen – no	No
	4. Is bitrophic exposure possible?	No	No	Yes (possible)	Yes (likely)?	Yes (possible)	?	Yes (possible)
	5. Are transgene product or metabolites detectable after feeding on plant tissue or secretion?	–	–	Yes?	?	Yes?	?	Yes?
	6. Does bitrophic exposure occur?	No	No	Yes (possible)	Yes (likely)?	Yes (possible)	?	Yes (possible)
TRITROPHIC EXPOSURE VIA HERBIVORE PRODUCTS	7. Does the predator feed on prey products/excretions (e.g. honeydew, frass, faeces)?	No	No	No	No	No	No	No

8. Do any of these herbivore products have detectable transgene products or metabolites?	–	–	–	–	–	–	–
9. Are herbivore products an important part of the predator diet?	–	–	–	–	–	–	–
10. Is tritrophic exposure via feeding on herbivore products possible?	No	No	No	No	No	No	No
11. Are transgene product or metabolites detectable in predator after feeding on herbivore products?	–	–	–	–	–	–	–
12. Does tritrophic exposure via feeding on herbivore products occur?	No	No	No	No	No	No	No
TRITROPHIC EXPOSURE VIA PREY							
13. Does the predator feed on prey that feed on the transgenic plant tissues? (See Table 7.2 and Chapter 6, this volume.)	Yes	Yes	Yes	Yes	Yes	Yes	Yes

Continued

Table 7.3. Continued

Selected species	O. javana	P. pseudoannulata	M. sexmaculatus	E. furcellata	P. fuscipes	I. scutellaris	O. indica
14. Is the prey likely to be exposed to transgene product or metabolites when eaten by the predator? (Chapter 6, this volume.)	Yes (various)	Yes (various)	Yes (thrips); yes? (low conc in aphids?)	Yes (lep larvae)	Yes?	Yes? (low conc in aphids?)	Yes (lep larvae)
15. Is this prey an important part of the predator's diet?	Yes	Yes	Yes	Yes	Yes	Yes	Yes
16. Is tritrophic exposure via feeding on prey possible?	Yes (likely)	Yes (likely)	Yes? (possible)	Yes (likely)	Yes (very likely)	Yes? (possible)	Yes (likely)
17. Are transgene product or metabolites detectable in natural enemy after feeding on prey?	Yes?	Yes?	Yes? (after feeding on aphids)	Yes?	?	?	Yes?
18. Does tritrophic exposure occur through prey?	Yes? (likely)	Yes? (likely)	Yes? (likely)	Yes (likely)	Yes (very likely)	Yes? (likely?)	Yes (likely)
HIGHER TROPHIC LEVEL EXPOSURE							
19. Does the predator cannibalize its own species or eat other intraguild foods (prey that are natural enemies themselves)?	Yes	Yes	Yes	Low	Yes	No	Yes

Question							
20. Is this species possibly exposed?	Yes	Yes	Yes	Likely	Yes	–	Yes
21. Are any of the intraguild foods significant food sources for the natural enemy?	Yes	Yes	No	No	No	–	No
22. Is higher trophic level exposure possible via cannibalism or intraguild feeding?	Yes – possible	Yes – possible	Yes – possible	Yes – low	Yes – possible	No	Yes – possible
23. Are transgene product or metabolites detectable in the natural enemy after cannibalism or intraguild feeding?	?	?	?	?	?	–	?
24. Does higher trophic level exposure occur via cannibalism or intraguild feeding?	Yes (possibly)	Yes (possibly)	Yes (possibly)	Yes (low)	Yes (possibly)	No	Yes (possibly)
BEHAVIOURAL MODIFICATION OF EXPOSURE 25. What feeding preferences or other behaviour could increase or decrease exposure?	Prey activity	Prey activity	None	Emigration	None	Prey density	None
26. Does natural enemy avoid eating exposed prey?	No	No	No	No	No	No	No

Continued

Table 7.3. Continued

Selected species	O. javana	P. pseudoannulata	M. sexmaculatus	E. furcellata	P. fuscipes	I. scutellaris	O. indica
27. Are behaviours likely to increase or decrease exposure?	?	?	No	?	No	Increase	No
EXPOSURE AFTER GENE FLOW — 28. Could the predator eat prey on plants that have received the transgene because of gene flow? (See Chapter 11, this volume.)			? Only on a small scale in some areas				
29. Is exposure via gene flow recipients possible?			? Only on a small scale in some areas				
30. Final assessment of possible exposure	Tritrophic exposure via prey – likely. Tritrophic exposure via intraguild predation – possible. Behavioural modification – possible.	Tritrophic exposure via prey – likely. Tritrophic exposure via intraguild predation – possible. Behavioural modification – possible.	Bitrophic exposure via pollen – possible. Tritrophic exposure via prey – likely? Tritrophic exposure via intraguild predation – possible.	Bitrophic exposure via plant fluids – unknown but likely. Tritrophic exposure via prey – likely. Behavioural modification – possible.	Bitrophic exposure via pollen – possible. Tritrophic exposure via prey – likely. Tritrophic exposure via intraguild predation – possible.	Bitrophic exposure via nectar – unknown. Tritrophic exposure via prey – likely? Behavioural modification – increase.	Bitrophic exposure via pollen – possible. Tritrophic exposure via prey – likely. Tritrophic exposure via intraguild predation – possible.

versus generalist (*Menochilus*, *Pardosa*), ground predator (*Pardosa*) versus plant-dwelling predators (*Eocanthecona*, *Menochilus*), possibly important biological control agents (all three species) and diversity of taxonomy (different classes/orders).

7.4. Identification of Potential Adverse Effects

An adverse effect results from direct and/or indirect exposure to the genetically modified plants (including via food chains). In this section, adverse effect pathways are developed that lead to risk hypotheses and scientifically testable hypotheses for pre-release experiments. The primary adverse effect pathway involves release of non-target pests from biological control, resulting in greater crop damage or higher plant disease incidence. The increase in damage or disease would result in reduced crop quality or quantity. The pathway begins with a decrease in the predator population and/or function and assumes that the predator reduces the abundance of the pest.

   M. sexmaculatus is primarily a predator of aphids and *A. gossypii* is one of the most important pests of cotton in Vietnam because it transmits cotton blue disease (CBD), one of the most important causes of yield loss (Chapter 2, this volume). If *M. sexmaculatus* densities are reduced, *A. gossypii* may become more abundant, resulting in higher incidence of CBD in cotton, with the consequence that *A. gossypii* may become an even larger pest problem than it is already (cf. Ma *et al.*, 2006). So far, the effects of different Bt endotoxins have had no or negligible effects on coccinellid beetles (e.g. Lundgren and Wiedenmann, 2005), but reduced nutritional quality of Bt-feeding prey and long-term exposure potentially may show effects on biology and behaviour (Zhang *et al.*, 2006b,c). As *M. sexmaculatus* also takes prey other than aphids, any reduction of alternative prey (cf. Parajulee *et al.*, 2006) could have the potential to affect coccinellid populations. So far, there are no indications that populations of herbivores other than lepidopteran larvae would be reduced strongly in Bt cotton (e.g. Whitehouse *et al.*, 2005, 2007); nevertheless, this hypothesis should be checked and falsified. Intraguild interactions have the potential to affect coccinellid beetles adversely (e.g. Srivastava *et al.*, 1987; Agarwala *et al.*, 2003) and an increase in density of intraguild predators, e.g. due to reduced insecticide applications in Bt cotton, may affect coccinellid populations in turn. A change in populations of coccinellid parasitoids, such as *Tetrastichus coccinellae* (Hymenoptera: Chalcidae; GSPP, 1991; Pham, 2002), potentially also may affect *M. sexmaculatus* densities.

   P. pseudoannulata is an important surface-active predator that has limited ability to climb plants. They may be important predators of *A. devastans* (Hemiptera: Cicadellidae), but are less important for aphids and thrips (Nguyen, 1996; Pham, 2002). If *P. pseudoannulata* densities are reduced, there may be increases in *A. devastans* densities, reducing cotton quality and quantity. Toxicity of Bt to spiders has not been reported so far; however, spiders have hardly been studied in this respect (Lövei and Arpaia, 2005; Ludy and Lang, 2006). Reduced quality of prey in Bt cotton, especially of lepidopteran larvae, may have adverse effects on arthropod generalist predators (Meissle *et al.*, 2005). Lepidopteran prey would decrease in Bt cotton as expected (Whitehouse *et al.*, 2007) and reduction of prey density potentially may affect spider numbers, which was perhaps the reason for an observed slight drop in spider abundance in Bt cotton in one study (Whitehouse *et al.*, 2005). If hymenopteran parasitoids specialized on spiders (Hymenoptera: Scelionidae, Baeini; Carey *et al.*, 2006) increase due to reduced insecticide application in Bt cotton, this potentially could lead to higher parasitation rates of wolf spider eggs.

E. *furcellata* is an important canopy-dwelling predator of *H. armigera* (Pham *et al.*, 1994b; Pham, 2002; Truong and Vu, 2004) and other lepidopteran and chrysomelid larvae. Because their main prey will be reduced on Bt cotton due to the action of the Bt, an effect of reduced densities of *E. furcellata* would be expected mainly in intercrops, where it may lead to a predation release of lepidopteran pests. With the exception of two heteropteran species (Ponsard *et al.*, 2002), Bt toxins seem not to harm predacious Heteroptera (e.g. Zwahlen *et al.*, 2000; Torres and Ruberson, 2006, 2008); however, to our knowledge, *E. furcellata* has not been tested in this respect. In Bt cotton, lepidopteran prey would decrease as expected (e.g. Whitehouse *et al.*, 2005), which could be a reason for observed reduced abundance of some predacious heteropterans in Bt cotton (Men *et al.*, 2003; Naranjo, 2005a). It is known that the eggs of heteropteran bugs on cotton serve as hosts for some parasitoid species belonging to the genus *Telenomus* (Hym.: Scelionidae) (Pham, 2002; Miayo, 2001). If these parasitoids increase due to a reduction of insecticide application, and fewer of their lepidopteran hosts are available due to the action of Bt cotton, parasitism of *E. furcellata* eggs may increase.

In general, all three species, *M. sexmaculatus*, *P. pseudoannulata* and *E. furcellata*, are also important predators of pests in the crops that are intercropped or grown near cotton in Vietnam, such as beans, vegetables, maize or rice. For example, *M. sexmaculatus* is an important predator of brown planthopper on rice.[1] If their populations are reduced, then their aphid, lepidopteran or other pest prey on the intercrops may become more abundant, resulting in higher losses to the intercrops.

7.5. Formulation of Risk Hypotheses

Combining the potential exposure pathways and adverse effects pathways, we note that cultivation of Bt cotton can interact negatively with predator populations in various ways (Fig. 7.1). In addition, it is important to consider that management of a field may have consequences for neighbouring crop fields and natural habitats if predators move from cotton to other crops or natural areas during their life cycle. For example, a decline in the density of *M. sexmaculatus* may also reduce biological control of aphids in crops intercropped

---

[1] http://www.knowledgebank.irri.org/Beneficials/Scientific_name_Menochilus_sexmaculatus_Fabricius.htm.

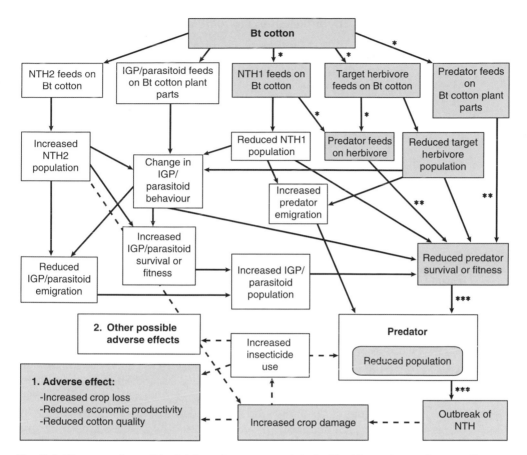

Fig. 7.1. Diagram of possible risk hypotheses associated with at least three adverse effect pathways for predators associated with Bt cotton. Arrows represent causal connections. Large 'predator' box includes the possible response of a predator (entity) to Bt cotton. Dashed lines indicate pathways that are not discussed in detail in this chapter. NTH = non-target herbivore; NTH1 and NTH2 are distinct non-target herbivore species. IGP = intraguild predator. Likely major pathways are shown in grey, key links are marked with stars.

with cotton, e.g. vegetables, maize, soybean, or a decline in the density of *P. pseudoannulata* could also result in lower biological control of leafhoppers or small lepidopteran larvae on crops intercropped with cotton, or reduced *E. furcellata* predation on lepidopteran larvae associated with the intercrops, which could increase in abundance and cause crop losses (both in intercrops and cotton). The increased pest populations could prompt farmers to change insecticide regimes, which might stimulate resurgence of other pests. In this section, we develop testable risk hypotheses and, in the following section, the likelihood of these hypotheses will be assessed (as outlined in Chapter 5, this volume).

Risk hypotheses for the Ladybird Beetle, *Menochilus sexmaculatus*

Risk hypothesis 1.1 (RH1.1)

Feeding on Bt cotton material and Bt-fed prey has an adverse effect on M. sexmaculatus abundance, leading to an outbreak of aphid populations and increased occurrence of CBD.

Adult feeding on (a) pollen and/or (b) *A. gossypii* and other prey may lead to lethal and sublethal effects, such as an increase of mortality of larvae or adults and/or decrease of longevity and fecundity of adults. It may also decrease fertility of eggs laid by adults, or obstruct maturation to the adult stage. Any of these and more could result in decreased densities of *M. sexmaculatus*. Consequently, this may lead to an increase in pests such as aphids, which may result in crop losses, and/or more rapid transmission of the pathogen that causes CBD, leading to crop losses.

Risk hypothesis 1.2 (RH1.2)

Early-season reduction of prey due to Bt cotton affects M. sexmaculatus negatively, thus leading to an outbreak of aphid populations and increased occurrence of CBD.

Effects of Bt cotton early in the season may cause reduction of the density of prey. Thrips are abundant on cotton in Vietnam and *M. sexmaculatus* feed on them (see Chapters 2 and 6, this volume; Nguyen Thi Hai, Nha Ho, October 2007, personal communication). Theoretically, a limited abundance of thrips prior to the appearance of *A. gossypii* could lead to lower survival and/or fitness or higher emigration of ladybird beetles due to reduced food resources. Consequently, population density of *M. sexmaculatus* would decrease, *A. gossypii* abundance would increase in the absence of the predator, CBD becomes more prevalent and greater crop losses occur.

Risk hypothesis 1.3 (RH1.3)

Due to a higher intraguild predation in Bt cotton aphids escape biological control, leading to an increased occurrence of CBD.

An increase in the density of geocorid or anthocorid predators, possibly mediated through an increase in their prey, such as thrips, or a change in insecticide use (see Chapter 6, this volume), can lead to higher intraguild predation (IGP) on ladybird beetle eggs. This could reduce densities of *M. sexmaculatus*, allowing *A. gossypii* to escape biological control, resulting in a higher transmission of CBD and greater crop losses.

Risk hypothesis 1.4 (RH1.4)

Increased parasitation of M. sexmaculatus in Bt cotton leads to aphid outbreaks and to higher incidence of CBD.

If parasitoids increase due to lower mortality under the changed insecticide regimes on Bt cotton, *M. sexmaculatus* populations may decline due to parasitism. As a result, aphids may increase, leading to a higher incidence of CBD and crop losses.

Risk hypotheses for the Wolf Spider, *Pardosa pseudoannulata*

Risk hypothesis 2.1 (RH2.1)
Foraging on Bt cotton-consuming prey reduces abundance of P. pseudo-annulata, *resulting in lower control of secondary pests.*

P. *pseudoannulata* may be exposed to Bt toxin through lepidopteran herbivore prey, which may increase mortality and development time, or decrease fecundity of the wolf spiders, leading to a decrease in population density. This could result in subsequent increases in populations of some main prey species, such as A. *devastans*, resulting in greater crop losses.

Risk hypothesis 2.2 (RH2.2)
Decrease of the prey H. armigera *due to Bt cotton reduces abundance of* P. pseudoannulata, *resulting in the increase of other pest organisms.*

Bt cotton reduces density of small *H. armigera* larvae, an important prey of P. *pseudoannulata*, which could lead to a reduction in the population density of P. *pseudoannulata* and result in an increase in population density of A. *devastans* and cotton yield loss.

Risk hypothesis 2.3 (RH2.3)
Increased parasitization of P. pseudoannulata *in Bt cotton leads to pest outbreaks.*

Reduction in insecticide use associated with the cultivation of Bt cotton may benefit populations of a wolf spider egg parasitoid, which in turn may parasitize more P. *pseudoannulata* eggs, reducing spider population density. This could result in subsequent increases in populations of some main prey species, such as A. *devastans*, and increased crop losses.

Risk hypotheses for the Predacious Bug, *Eocanthecona furcellata*

Risk hypothesis 3.1 (RH3.1)
Feeding on Bt cotton and Bt-fed prey has an adverse effect on E. furcellata, *leading to a predation relase of pest species in intercrops.*

If *E. furcellata* is exposed through plant feeding on cotton and/or by feeding on sublethally or unaffected lepidopteran larvae on Bt cotton, this may lead to increased mortality and development time and decreased fecundity of E. *furcellata*, which results in a reduction in its population density. In turn, this may lead to reduced E. *furcelluta* predation on lepidopteran larvae associated with the intercrops, which could increase in abundance and cause crop losses.

Risk hypothesis 3.2 (RH3.2)
Reduced numbers of H. armigera *in Bt cotton lead to lower abundance of* E. furcellata *and reduced biological control in intercrops.*

Bt cotton reduces numbers of lepidopteran larvae on cotton that are well controlled by the Bt toxin, such as *H. armigera*, the preferred prey of E. *furcellata*, which leads to lower fitness or higher emigration of the bugs due to limited food resources. Consequently, the population density of *E. furcellata* decreases, subsequently the abundance of lepidopteran larvae on the intercrops increases and intercrop losses increase.

Risk hypothesis 3.3 (RH3.3)
Increased parasitism of E. furcellata *in Bt cotton leads to reduced biological control in intercrops.*

If the heteropteran egg parasitoid *Telenomus* increases due to lower mortality under the changed insecticide regimes on Bt cotton, there may be increased parasitism of heteropteran eggs, reduced *E. furcellata* density, lower biological control of lepidopteran larvae on the intercrops and greater crop losses for the intercrops.

7.6. Prioritization of Risk Hypotheses and Analysis Plan

The likelihood and seriousness of a possible risk is determined from the combination of the likelihood of the potential exposure pathway and the likelihood of the adverse effect pathway. Risk hypotheses with higher relative likelihoods of exposure and adverse effects can be prioritized over the others, in order to focus effort on the greatest risks. In addition, factors contributing to the seriousness of the consequences could be considered to prioritize the risk hypotheses. Such factors would include the potential irreversibility, spatial scale and severity of the consequences (OGTR, 2005). The higher priority risk hypotheses for Bt cotton and predators (Table 7.4) are *A. gossypii* and/or CBD problems caused by decreased *M. sexmaculatus* populations on Bt cotton, resulting from bitrophic exposure (RH1.1a) and tritrophic exposure via aphid prey (RH1.1b) or reduction of other prey (RH1.2), pest problems caused by decreased *P. pseudoannulata* populations on Bt cotton, due to tritrophic exposure via prey (RH2.1), and lepidopteran pest problems caused by decreased *E. furcellata* populations on Bt cotton, resulting from feeding on exposed but unaffected (or sublethally affected) lepidopteran larvae (RH3.1, experimental protocols in Yasuda and Wakamura, 1992). The other, lower priority risk hypotheses can be considered later as necessary.

The next section describes approaches to testing the highest priority risk hypotheses for *M. sexmaculatus*. This species was chosen as an indicator of risk because it is believed to be one of the most significant beneficial species in Bt cotton and associated crops and is linked most closely to the potential serious problem of an increase in CBD on Bt cotton (mediated by a possible predation release of the CBD-transmitting aphids if the aphid-feeding *M. sexmaculatus* decrease in abundance).

7.7. Experimental Designs to Test Risk Hypotheses for *M. sexmaculatus*

Knowledge gaps

In the following text, we develop an analysis plan based on RH1.1 for *M. sexmaculatus*. During the formulation of the risk hypotheses, it became obvious that

Table 7.4. Prioritization of risk hypotheses. Rank values of exposure pathway and adverse effects pathway: 1 = highest rank, 2 = intermediate rank, 3 = lowest rank. Likelihood product is the multiplication product of the two preceding rank numbers (rank values: 1–2 = high likelihood, 3–4 = medium likelihood, 6–9 = low likelihood). Ranks of exposure pathway and adverse effects pathway were determined based on Association and Functional Significance criteria (Table 7.1), exposure criteria (Table 7.3) and available literature, and finally assessed and designated by expert consultation and discussion.

High priority predator		Risk hypothesis	Relative likelihood		
			Exposure pathway	Adverse effect pathway	Likelihood product
M. sexmaculatus	RH1.1a	Feeding on Bt cotton material has an adverse effect on *M. sexmaculatus* abundance, leading to an outbreak of aphid populations and increased occurrence of CBD.	1	1	1 (high)
M. sexmaculatus	RH1.1b	Feeding on Bt cotton consuming prey has an adverse effect on *M. sexmaculatus* abundance, leading to an outbreak of aphid populations and increased occurrence of CBD.	1	1	1 (high)
M. sexmaculatus	RH1.2	Early season reduction of prey due to Bt cotton reduces abundance of *M. sexmaculatus*, thus leading to an outbreak of aphid populations and increased occurrence of CBD.	2	1	2 (high)
M. sexmaculatus	RH1.3	Due to a higher intraguild predation in Bt cotton aphids escape biological control, leading to an increased occurrence of CBD.	2	2	4 (medium)
M. sexmaculatus	RH1.4	Increased parasitization of *M. sexmaculatus* in Bt cotton leads to aphid outbreaks and to higher CBD incidence.	2	2	4 (medium)

Continued

Table 7.4. *Continued*

High priority predator		Risk hypothesis	Exposure pathway	Adverse effect pathway	Likelihood product
P. pseudoannulata	RH2.1	Foraging on Bt cotton consuming prey reduces abundance of *P. pseudoannulata*, resulting in lower control of secondary pests.	1	1	1 (high)
P. pseudoannulata	RH2.2	Decrease of the prey *H. armigera* due to Bt cotton reduces abundance of *P. pseudoannulata*, resulting in the increase of other pest organisms.	2	2	4 (medium)
P. pseudoannulata	RH2.3	Increased parasitization of *P. pseudoannulata* in Bt cotton leads to pest outbreaks.	2	2	4 (medium)
E. furcellata	RH3.1	Feeding on Bt cotton material and Bt-fed prey has an adverse effect on *E. furcellata*, leading to a predation release of pest species in intercrops.	1	1	1 (high)
E. furcellata	RH3.2	Reduced numbers of *H. armigera* in Bt cotton lead to lower abundance of *E. furcellata* and reduced biological control in intercrops.	1	2	2 (high)
E. furcellata	RH3.3	Increased parasitization of *E. furcellata* in Bt cotton leads to reduced biological control in intercrops.	2	3	6 (low)

Relative likelihood

quite a few information gaps existed, which made it difficult to characterize the actual risk that an effect on *M. sexmaculatus* would affect Bt cotton cultivation or cultivation of intercrops negatively. The missing information created uncertainty in the assessment and more research is called for to fill these gaps. Examples for the main sources of uncertainty associated with *M. sexmaculatus* are:

- What are the pests under biological control by *M. sexmaculatus* or other common coccinellids in the cotton intercropping systems in Vietnam?

Is there history or experience with insecticide-induced pest flare-ups that could point toward the existence of effective biological control?

- How large a reduction in *M. sexmaculatus* populations is necessary to allow cotton aphid populations to spread CBD more rapidly? This problem can be considered in greater detail: What is the quantitative relationship between *M. sexmaculatus* population density, cotton aphid population density and movement and transmission of CBD?

- What are the intraguild predators of *M. sexmaculatus* and does IGP of eggs or pupal parasitism reduce *M. sexmaculatus* population density? How much greater would IGP on eggs or pupal predation have to be to suppress *M. sexmaculatus* population density? The scientific literature suggests that IGP and parasitism suppress population density of aphidophagous coccinellids occasionally, but the effect is transient (RH1.3 and RH1.4).

- What are the important early season prey species of *M. sexmaculatus* and are any of them reduced by Bt cotton? These species will influence mid and late-season interactions greatly (RH1.2).

Laboratory experiments

To assess RH1.1b for *M. sexmaculatus*, we note that the important causal links are indicated by 'stars' in Fig. 7.1. The two links with *** are key links because, if either is not significant, then there will be no consequent adverse effect for any of the risk hypotheses (RH1.1, RH1.2, RH1.3 and RH1.4). The first link is between beetle survival/fitness and beetle population size, and the second link is between beetle population size and pest population outbreaks and resultant crop losses. Considerable research on other coccinellids has demonstrated that these links can be insignificant, i.e. reduced survival of coccinellids may not lead necessarily to decreased population sizes, and reduced coccinellid population sizes may not result automatically in pest outbreaks; so, either link might merit investigation. The two ** links (Fig. 7.1) may be good assessment targets because there are only two links to falsify RH1.1. These links are between bitrophic exposure and beetle survival/fitness and tritrophic exposure and beetle survival/fitness. We do not know if the Bt protein or its derivatives are toxic to *M. sexmaculatus*. Only the second link need be tested for RH1.1b.

There are two options to test the ** links, i.e. the links between bitrophic exposure and beetle survival/fitness and tritrophic exposure and beetle survival/fitness: using purified transgene protein or using the Bt plant. Although both options can be used, this could increase the cost of the assessment.

Option 1. Laboratory feeding trial using purified Bt toxin

Feed adult and larval *M. sexmaculatus* with purified toxin produced by Bt cotton. This experiment determines if the transgene product affects beetle survival and fitness components, but does not evaluate Bt cotton as a whole.

Administer toxin via treatment of food. Two treatments are needed: food with Bt toxin and food without Bt toxin (control). Ensure that the food with Bt toxin has at least 50× the concentration of the typical food consumed in the field (a concentration higher than would be anticipated in the field is a safety factor typically used in ecotoxicological studies; often even higher factors such as 100× and more are used). For adults, use newly hatched, unfed, laboratory-reared adult beetles (equal male–female ratio). Standardize food supply of beetles, renew food/prey daily and note that the beetles consume the treated foods, record daily food consumption, number of eggs laid and day of death. In addition, measure body weight of beetles before treatment and periodically thereafter (every other day or every third day). Upon death, weigh immediately and assay the adult for presence of Bt (using ELISA). Aim for a sample size of 40 beetles per treatment, replicated twice. For larvae, follow the same procedure as outlined above for adults, starting with neonates (sometimes it is best to let the neonates leave the egg mass before using them). It may be necessary to rear larvae individually. Record daily food consumption, the time of death, time of each moult and body weight (every other day or every third day). Survival statistics, *t*-tests or repeated measures ANOVA can be used to analyse data. If any adverse effect is observed, repeat the experiment, using foods with multiple concentrations of Bt toxin to quantify a NOEL (no observed effect level) or benchmark dose and LC_{50} or EC_{50}. At least five concentrations plus the control should be used. Record similar data for adults and larvae as above, aiming for the same sample size as above for each concentration.

Option 2. Laboratory trials using the Bt plant
Feed adults and larvae with prey reared on Bt cotton to test RH1.1b or pollen from Bt plants to test RH1.1a. This experiment determines if Bt cotton affects survival or fitness of predators and tests for effects of Bt toxin at concentrations typical in the field, but at higher levels of exposure than expected in the field.

Grow Bt and non-Bt cotton in a greenhouse that will not allow the transgene to escape. Several different lines of non-transgenic cotton should be tested in order to compare the effects of Bt cotton with the mean effects of non-transgenic plants (the latter demonstrating the average background effect of non-transgenic cotton on the natural enemy). Grow cotton aphids on these plants. Confirm that the aphids contain Bt toxin. Feed adults or larvae aphids *ad libitum* from either the Bt or non-Bt plant, recording similar data as in Option 1 (Omkar *et al.*, 2005a). If this experiment shows an effect, a field study is compulsory in order to determine if the effect also occurs under more natural conditions. Even if no adverse effects are detected in laboratory experiments, additional (semi-)field tests should not be excluded automatically (see below for discussion on field studies).

Option 1 tests both RH1.1a and RH1.1b at the same time, but requires quantities of purified transgene product, methods to quantify protein concentrations and at least one semi-field trial to follow up the results. For some predators, it may be difficult to simulate exposure appropriately using this option.

Option 2 simulates exposure appropriately, does not require purified transgene product and may not require a field experiment, but does require sufficient greenhouse space to grow enough transgenic plants and adequate labour to conduct the experiments.

Field experiments

Both of the above laboratory-based options may be an insufficient test of RH1.1, even when they show no effect ('no-effect results'). This is because the extrapolations from the laboratory experiment to the whole ecosystem may be inaccurate. Laboratory experiments can be conducted for only a short period of time, for a restricted set of beetle traits, behaviours and population parameters and small spatial scales, so some responses to the transgenic crop may be missed (e.g. the potential effects of simultaneous and long-term combinations of bi- and tritrophic exposure of *M. sexmaculatus*). An example for the uncertainty of 'no-effect results' may be the case of the Monarch butterfly, *Danaus plexippus*. Monarch larvae (Lepidoptera) were fed with amounts of Mon810 Bt maize pollen several times higher than would be expected under natural conditions and the larvae showed no negative response (e.g. Hellmich *et al.*, 2001; Sears *et al.*, 2001). However, the conclusion that Mon810 maize pollen feeding would have no effect on the larvae would have been wrong because, in the field, larvae were affected negatively (Dively *et al.*, 2004), possibly because laboratory studies often do not account for a continuous chronic pollen exposure and additional environmental stressors mediating the effect.

Another criterion for deciding whether or not to conduct field studies can be the seriousness of the potential adverse effect. For example, the occurrence and increase of CBD is considered to be very detrimental and of high agricultural and economic importance in Vietnam. Thus, field experiments assessing this risk can be justified and are recommended to measure the magnitude and likelihood of this risk directly (see below). The authors agree with Torres and Ruberson (2005) in that population level and large-scale effects should be the ultimate end points in risk assessment trials, evaluated over sufficiently long periods to consider environmental variability – despite the challenges, limitations and costs of field work.

As an addition to the above laboratory-based Options 1 and 2, we propose, therefore, a field experiment to test one of the *** key links of Fig. 7.1; that is, the relationship between predators, aphids and the occurrence of CBD. This experiment can be conducted with or without the Bt crop. To determine if a reduction in *M. sexmaculatus* populations (and other predators) will allow *A. gossypii* populations to outbreak and result in greater CBD, it is necessary to remove the natural enemy fauna selectively. This can be done using selective insecticides as a positive control which remove most (but not all) predators, while not affecting aphids. For example, methoxyclor is known to eliminate aphid natural enemies and not harm aphids. A field experiment with non-Bt cotton and two or three treatments (no methoxyclor, low rate of methoxyclor and normal rate of methoxyclor) should be replicated at least four times using sufficiently sized plots. Estimate the density of aphids and all aphid natural enemies before applying insecticide. After a suitable re-entry period has passed, estimate the density of aphids and all aphid natural enemies twice a week, until aphid populations begin to decline. Estimate the prevalence of CBD in cotton at a convenient time when symptoms should be evident. If there is no aphid outbreak, or no increase in the prevalence of CBD, the natural enemies are not important biological control agents and RH1.1 and RH1.2 (and possibly RH1.3 and RH1.4) are falsified. If these hypotheses are not falsified, additional experiments are required to clarify the relative contribution of coccinellids to aphid mortality. If different densities of natural enemies result in different degrees of severity of CBD, the results of these experiments can be used to figure out a threshold value for the adverse effect as to what reduction in a predator density would be still acceptable.

We are not arguing that field experiments can and should replace laboratory trials. Laboratory studies are crucial to reveal and investigate further single and separated effect–cause relationships, thus supporting results from field experiments, which may often be difficult to interpret without supporting laboratory data. On the other hand, we make the case that field experiments should not be excluded in principle simply because early tier tests indicate no harmful effects (Lang *et al.*, 2007), as is sometimes suggested, because if laboratory experiments do not show an adverse impact, impact still may occur under semi-field or field conditions. Therefore (semi-)field studies should always be done in order to confirm or refute the laboratory results.

All field experiments should be designed, conducted and analysed in a way that the probability of detecting potential effects is reasonably high (Marvier, 2002; Box 7.2). An 80% likelihood (power) of detecting a GMO effect generally is considered sufficient (e.g. Bourguet *et al.*, 2002). Therefore, we suggest that experiments are designed to try to attain a statistical power of 80% to detect a GMO effect of 30%; for example, 80% power to detect a reduction of the population density of *M. sexmaculatus* by 30%. A prospective power analysis prior to the field tests should be obligatory in order to calculate the necessary sample size (e.g. Andow, 2003; Perry *et al.*, 2003; Lang, 2004). The variance of the field data needed for the power calculation may be either obtained from the literature or derived from a preliminary monitoring trial (e.g. Bourguet *et al.*, 2002; Meissle and Lang, 2005). If resources are limited, it is preferable to reduce the number of sample dates (effort per replication) rather than number of replicate plots/fields (e.g. Naranjo, 2005a). A satisfactory field study would also include longer-term study periods, including several generations of the focus species, the exact duration of 'long-term' depending on the focus non-target organism, the crop and the receiving environment (Andow and Hilbeck, 2004). It is particularly important to conduct studies in fields of adequate size (i.e. commercial field size) because, if plot sizes are too small, existing effects can be masked and remain undetected (e.g. Duffield and Aebischer, 1994; Prasifka *et al.*, 2005). Likewise, if the different treatment plots or fields are too close to each other, existing effects may not be detected for very mobile organisms because the populations of the different treatments can mix.

Box 7.2. Recommendations for field studies of invertebrate predators for Bt cotton assessment.

No.	Recommendation
1	Determine variance of field abundance of the selected predator species, preferably from a preliminary field trial (or alternatively by obtaining the data from the literature).
2	Calculate necessary sample size (number of plots/fields) with a prospective power analysis. Experiments should have sufficient power to detect an effect of a 30% reduction in abundance with 80% likelihood.
3	Use plots of sufficient size, i.e. of common commercial field size.
4	Ensure that distances between different treatments are adequate, i.e. the difference between a Bt cotton field and the control should be larger than the average dispersal distance of the selected species, or make sure that movement of animals is prevented, for example, by establishing dispersal borders.
5	Conduct the study over several seasons, i.e. a minimum of 3 years.
6	Make sure to establish a control that shows a 'pure' GMO effect, e.g. in the case of Bt cotton, an unsprayed conventional non-Bt cotton. If current or planned agricultural practices are not equal to the 'pure GMO effect control', include additional controls representative of current practice, e.g. a sprayed conventional non-Bt cotton or an organically-grown cotton. Make sure to account for controls representing future and/or political management goals, such as integrated pest management, organic farming. Consider testing several different lines of non-transgenic cotton in order to compare the effects of Bt cotton with the average effects of non-transgenic cultivars.

7.8. Conclusions

It is necessary to assess potential environmental risks and benefits of genetically modified plants on arthropod predators before introducing them on a large scale in Vietnam in order to support sustainable agriculture. We considered the possible effects of Bt cotton on predatory invertebrate species, which are crucial for natural biological control of pest organisms in agriculture. The selection method ranked the predators according to their association with cotton and their functional significance in the agroecosystem, and proved to be effective to prioritize species for the risk assessment. Starting from a pool of 69 species, we prioritized 14 predacious species which we considered to be important for testing the potential risk caused by cultivation of Bt cotton in Vietnam (Table 7.1). We used additional selection criteria to focus on seven species. As case examples, detailed risk hypotheses describing specific Bt cotton effects were developed for three species (*P. pseudoannulata*, *E. furcellata*, *M. sexmaculatus*). Ten different risk hypotheses were developed for these three predator species, of which four were prioritized for detailed investigation. As an example, four experimental studies were developed which provided alternative strategies for testing one of the prioritized risk hypotheses. These proposed experiments, both laboratory and field, centre around the coccinellid *M. sexmaculatus* because it is an important predator of aphids, which are vectors for the CBD pathogen. Because of the high economic implications of CBD, it was considered to be of high relevance to check what influence the cultivation of Bt cotton could have on the coccinellid–aphid–CBD system.

References

Agarwala, B.K., Bardhanroy, P., Yasuda, H. and Takizawa, T. (2001) Prey consumption and oviposition of the aphidophagous predator *Menochilus sexmaculatus* (Coleoptera: Coccinellidae) in relation to prey density and adult size. *Environmental Entomology* 30(6), 1182–1187.

Agarwala, B.K., Bardhanroy, P., Yasuda, H. and Takizawa, T. (2003) Effects of conspecific and heterospecific competitors on feeding and oviposition of a predatory ladybird: a laboratory study. *Entomologia Experimentalis et Applicata* 106, 219–226.

Ambrosino, M.D., Luna, J.M., Jepson, P.C. and Wratten, S.D. (2006) Relative frequencies of visits to selected insectary plants by predatory hoverflies (Diptera: Syrphidae), other beneficial insects, and herbivores. *Environmental Entomology* 35(2), 394–400.

Andow, D.A. (2003) Negative and positive data, statistical power, and confidence intervals. *Environmental Biosafety Research* 2(2), 1–6.

Andow, D.A. and Hilbeck, A. (2004) Science-based risk assessment for non-target effects of transgenic crops. *BioScience* 54(7), 637–649.

Behura, B.K. and Parida, G.B. (1979) Observation on the life history of *Menochilus sexmaculatus* (Fabr.) (Coleoptera, Coccinellidae). *Symposium on Recent Trends Aphidology Studies Bhubaneswar*, Sup.1, p. 31.

Birch, A.N.E., Wheatley, R.E., Anyango, B., Arpaia, S., Capalbo, D., Getu Degaga, E., Fontes, E., Kalama, P., Lelmen, E., Lövei, G., Melo, I.S., Muyekho, F., Ngi-Song, A., Ochieno, D., Ogwang, J., Pitelli, R., Schuler, T., Sétamou, M., Sithanantham, S., Smith, J., Van Son, N., Songa, J., Sujii, E., Tan, T.Q., Wan, F.-H. and Hilbeck, A. (2004) Biodiversity and non-target impacts: a case study of Bt maize in Kenya. In: Hilbeck, A. and Andow, D.A. (eds) *Environmental Risk Assessment of Genetically Modified Organisms, Volume 1: A Case Study of Bt Maize in Kenya.* CAB International, Wallingford, UK, pp. 117–186.

Bourguet, D., Chaufaux, J., Micoud, A., Delos, M., Naibo, B., Bombarde, F., Marque, G., Eychenne, N. and Pagliari, C. (2002) *Ostrinia nubilalis* parasitism and the field abundance of non-target insects in transgenic *Bacillus thuringiensis* corn (*Zea mays*). *Environmental Biosafety Research* 1(1), 49–60.

Carey, D., Murphy, N.P. and Austin, A.D. (2006) Molecular phylogenetics and the evolution of wing reduction in the Baeini (Hymenoptera: Scelionidae): parasitoids of spider eggs. *Invertebrate Systematics* 20(4), 489–501.

Coll, M. and Guershon, M. (2002) Omnivory in terrestrial arthropods: mixing plant and prey diets. *Annual Review of Entomology* 47, 267–297.

Cui, J.J. and Xia, J.Y. (2000) Effects of Bt (*Bacillus thuringiensis*) transgenic cotton on the dynamics of pest population and their enemies. *Acta Phytophylacica* 27, 141–145 (in Chinese with English abstract).

DeBach, P. (1975) *Biological Control by Natural Enemies.* Cambridge University Press, New York.

DeClercq, P. (2000) Predaceous stinkbugs (Pentatomidae: Asopinae). In: Schaefer, C.W. and Panizzi, A.R. (eds) *Heteroptera of Economic Importance.* CRC Press LLC, Boca Raton, Florida, pp. 737–789.

Dively, G.P., Rose, R., Sears, M.K., Hellmich, R.L., Stanley-Horn, D.E., Calvin, D.D., Russo, J.M. and Anderson, P.L. (2004) Effects on monarch butterfly larvae (Lepidoptera: Danaidae) after continuous exposure to Cry1Ab-expressing corn during anthesis. *Environmental Entomology* 33(4), 1116–1125.

Duffield, S.J. and Aebischer, N.J. (1994) The effect of spatial scale of treatment with dimethoate on invertebrate population recovery in winter wheat. *Journal of Applied Ecology* 31(2), 263–281.

Faria, M.R., Lundgren, J.G., Fontes, E.M.G., Fernandes, O.A., Schmidt, F., Tuat Nguyen Van and Andow, D.A. (2006) Assessing the effects of Bt cotton on generalist arthropod predators. In: Hilbeck, A., Andow, D.A. and Fontes, E.M.G. (eds) *Environmental Risk Assessment of Genetically Modified Organisms, Volume 2: Methodologies for Assessing Bt Cotton in Brazil.* CAB International, Wallingford, UK, pp. 175–199.

GSPP (General Station of Plant Protection) (1991) *A List of Natural Enemies of Chinese Rice Pests.* General Station of Plant Protection, Ministry of Agriculture, P.R. China (in Chinese).

Guo, J.-Y., Wan, F.-H. and Dong, L. (2004) Survival and development of immature *Chrysopa sinica* and *Propylea japonica* feeding on *Bemisia tabaci* propagated on transgenic Bt cotton. *Chinese Journal of Biological Control* 20, 164–169 (in Chinese with English abstract).

Ha Quang Hung (1995) Study on cotton insect pests and their parasite insects in cotton farm at To Hieu – Son La in 1994. *Plant Protection Journal of Vietnam* 2, 11–17 (in Vietnamese).

Hagen, K.S., van den Bosch, R. and Dahlsten, D.L. (1971) The importance of naturally-occurring biological control in the Western USA. In: Huffaker, C.B. (ed.) *Biological Control.* Plenum Press, New York, London, pp. 253–293.

Harwood, J.D., Wallin, W.G. and Obrycki, J.J. (2005) Uptake of Bt endotoxins by non-target herbivores and higher order arthropod predators: molecular evidence from a transgenic corn agroecosystem. *Molecular Ecology* 14, 2815–2823.

Head, G., Moar, W., Eubanks, M., Freeman, B., Ruberson, J., Hagerty, A. and Turnipseed, S. (2005) A multiyear, large-scale comparison of arthropod populations on commercially managed Bt and non-Bt cotton fields. *Environmental Entomology* 34, 1257–1266.

Heinrichs, E.A. and Mochida, O. (1984) From secondary to major pest status: the case of insecticide-induced rice brown planthopper, *Nilaparvata lugens* resurgence. *Protection Ecology* 7, 191–218.

Hellmich, R.L., Siegfried, B.D., Sears, M.K., Stanley-Horn, D.E., Daniels, M.J., Mattila, H.R., Spencer, T., Bidne, K.G. and Lewis, L.C. (2001) Monarch larvae sensitivity to *Bacillus thuringiensis*-purified proteins and pollen. *Proceedings of the National Academy of Sciences of the USA* 98(21), 11925–11930.

Heong, K.L. and Schoenly, K.G. (1998) Impact of insecticides on herbivore-natural enemy communities in tropical rice ecosystems. In: Haskell, P.T. and McEwen, P. (eds) *Ecotoxicology: Pesticides and Beneficial Organisms.* Chapman & Hall Press, London, pp. 381–403.

Hilbeck, A., Andow, D.A., Arpaia, S., Birch, A.N.E., Fontes, E.M.G., Lövei, G.L., Sujii, E., Wheatley, R.E. and Underwood, E. (2006) Methodology to support non-target and biodiversity risk assessment. In: Hilbeck, A., Andow, D.A. and Fontes, E.M.G. (eds) *Environmental Risk Assessment of Genetically Modified Organisms, Volume 2: Methodologies for Assessing Bt Cotton in Brazil.* CAB International, Wallingford, UK, pp. 108–132.

Kenmore, P.E. (1979) Limits of the brown planthopper problem: implications for integrated pest management. Paper presented at a Saturday Seminar, 30 June 1979. International Rice Research Institute, Laguna, Philippines.

Lang, A. (2004) Monitoring the impact of *Bt* maize on butterflies in the field: estimation of required sample sizes. *Environmental Biosafety Research* 3, 55–66.

Lang, A., Lauber, E. and Darvas, B. (2007) Early-tier tests insufficient for GMO risk assessment. *Nature Biotechnology* 25, 35–36.

Liu, W.X., Wan, F.-H. and Guo, J.-Y. (2002) Structure and seasonal dynamics of arthropods in transgenic Bt cotton field. *Acta Entomologica Sinica* 22, 729–735 (in Chinese with English abstract).

Liu, W.X., Wan, F.-H., Guo, J.-Y. and Lövei, G.L. (2004) Spiders and their seasonal dynamics in transgenic Bt versus conventionally managed cotton fields in north-central China. In: Samu, F. and Szinetár, C. (eds), *Proceedings of the 20th European Colloquium of Arachnology, Szombatheley, 22–26 July 2002*. Plant Protection Institute & Berzsenyi College, Budapest, pp. 337–342.

Lövei, G.L. and Arpaia, S. (2005) The impact of transgenic plants on natural enemies: a critical review of laboratory studies. *Entomologia Experimentalis et Applicata* 144, 1–14.

Ludy, C. (2004) Intentional pollen feeding in the spider *Araneus diadematus* Clerck, 1757. *Newsletter of the British Arachnological Society* 101, 4–5.

Ludy, C. and Lang, A. (2006) Bt maize pollen exposure and impact on the garden spider, *Araneus diadematus*. *Entomologia Experimentalis et Applicata* 118, 145–156.

Lundgren, J.G. and Wiedenmann, R.N. (2005) Tritrophic interactions among Bt (CryMb1) corn, aphid prey, and the predator *Coleomegilla maculata* (Coleoptera: Coccinellidae). *Environmental Entomology* 34(6), 1621–1625.

Lundgren, J.G., Razzak, A.A. and Wiedenmann, R.N. (2004) Population responses and food consumption by predators *Coleomegilla maculata* and *Harmonia axyridis* (Coleoptera: Coccinellidae) during anthesis in an Illinois cornfield. *Environmental Entomology* 33(4), 958–963.

Lundgren, J.G., Huber, A. and Wiedenmann, R.N. (2005) Quantification of consumption of corn pollen by the predator *Coleomegilia maculata* (Coleoptera: Coccinellidae) during anthesis in an Illinois cornfield. *Agricultural and Forest Entomology* 7(1), 53–60.

Ma, X.M., Liu, X.X., Zhang, Q.W., Zhao, J.Z., Cai, Q.N., Ma, Y.A. and Chen, D.M. (2006) Assessment of cotton aphids, *Aphis gossypii*, and their natural enemies on aphid-resistant and aphid-susceptible wheat varieties in a wheat–cotton relay intercropping system. *Entomologia Experimentalis et Applicata* 121, 235–241.

Marvier, M. (2002) Improving risk assessment for non-target safety of transgenic crops. *Ecological Applications* 12(4), 1119–1124.

Marvier, M., McCreedy, C., Regetz, J. and Kareiva, P. (2007). A meta-analysis of effects of Bt cotton and maize on non-target invertebrates. *Science* 316, 1475–1477.

Meissle, M. and Lang, A. (2005) Comparing methods to evaluate the effects of Bt maize and insecticide on spider assemblages. *Agriculture Ecosystems and Environment* 107, 359–370.

Meissle, M., Vojtech, E. and Poppy, G.M. (2005) Effects of Bt maize-fed prey on the generalist predator *Poecilus cupreus* L. (Coleoptera: Carabidae). *Transgenic Research* 14(2), 123–132.

Men, X., Ge, F., Liu, X. and Yardim, E.N. (2003) Diversity of arthropod communities in transgenic Bt cotton and non-transgenic cotton agroecosystems. *Environmental Entomology* 32(2), 270–275.

Men, X., Ge, F., Edwards, C.A. and Yardim, E.N. (2004) Influence of pesticide applications on pest and predators associated with transgenic Bt cotton and non-transgenic cotton plants. *Phytoparasitica* 32, 246–254.

Miayo, E. (2001) Biology and mass production of assassin bug, *Eocanthecona furcellata* Wolf, as biocontrol agent for major insect pests of pechay and tomato. Research Highlights 2000, Bureau of Agricultural Research – Department of Agriculture, Quezon City, Philippines.

Naranjo, S.E. (2005a) Long-term assessment of the effects of transgenic Bt cotton on the abundance of non-target arthropod natural enemies. *Environmental Entomology* 34, 1193–1210.

Naranjo, S.E. (2005b) Long-term assessment of the effects of transgenic Bt cotton on the function of the natural enemy community. *Environmental Entomology* 34, 1211–1223.

Nguyen Thi Hai (1996) Cotton pests and their natural enemies. In: *Selected Scientific Reports of Research on Cotton during 1976–1996*. Nhaho Cotton Research Centre, Agricultural Publishing House, Ho Chi Minh City, Vietnam, pp. 108–120 (in Vietnamese).

Nguyen Van Huynh, Huynh Quang Xuan and Luu Ngoc Hai (1980) Preliminary study on key natural enemies of brown planthopper. In: *Report of Research on Brown Planthopper in the MeKong Delta during 1977–1979*. Agricultural Publishing House, Ha Noi, Vietnam, pp. 134–142 (in Vietnamese).

OGTR (2005) *Risk Analysis Framework*. Office of the Gene Technology Regulator, Australia. www.ogtr.gov.au/pubform/riskassessments.htm (accessed May 2007).

Omkar, A.P. and Singh, S.K. (2005) Development and immature survival of two aphidophagous ladybirds, *Coelophora biplagiata* and *Micraspis discolor. Insect Science* 12, 375–379.

Omkar, G.M., Srivastava, S., Gupta, A.K. and Singh, S.K. (2005a) Reproductive performance of four aphidophagous ladybirds on cowpea aphid, *Aphis cracciuora* Koch. *Journal of Applied Entomology* 129(4), 217–220.

Omkar, A.P., Mishra, G., Srivastava, S., Singh, S.K. and Gupta, A.K. (2005b) Intrinsic advantages of *Cheilomenes sexmaculata* over two coexisting *Coccinella* species (Coleoptera: Coccinellidae). *Insect Science* 12(3), 179–184.

Parajulee, M.N., Shrestha, R.B. and Leser, J.F. (2006) Influence of tillage, planting date, and Bt cultivar on seasonal abundance and within-plant distribution patterns of thrips and cotton fleahoppers in cotton. *International Journal of Pest Management* 52(3), 249–260.

Perry, J.N., Rothery, P., Clark, S.J., Heard, M.S. and Hawes, C. (2003) Design, analysis and statistical power of the farm-scale evaluations of genetically modified herbicide-tolerant crops. *Journal of Applied Ecology* 40(1), 17–31.

Pham Van Lam (1993) Preliminary results of identification of natural enemies of cotton pests. *Journal of Plant Protection of Vietnam* 5, 2–5 (in Vietnamese).

Pham Van Lam (1996a) Contributions to the study on fauna of Hymenopterous parasitoids in Vietnam. In: *Selected Scientific Reports on Biological Control of Pests and Weeds during 1990–1995*. Agricultural Publishing House, Ha Noi, pp. 95–103 (in Vietnamese).

Pham Van Lam (1996b) The influence of plant protection measures on natural enemies in cotton fields. *Journal of Plant Protection of Vietnam* 3, 21–25 (in Vietnamese).

Pham Van Lam (2000) *A List of Rice Arthropod Pests and Their Natural Enemies in Vietnam*. Agricultural Publishing House, Ha Noi (in Vietnamese).

Pham Van Lam (2002) Findings on collecting and identifying natural enemies of key pests on economic crops in Vietnam. In: Pham Van Lam (ed.) *Natural Enemy-Resources of Pests: Studies and Implementation*. Agricultural Publishing House, Ha Noi, pp. 7–57 (in Vietnamese).

Pham Van Lam (2004) On the biology of the ladybeetle *Menochilus sexmaculatus* Fabr. (Coleoptera: Coccinellidae). *Journal of Plant Protection of Vietnam* 3, 22–27 (in Vietnamese).

Pham Van Lam, Nguyen Thi Hai and Ngo Trung Son (1993) Some preliminary findings on natural enemies in cotton fields. *Abstracts. Scientific Conference on Plant Protection*, 24–25 March 1993. Agricultural Publishing House, Ha Noi, pp. 40–41 (in Vietnamese).

Pham Van Lam, Luong Thanh Cu and Nguyen Thi Diep (1994a) Laboratory studies on biology of predacious bug *Eocanthecona furcellata* Wolff (Hemiptera: Pentatomidae). *Journal of Plant Protection of Vietnam* 1, 5–9.

Pham Van Lam, Luong Thanh Cu and Nguyen Thi Diep (1994b) Prey range and voracity of predacious bug, *Eocanthecona furcellata* Wolff (Hemiptera: Pentatomidae). *Journal of Plant Protection of Vietnam* 2, 1–4 (in Vietnamese).

Polis, G.A. and Holt, R.D. (1992) Intraguild predation: the dynamics of complex trophic interactions. *Trends in Ecology and Evolution* 7, 151–154.

Ponsard, S., Gutierrez, A.P. and Mills, N.J. (2002) Effect of Bt-toxin (Cry1Ac) in transgenic cotton on the adult longevity of four heteropteran predators. *Environmental Entomology* 31(6), 1197–1205.

Prasifka, J.R., Hellmich, R.L., Dively, G.P. and Lewis, L.C. (2005) Assessing the effects of pest management on non-target arthropods: the influence of plot size and isolation. *Environmental Entomology* 34(5), 1181–1192.

Reed, G.L., Jensen, A.S., Riebe, J., Head, G. and Duan, J.J. (2001) Transgenic Bt potato and conventional insecticides for Colorado potato beetle management: comparative efficacy and non-target impacts. *Entomologia Experimentalis et Applicata* 100, 89–100.

Sears, M.K., Hellmich, R.L., Stanley-Horn, D.E., Oberhauser, K.S., Pleasants, J.M., Mattila, H.R., Siegfried, B.D. and Dively, G.P. (2001) Impact of Bt corn pollen on monarch butterfly populations: a risk assessment. *Proceedings of the National Academy of Sciences of the USA* 98(21), 11937–11942.

Sharma, H.C., Arora, R. and Pampapathy, G. (2007) Influence of transgenic cottons with *Bacillus thuringiensis* cry1Ac gene on the natural enemies of *Helicoverpa armigera*. *Biocontrol* 52(4), 469–489.

Sih, A., Englund, G. and Wooster, D. (1998) Emergent impacts of multiple predators on prey. *Trends in Ecology and Evolution* 13, 350–355.

Sisterson, M.S., Biggs, R.W., Olson, C., Carrière, Y., Dennehy, T.J. and Tabashnik, B.E. (2004) Arthropod abundance and diversity in Bt and non-Bt cotton fields. *Environmental Entomology* 33, 921–929.

Srivastava, A.S., Katiyar, R.R., Upadhyay, D. and Singh, S.V. (1987) *Canthecona furcellata* Wolff (Hemiptera: Pentatomidae) predating on *Menochilus sexmaculata* (F.) (Coleoptera: Coccinellidae). *Indian Journal of Entomology* 49, 558.

Torres, J.B. and Ruberson, J.R. (2005) Canopy- and ground-dwelling predatory arthropods in commercial Bt and non-Bt cotton fields: patterns and mechanisms. *Environmental Entomology* 34, 1242–1256.

Torres, J.B. and Ruberson, J.R. (2006) Interactions of Bt-cotton and the omnivorous big-eyed bug *Geocoris punctipes* (Say), a key predator in cotton fields. *Biological Control* 39, 47–57.

Torres, J.B. and Ruberson, J.R. (2007) Abundance and diversity of ground-dwelling arthropods of pest management importance in commercial Bt and non-Bt cotton fields. *Annals of Applied Biology* 150, 27–39.

Torres, J.B. and Ruberson, J.R. (2008) Interactions of *Bacillus thuringiensis* Cry1Ac toxin in genetically engineered cotton with predatory heteropterans. *Transgenic Research.* DOI 10.1007/s11248–007–9109–8.

Torres, J.B., Ruberson, J.R. and Adang, M.J. (2006) Expression of *Bacillus thuringiensis* Cry1Ac protein in cotton plants, acquisition by pests and predators: a tritrophic analysis. *Agricultural and Forest Entomology* 8(3), 191–202.

Truong Xuan Lam and Vu Quang Con (2004) *Predacious Bugs on Some Crops in the North of Vietnam.* Agricultural Publishing House, Hanoi, (in Vietnamese).

van den Bosch, R., Leigh, T.F., Falcon, L.A., Stern, V.M., Gonzales, D. and Hagen, K.S. (1971) The developing program of integrated control of cotton pests in California. In: Huffaker, C.B. (ed.) *Biological Control.* Plenum Press, New York, London, pp. 377–394.

Vogelei, A. and Greissl, R. (1989) Survival strategies of the crab spider *Thomisus onustus* Walckenaer 1806 (Chelicerata, Arachnida, Thomisidae). *Oecologia* 80, 513–515.

Vu Quang Con, Pham Huu Nhuong and Nguyen Thi Hai (1994) Some preliminary results of study on biological characters of predator bugs *Eocanthecona furcellata* Wolff in Nha Ho (Ninh Thuan Province). *Journal of Plant Protection of Vietnam* 4, 17–19 (in Vietnamese).

Wan, F.-H., Liu, W.X. and Guo, J.-Y. (2002) Comparison analyses of the functional groups of natural enemy in transgenic Bt cotton field and non-transgenic cotton fields with IPM, and chemical control. *Acta Entomologica Sinica* 22, 935–942 (in Chinese with English abstract).

Wang, C.Y. and Xia, J.Y. (1997) Differences of population dynamics of boll-worms and of population dynamics of major natural enemies between Bt transgenic cotton and conventional cotton. *China Cotton* 24, 13–15 (in Chinese with English abstract).

Wang, S., Just, D.R. and Pinstrup-Andersen, P. (2006) Tarnishing silver bullets: Bt technology adoption, bounded rationality and the outbreak of secondary pest infestations in China. Selected Paper prepared for presentation at the American Agricultural Economics Association Annual Meeting, Long Beach, California, 22–26 July 2006. www.grain.org/research/btcotton.cfm?id=374 (accessed 17 July 2007).

Whitehouse, M.E.A., Wilson, L.J. and Fitt, G.P. (2005) A comparison of arthropod communities in transgenic Bt and conventional cotton in Australia. *Environmental Entomology* 34, 1224–1241.

Whitehouse, M.E.A., Wilson, L.J. and Constable, G.A. (2007) Target and non-target effects on the invertebrate community of Vip cotton, a new insecticidal transgenic. *Australian Journal of Agricultural Research* 58, 273–285.

Wilson, W.D., Flint, H.M., Daeton, R W., Fuschhoff, D.A., Perlak, E.J., Armstrong, T.A., Fuchs, R.L., Berberich, S.A., Parks, N.J. and Stapp, B.R. (1992) Resistance of cotton lines containing a *Bacillus thuringiensis* toxin to pink bollworm (Lepidoptera: Noctuidae) and other insects. *Journal of Entomological Science* 34, 415–425.

Wu, K., Li, W., Feng, H. and Guo, Y. (2002) Seasonal abundance of the mirids *Lygus lucorum* and *Adelphocoris* spp. (Hemiptera: Miridae) on Bt cotton in northern China. *Crop Protection* 21, 997–1002.

Yasuda, T. and Wakamura, S. (1992) Rearing of the predatory stink bug, *Eocanthecona furcellata* (Wolff) (Heteroptera: Pentatomidae), on frozen larvae of *Spodoptera litura* (Fabricius) (Lepidoptera: Noctuidae). *Japanese Journal of Applied Entomology and Zoology* 27(2), 303–305.

Zhang, G.-F., Wan, F.-H., Lövei, G.L., Liu, W.X. and Guo, J.-Y. (2006a) Transmission of Bt toxin to the predator *Propylea japonica* (Coleoptera: Coccinellidae) through its aphid prey feeding on transgenic Bt cotton. *Environmental Entomology* 35(1), 143–150.

Zhang, G.-F., Wan, F.-H., Liu, W. X. and Guo, J.-Y. (2006b) Early instar response to plant-derived Bt-toxin in a herbivore (*Spodoptera litura*) and a predator (*Propylea japonica*). *Crop Protection* 25, 527–533.

Zhang, S.-Y., Li, D.-M., Cui, J. and Xie, B.-Y. (2006c) Effects of Bt-toxin Cry1Ac on *Propylaea japonica* Thunberg (Col., Coccinellidae) by feeding on Bt-treated Bt-resistant *Helicoverpa armigera* (Hübner) (Lep., Noctuidae) larvae. *Journal of Applied Entomology* 130(4), 206–212.

Zwahlen, C. and Andow, D.A. (2005) Field evidence for the exposure of ground beetles to Cry1Ab from transgenic corn. *Environmental Biosafety Research* 4, 113–117.

Zwahlen, C., Nentwig, W., Bigler, F. and Hilbeck, A. (2000) Tritrophic interactions of transgenic *Bacillus thuringiensis* corn, *Anaphothrips obscurus* (Thysanoptera: Thripidae), and the predator *Orius majusculus* (Heteroptera: Anthocoridae). *Environmental Entomology* 29(4), 846–850.

8 Potential Effects of Transgenic Cotton on Non-target Insect Parasitoids in Vietnam

Nguyễn Văn Huỳnh, A. Nicholas E. Birch, Josephine Songa, Nguyễn Văn Đĩnh and Trần Thế Lâm

Insect parasitoids are considered to be an important component of biodiversity because of their species richness, population abundance and their role in insect pest control (Price, 1997). In agricultural ecosystems, their activities, together with predators, help to maintain herbivores or insect pests below economic thresholds. Parasitoids differ from predators in that they do not kill their host immediately. Hosts are killed slowly, as the parasitoid consumes it. Adult parasitoids are dependent on non-host food sources, such as floral nectar, pollen and honeydew (Leius, 1960; Powell, 1986; Jervis et al., 1992, 1993). This characteristic of high intimacy with the host, compared with most predators, may lead to a more profound reduction of their functional quality if their host quality and abundance is affected by any environmental factor, such as the introduction of an insecticidal transgenic crop (Schuler et al., 2001; Groot and Dicke, 2002; Andow and Hilbeck, 2004; Poppy and Sutherland, 2004; Birch and Wheatley, 2005; Lövei and Arpaia, 2005).

Therefore, the cultivation of Bt cotton potentially could cause adverse effects on parasitoids, either when they feed as adults on nectar and pollen of Bt cotton (Greenplate, 1997, 1999; Chapter 9, this volume), or on honeydew excreted by sap-sucking insect pests, or when they parasitize herbivorous insect pests feeding on Bt cotton (Ramirez-Romero et al., 2007). Bt toxins in cotton are assumed to affect mainly larvae of the target lepidopteran insects (Adamczyk and Meredith, 2004) and are considered unlikely to affect parasitoids. However, Bt and other insecticidal transgenic crops have affected parasitoids adversely in 39.8% of examined cases (Lövei and Arpaia, 2005), so effects on parasitoids do merit testing until more is known about the mode of action of Bt toxins and their metabolites in non-target insects (Hilbeck and Schmidt, 2006). In addition to any disruption to host–parasitoid interactions, the cultivation of Bt cotton may suppress the major lepidopteran pests targeted by the Bt trait and thus may cause an outbreak of secondary pests that are released from competition (see Chapter 6, this volume). Compared to the potential effects of Bt crops on predators, the effects on parasitoids are likely to be more subtle but profound and act over multiple generations. In addition, parasitoids move between crops searching for suitable hosts, so disruption of parasitoids on a Bt crop could reduce biological control of pests on other crops or on natural vegetation in the surrounding area.

The framework for analysis of non-target impacts by transgenic crops suggested by the GMO ERA Project (Birch *et al.*, 2004; Pallini *et al.*, 2006) was followed for the case of a Bt cotton introduced for cultivation in Vietnam. The steps of an environmental risk assessment model (Chapter 5, this volume) are followed in this chapter for parasitoids.

1. Identify the relevant functional categories of biodiversity to be analysed and decide the most important functions for risk assessment (e.g. important natural enemy groups like parasitoids of cotton pests).
2. List non-target parasitoid species on cotton and prioritize them using a Selection Matrix.
3. Assess potential direct and indirect exposure pathways to transgenic product or metabolites in the transgenic plant for the high priority species selected in Step 2.
4. Identify potential adverse effects pathways, including pathways for each of the likely exposure pathways from Step 3, and 'knock-on' pathways.
5. Formulate research hypotheses based on risk hypotheses and design experiments for risk assessment based on the exposure pathways and the adverse effects pathways, to confirm or refute the risk hypothesis.
6. Formulate improved risk hypotheses and design laboratory, greenhouse and field experiments to test these hypotheses.

8.1. Prioritization of Parasitoid Species on Cotton in Vietnam

A list of 25 taxa of insect parasitoids of hosts on cotton was prepared and confirmed as common and abundant in cotton in Vietnam (Pham, 1996, 2003; Phạm Văn Lầm, Ho Chi Minh City, 2004, personal communication). All but four of these taxa are parasitoids of the major lepidopteran pests of cotton, which include *Helicoverpa armigera* Hübner, *Spodoptera litura* (F.), *S. exigua* (Hübner) (all Lepidoptera: Noctuidae) and *Pectinophora gossypiella* Saunders (Lepidoptera: Gelechiidae) and the less damaging but ubiquitous *Earias fabia* (Stoll) (synonym: *E. vitella* F.), *Anomis flava* F. (both Lepidoptera: Noctuidae) and *Sylepta derogata* F. (Lepidoptera: Pyralidae) (Nguyen and Sen, 2003). Three taxa are parasitoids of nymphs of the generalist aphid, *Aphis gossypii* Glover, which transmits viral diseases including cotton blue disease (CBD), and one taxon is an egg parasitoid of the sucking pest *Cletus* sp. (Hemiptera: Coreidae). All of these pests, except *P. gossypiella*, are also polyphagous on many crops that are grown near cotton and on weeds and wild plants (Nguyen and Sen, 2003). See Chapters 2 and 6, this volume, for more information on these pests.

A Selection Matrix was used to rank the relative importance of these parasitoids in relation to their biological control function and their potential

Table 8.1. Selection Matrix for parasitoid taxa associated with cotton in Vietnam, showing scores for six association criteria and four significance criteria. High rank = 1, medium rank = 2, low rank = 3. Association with Crop criteria: A1 = geographic distribution; A2 = habitat specialization: degree of association with cotton habitat; A3 = prevalence: proportion of suitable cotton habitat occupied; A4 = abundance on cotton crop; A5 = phenology: degree of overlap of host life history with cotton; A6 = phenology: proportion of cotton growing season when species is present (E = early, M = middle, L = late; 1 = all, 2 = E–M; 3 = M or ML). Functional Significance criteria: BC = significance as biological control agent in cotton; OC = significance in association with other crops; NT = significance in natural areas; FN = significance as a food for other natural enemies. Mean is the average for the association or significance criteria: MA = mean of Association with Crop criteria; MS = mean of Functional Significance criteria. Sum is the sum of the two averages.

Species/taxon	Order and Family	Host species	Life stage parasitized	\multicolumn{7}{c}{Association with crop}							\multicolumn{5}{c}{Functional significance}					Sum
				A1	A2	A3	A4	A5	A6	MA	BC	OC	NT	FN	MS	
Chetogena spp.	Diptera: Tachinidae	*Helicoverpa armigera*	Larva	1	1	3	1	2	1	1.50	3	3	2	1	2.25	3.75
Charops sp.	Hymenoptera: Ichneumonidae	*Spodoptera litura*	Larva	2	2	3	2	1	2	2.00	3	3	1	2	2.25	4.25
Eriborus vulgaris Morley	Hymenoptera: Ichneumonidae	*H. armigera*	Larva	2.5	3	3	3	3	1	2.58	?	?	?	?	?	?
Pristomerus sp.	Hymenoptera: Ichneumonidae	*S. litura* & *S. exigua*	Larva	2.5	3	3	3	2	2	2.58	?	?	?	?	?	?
Temelucha sp.	Hymenoptera: Ichneumonidae	*S. exigua*	Larva	1	1	2	2	1.5	2	1.58	3	3	1	2	2.25	3.83
Trathala flavorbitalis Cam.	Hymenoptera: Ichneumonidae	*S. exigua*	Larva	2.5	3	3	3	1	2	2.42	?	?	?	?	?	?
Xanthopimpla flavolineata Cameron	Hymenoptera: Ichneumonidae	*H. armigera, Anomis flava*	Larva	2	1	2	1	2	1	1.50	3	3	2	2	2.50	4.00
Xanthopimpla enderleini Krieger	Hymenoptera: Ichneumonidae	*H. armigera, A. flava*	Larva	3	1	3	2	1.5	1	1.92	?	?	?	?	?	?
Apanteles sp. 1 + 2	Hymenoptera: Braconidae	*H. armigera, A. flava*	Larva	1	1	1	1	2	1	1.17	1	1	2	1	1.25	2.42

Species	Family	Host	Stage							Mean					Mean	Total
Bracon sp.	Hymenoptera: Braconidae	H. armigera	Larva	2	1	3	1	1	1	1.50	3	3	2	2	2.50	4.00
Chelonus sp. 1	Hymenoptera: Braconidae	H. armigera, S. exigua	Larva	2	1	2	1.5	3	1	1.75	3	3	2	2	2.50	4.25
Chelonus sp. 2	Hymenoptera: Braconidae	Pectinophora gossypiella	Larva	3	1	3	2	2	3	2.33	?	?	?	?	?	?
Rogas sp.	Hymenoptera: Braconidae	Earias vitella	Larva	?	?	?	?	1	3	2.00	?	?	?	?	?	?
Microplitis sp.	Hymenoptera: Braconidae	A. flava, S. litura	Larva	2	1	3	3	1	2	2.00	?	?	?	?	?	?
Copidosoma sp.	Hymenoptera: Encyrtidae	A. flava & others	Larva	2	1	3	2	1	2	1.83	?	?	?	?	?	?
Brachymeria lasus Walker	Hymenoptera: Chalcicidae	A. flava & others	Larva	2	1	3	2	1	2	1.83	2	1	2	2	1.75	3.58
Brachymeria sp.	Hymenoptera: Chalcicidae	Sylepta derogata	Larva	3	1	3	3	1	3	2.33	?	?	?	?	?	?
Euplectrus sp.	Hymenoptera: Eulophidae	A. flava	Larva	3	1	3	3	1.5	2	2.25	?	?	?	?	?	?
Elasmus sp.	Hymenoptera: Eulophidae	S. derogata & others	Larva	3	1	3	2	1	3	2.17	?	?	?	?	?	?
Trichogramma chilonis Ishii	Hymenoptera: Trichogrammatidae	H. armigera, A. flava, S. derogata, E. vitella	Egg	1	3	1	1	3	1	1.67	1	1	1	3	1.50	3.17
Gryon sp.	Hymenoptera: Scelionidae	Cletus sp.	Egg	2	3	3	3	1	3	2.50	?	?	?	?	?	?
Aphelinus sp.	Hymenoptera: Aphelinidae	Aphis gossypii	Nymph	2	3	1	1	2	3	2.00	3	1	3	1	2.00	4.00
Lysiphlebus sp.	Hymenoptera: Aphidiidae	A. gossypii	Nymph	3	3	3	3	2	2	2.67	?	?	?	?	?	?

association with cotton (Table 8.1). Regarding crop phenology and how this affects interactions with parasitoids, we assessed during which parts of the cotton growing season each parasitoid taxon was present; 'early', 'middle', 'late' and 'all season' descriptors were used. The cropping pattern of cotton with other crops was also considered relevant to parasitoid ecology: in the central region of Vietnam, cotton is cultivated mostly together with a diverse range of intercropped species, while in the south, cotton may be intercropped with cucumber, watermelon or maize. Such cropping patterns may affect herbivore feeding preference and abundance, thereby affecting the parasitoids that depend on them as hosts.

Ranks for the association with the crop were averaged across the five criteria and the significance ranks were averaged across the four criteria (see Table 8.1). The two means were summed and ranked to provide a preliminary risk estimate.

8.2. Knowledge Gaps and Uncertainty Associated with Parasitoid Species on Cotton in Vietnam

The Selection Matrix exercise highlighted significant knowledge gaps and uncertainty associated with cotton parasitoids in Vietnam. Of the 23 taxa listed as parasitizing cotton pests, 17 have not been identified to species level. It was therefore not possible to rank 13 taxa for functional significance and these species were given a ? to indicate uncertainty. A clearer knowledge of taxonomy, host specificity, behaviour and ecology of these taxa is essential for risk assessment and we recommend further research and examination of the literature to fill these knowledge gaps in Vietnam. It is also probable that, with more research, additional parasitoid species can be found on hosts on cotton in Vietnam.

In order to illustrate how a risk assessment can proceed once these knowledge gaps have been filled, we selected three taxa that represented a wide taxonomic, ecological and behavioural diversity. The three selected taxa were:

1. *Apanteles* sp. (Hymenoptera: Braconidae) is a widely distributed larval parasitoid of *H. armigera*, *A. flava* and *E. vitella* (Pham, 1996, 2003). Both the hosts and the parasitoid are considered easy to mass rear and use in experiments. In addition to ecological considerations, ease of laboratory rearing is also an important consideration, since such facilities are limited in scope in Vietnam.
2. *Trichogramma chilonis* Ishii (Hymenoptera: Trichogrammatidae) is a common egg parasitoid of *H. armigera* and *A. flava* in Vietnam (Nguyen and Nguyen, 1982) and *Earias* spp. in India (Sekhon and Varma, 1983). Mass rearing and commercial release of this parasitoid for augmentative biocontrol have been widely reported (Waage and Greathead, 1986) and the species is used in Vietnam.
3. *Aphelinus* sp. (Hymenoptera: Aphelinidae) is rather common and has been observed attacking aphid species of major crop plants in Vietnam. This parasitoid was selected with the objective of controlling the virus vector *A. gossypii* (as a part of an IPM system). It is also easy to mass rear (Takada, 2002).

For the rest of this chapter, we focus on these three taxa as illustrative of the general processes and steps involved. This selection does not imply that these are the most important parasitoids to investigate on Bt cotton in Vietnam. We recommend that risk hypotheses for more taxa are evaluated once knowledge gaps have been filled, to ensure that the potential risks are considered.

8.3. Identifying Exposure Pathways

There are many ecological exposure pathways through which Bt cotton could interact with parasitoids (Fig. 8.1).

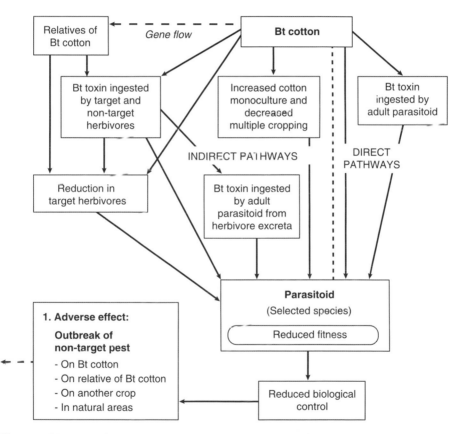

Fig. 8.1. Diagram of possible risk hypotheses associated with the adverse effect scenarios leading to crop losses (assessment end points) for parasitoids associated with Bt cotton. Arrows represent causal connections. Large 'parasitoid' box includes the possible responses (e.g. measurement end points) of a parasitoid (entity) to Bt cotton (stressor). Arrows leading into the 'parasitoid' box may affect any of the measurement end points in the box. Dashed lines indicate pathways that are not discussed in detail in this chapter. Dotted line separates indirect pathways from direct pathways.

Bitrophic exposure

Bt toxin and active metabolites are present in pollen of Bt cotton varieties that produce Cry1Ac and VIP3A proteins (Chapter 4, this volume). They are less likely to be present in floral nectar, because it is normally composed only of plant sugars. Extra-floral nectar also originates from the phloem, so will have broadly similar composition to floral nectar, but its concentration and abundance (particularly on cotton leaves) can vary widely in response to herbivore damage (Röse *et al.*, 2006). Nearly all parasitoids consume both floral and extra-floral nectar (Jervis *et al.*, 1993). Some parasitoid species also eat pollen, and parasitoids feeding on floral nectar might imbibe some pollen incidentally (Jervis *et al.*, 1993). Therefore, adult parasitoids could acquire Bt toxin by feeding on pollen; they possibly might acquire it through feeding on floral and extra-floral nectar. It is also possible that parasitoids are exposed to Bt toxin or metabolites through consumption of plant guttation fluids.

Tritrophic exposure through herbivore excretion products

Bt toxin may be present in honeydew secreted by aphids or leafhoppers feeding on Bt cotton (e.g. Bernal *et al.*, 2002; Zhang *et al.*, 2006) and adult parasitoids may acquire it by feeding on the honeydew.

Tritrophic exposure through host herbivore

Bt toxin may be present in the herbivore host life stage in which the immature parasitoid develops and the immature stage may acquire the Bt toxin from the host. For *Trichogramma* and *Gryon*, this is host eggs and for the other species, this is host larvae or nymphs. It is likely that lepidopteran larvae feeding on Bt cotton will contain significant amounts of Bt toxins in their guts (Chapter 6, this volume). *A. gossypii* may take up Bt toxins in trace amounts (Zhang *et al.*, 2006), but it is not known how much Bt is present in cotton phloem. *Aphelinus* adults feed on the fluids of their hosts (Takada and Tokumaru, 1996), which would also expose this life stage if the Bt were present in the aphids. It is not known if the eggs of lepidopteran or hemipteran herbivores feeding on Bt crops contain Bt toxins (Chapter 5, this volume), so tritrophic exposure of egg parasitoids through the egg is uncertain.

Higher trophic level exposure

Although not considered explicitly in this chapter, exposure of parasitoids at higher trophic levels could occur via hyperparasitoids (parasitoids of parasitoids), or heteronomous parasitoids (sibling parasitism) (Sullivan and Völkl, 1999) and parasitoids of predators. The consequences of such exposure are difficult to predict.

A systematic assessment of potential exposure pathways for the selected species (Table 8.2) indicates that some kind of exposure is possible for all three

Table 8.2. Potential exposure assessment for Bt cotton. Assessment is based primarily on the literature on Cry1Ac cotton (Chapter 4, this volume), and '?' indicates uncertainty.

	Selected species or taxon	*Apanteles* sp.	*T. chilonis*	*Aphelinus* sp.
BITROPHIC EXPOSURE	Does parasitoid feed on plant tissues or products that contain the transgene product?	Yes	Yes	Yes
	Is this feeding important for the parasitoid?	Yes	Yes	Yes
	Is bitrophic exposure possible?	Yes	Yes	Yes
	Have the transgene product or metabolites been detected in the species after bitrophic feeding?	?	?	?
	Does bitrophic exposure occur?	?	?	?
TRITROPHIC EXPOSURE (HERBIVORE PRODUCTS)	Does the parasitoid feed on animal products or excretions that contain the transgene product or metabolites?	Yes	Yes	Yes
	Are herbivore products an important part of the parasitoid diet?	?	Yes	Yes
	Is tritrophic exposure via feeding on herbivore products possible?	Yes?	Yes?	?
	Have the transgene product or metabolites been detected in the parasitoid after this kind of feeding?	?	?	?
	Does tritrophic exposure via feeding on herbivore products occur?	?	?	?
TRITROPHIC EXPOSURE (PARASITIZED HOST)	Are any of the hosts which the species parasitizes exposed when parasitized?	Yes	Yes?	Yes?
	Is tritrophic exposure via the host possible?	Yes	?	?
	Have the transgene product or metabolites been detected in the parasitoid?	?	?	?

Continued

Table 8.2. *Continued*

	Selected species or taxon	*Apanteles* sp.	*T. chilonis*	*Aphelinus* sp.
	Does tritrophic exposure occur through host?	Yes?	?	?
HIGHER TROPHIC EXPOSURE (CANNIBALISM/ INTRAGUILD FEEDING)	Does the parasitoid superparasitize or multiparasitize or show heteronomous hyperparasitism?	Yes	Yes	Yes
	Is higher trophic exposure or exposure possible?	Yes	Yes	Yes
EXPOSURE VIA GENE FLOW RECIPIENTS	Can gene flow occur?	Only on small scale in some regions		
	If yes, could the parasitoid feed on plants or parasitize hosts on plants that have received the transgene through gene flow?	Yes?	?	?
	Is exposure via gene flow possible?	? On small scale	? On small scale	? On small scale
EXPOSURE CHANGED BY BEHAVIOUR	What change in parasitoid behaviour might increase or decrease exposure?	Searching	Searching	Searching
	Are behaviours likely to increase or decrease exposure?	?	?	?
SUMMARY	IS EXPOSURE POSSIBLE?	Yes	Yes	Yes
	DOES EXPOSURE OCCUR?	Yes?	?	?

parasitoid species, but no evidence is available that any exposure actually occurs for any of the species. Some gene flow of Bt transgenes to cultivated or unattended cotton present in the area is possible, constituting a possible 'knock-on' effect in ecosystem interactions involving parasitoids. However, gene flow is likely to be restricted in Vietnam (Chapter 11, this volume); therefore, exposure of parasitoids to Bt through gene flow recipients will be small. Finally, for the *Apanteles* sp., behavioural reactions associated with Bt cotton may reduce exposure. This high degree of uncertainty suggests that some of the following research hypotheses on direct or indirect exposure pathways of parasitoids should be investigated.

8.4. Identifying Adverse Effects Pathways

The main adverse effect is an outbreak of a non-target pest because biological control by a parasitoid is reduced (Fig. 8.1). An adverse effect on a parasitoid must translate into reduced biological control, increased pest attack and crop or biodiversity losses. An adverse effect on a parasitoid is necessary but insufficient in itself to cause this cascade. In other words, examination of the effects on parasitoids can tell us that an adverse effect is or is not possible, but it cannot tell us if an adverse effect does occur. Thus, it is appropriate to consider adverse effects pathways that begin with ecological effects on parasitoid fitness (important fitness parameters for parasitoids include survival, larval development time, adult longevity, adult dispersal, adult host finding, adult fecundity, adult sex ratio, adult weight, adult emergence rate), which could affect parasitoid population dynamics adversely.

'Knock-on' pathways

- The lepidopteran herbivore hosts killed by Cry1Ac Bt cotton, particularly *H. armigera*, *P. gossypiella*, *A. flava*, *E. vitella* and *S. derogata*, will be less abundant for parasitoids on Bt cotton and parasitoids may therefore alter their foraging and oviposition behaviours, resulting in changes in parasitism.

- The herbivore hosts that are not killed by Bt cotton could be less suitable and/or sublethally affected by Bt cotton (especially the *Spodoptera* spp., see Chapter 6, this volume). This could affect parasitoid development inside these hosts adversely. Baur and Boethel (2003) found longer development time, decreased adult longevity and lower fecundity of two *Cotesia* species that parasitized a sublethally affected host (soybean looper, *Pseudoplusia includens*) that was fed on Cry1Ac cotton tissue. Liu *et al.* (2005a,b) tested the effects on the braconid endoparasitoid *Microplitis mediator* on *H. armigera* larvae that were given a diluted dose of Bt cotton tissue so that they were sublethally affected. The parasitoids reared on the sublethally affected hosts showed longer development time, lower emergence rate and lower adult weight. The parasitoids emerged from the Bt-fed larvae in their third instar, whereas they emerged from fourth-instar larvae fed on the control diet. An ichneumonid endoparasitoid was also affected adversely by sublethally affected *H. armigera* hosts (Liu *et al.*, 2005c).

- Bt cotton may have a changed metabolic profile compared to non-Bt cottons, including changes in volatile production. Yan *et al.* (2004) reported that *H. armigera* responded to seven compounds and two minor unknown compounds in Bt cotton volatiles, among which concentrations of α-pinene and β-pinene were higher in Bt cotton than in normal cotton. This could cause changes in parasitoid behaviour.

- Sublethal or other effects of Bt cotton on herbivore hosts may affect their production of volatiles and changes in their feeding behaviour may affect the response of the plant to herbivore attack. These changes could affect parasitoid behaviour.

- Use of Bt cotton is expected to result in changes in pesticide use (e.g. different application frequency, rates, timing or products), which may affect non-target pests on cotton and alternative host herbivores in the locality. These changes may affect the parasitoids on target and non-target pests, resulting in pest outbreaks.

•   Use of Bt cotton may alter conventional cropping systems, for example, possibly stimulating a switch from intercropping to monocultures. Bt cotton is likely to be grown in central Vietnam as an irrigated crop in the dry season in relay or intercrop with many other crops, such as soybean, mung bean, maize, or jute (Chapter 2, this volume). If Vietnamese farmers adopt Bt cotton widely, it may end up being grown in large monocultures and reducing crop and habitat diversity that support parasitoids. This could lead to reduced parasitoid populations and pest outbreaks of non-target pests on cotton or target or non-target pests, including *H. armigera*, on other crops nearby.

8.5. Formulation of Risk Hypotheses

In the following text, *Apanteles* sp. is used as the case example. The following risk hypotheses concerning the adverse effects of Bt cotton on *Apanteles* sp. were developed based on the potential exposure pathways and adverse effects pathways (Fig. 8.2).

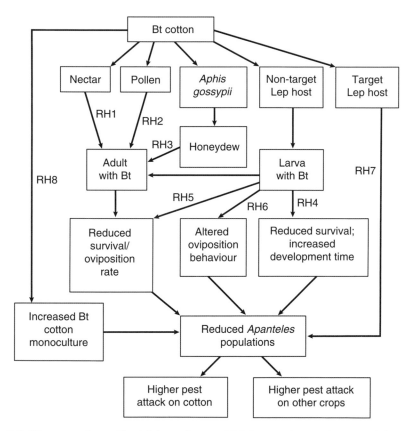

Fig. 8.2. Diagram of specific risk hypotheses (RH, for numbers refer to text) for potential adverse effects on *Apanteles* sp. associated with Bt cotton in Vietnam.

Risk hypothesis 1 (RH1, Bitrophic)

Floral or extra-floral nectar from Bt cotton contains Bt toxin or associated chemicals, which is taken up by adult *Apanteles* sp. This reduces adult survival or oviposition rate, leading to lower *Apanteles* sp. population size, which leads to pest outbreaks on cotton or other crops.

Risk hypothesis 2 (RH2, Bitrophic)

Pollen from Bt cotton contains Bt toxin or associated chemicals, which is taken up by adult *Apanteles* sp. This reduces adult survival or oviposition rate, leading to lower *Apanteles* sp. population size, which leads to pest outbreaks on cotton or other crops.

Risk hypothesis 3 (RH3, Tritrophic)

Honeydew excreted by *A. gossypii* feeding on Bt cotton contains Bt toxin or associated chemicals, which is consumed by adult *Apanteles* sp. This reduces adult survival or oviposition rate, leading to lower *Apanteles* sp. population size, which leads to pest outbreaks on cotton or other crops.

Risk hypothesis 4 (RH4, Tritrophic)

Lepidopteran larvae feeding on Bt cotton contain Bt toxin or associated chemicals and are unaffected by it (because they are either naturally insensitive or have acquired resistance). Larval *Apanteles* sp. developing inside these hosts acquire the toxin and have lower survival, longer development time or smaller size and/or fecundity of adult *Apanteles* is reduced, leading to lower *Apanteles* sp. population size, which leads to pest outbreaks on cotton or other crops.

Risk hypothesis 5 (RH5, Tritrophic)

S. exigua (or *S. litura*) feeding on Bt cotton are affected sublethally: they develop more slowly and have lower larval body weight and size. Larval *Apanteles* sp. developing inside these hosts also have a longer development time, lower emergence rate and/or lower adult weight, and the longevity or fecundity of adult *Apanteles* sp. is reduced, leading to lower *Apanteles* sp. population size, which leads to pest outbreaks on cotton or other crops.

Risk hypothesis 6 (RH6, 'Knock-on' effect)

S. exigua (or *S. litura*) feeding on Bt cotton release an altered volatile pattern in their frass, or the plant volatiles released by Bt cotton in response to

Spodoptera damage are altered, either due to changed feeding patterns of the sublethally affected larvae, or due to unintended changes in the Bt cotton plant. Host finding and oviposition behaviour by adult *Apanteles* sp. is altered. This altered behaviour results in lower *Apanteles* sp. population size, which leads to pest outbreaks on cotton or other crops.

Risk hypothesis 7 (RH7, 'Knock-on' effect)

Bt cotton kills the target lepidopteran pest, *H. armigera*, depleting hosts and reducing the population of *Apanteles* sp. that would have reproduced on *H. armigera*, resulting in outbreaks of pests which formerly were controlled by this parasitoid on cotton or other crops.

Risk hypothesis 8 (RH8, 'Knock-on' effects)

The introduction of Bt cotton management systems into relay (cotton/cucumber) or intercropped (with maize, soybean, mung bean) systems may result in altered pesticide use, altered fertilizer use, or changes in the cropping system itself, which affects *Apanteles* sp. adversely, resulting in pest outbreaks on cotton or other crops.

Risk hypothesis 9 (RH9, gene flow effects involving cultivated and volunteer cotton)

Gene flow could allow a relative of Bt cotton to express Bt toxin. Host herbivores feeding on this novel hybrid plant could have either a direct or indirect adverse effect on a parasitoid such as *Apanteles* sp. if such a hybrid were widespread in the regional agroecosystem.

8.6. Prioritization of Risk Hypotheses and Analysis Plan

Prior to testing the risk hypotheses, it is essential to specify the pest risk at the end of the risk hypotheses (terminal boxes in Fig. 8.2). All of these risk hypotheses result in reduced *Apanteles* sp. adult populations emerging from hosts on cotton. Thus, the hypothesized pest outbreaks must occur after this time. It is essential that the potential outbreak pest species and associated crops be identified clearly before any experimentation begins. This is essential because all of the experiments should be done to determine the risk if the identified pest species were to be released from biological control by *Apanteles* sp. and cause damage to a crop. If no potential outbreak pest species can be identified, then there is no need to conduct any of the experiments.

In addition, the identity and distinguishing species characteristics of *Apanteles* sp. need to be clarified sufficiently. This is necessary so that researchers can be confident that all experiments are being conducted on the same species and that the selected species is one that is significant in the field. It would be best to name the species officially, but if this is not possible to do in a timely manner, it may be appropriate to proceed if diagnostic morphological traits are used to identify the species.

Assuming that a potential outbreak pest species and an appropriate *Apanteles* species can be identified, then Risk Hypotheses 1–6 could be evaluated in the laboratory or greenhouse (Experiments 1 and 2). RH2 is considered more likely than the others because Bt cotton pollen is known to express Bt toxins (Chapter 4, this volume) and RH5 is considered likely because non-target host Lepidoptera feeding on Bt crops are often poorer hosts (Table 8.3). Some protocols for assessing these risk hypotheses are provided in the next section. Risk Hypotheses 1–3 can be falsified quickly if nectar, pollen or honeydew does not contain Bt toxin or an active metabolite; these biochemical tests should be the first experiments conducted (see Experiment 1). Risk can be estimated first by considering worst-case possibilities for the exposure pathways and adverse effects pathways, by providing constant, high exposure to the adult or larval parasitoids and estimating parasitoid fitness parameters (Experiment 1). Using these (or more extreme exposures), it is possible to calculate the maximum possible effect on the parasitoid population size in cotton and project how much this will reduce parasitism of the identified potential outbreak pest species. If the estimated worst-case risk is unacceptably high, then definitive experiments would need to be conducted to estimate expected exposure and release from biological control.

RH4 distinguishes a possible effect on the parasitoid caused by contact with the Bt toxin from any effect on the parasitoid caused by sublethal effects on the host. The only way to test RH4 is if a lepidopteran species is available that largely is unaffected by the Bt and is also a suitable host for the parasitoid. If a Cry1Ac resistant laboratory strain of a host species (e.g. *H. armigera*) is available, the laboratory experimental protocol of Schuler *et al.* (2004) can be used. In nature, sublethal Bt effects on the host may be confounded with a Bt effect on the parasitoid larvae developing in that host, which will be very difficult, if not impossible, to distinguish. However, it may not be necessary to make this distinction between Risk Hypotheses 4 and 5 within the risk assessment to fulfil the main aim of finding out whether or not the biocontrol capacity of that parasitoid in cotton or other crops is affected adversely.

RH7 is a population level effect on a parasitoid that can best be tested in small- and large-scale field trials.

RH8 is a complex hypothesis and will be difficult to assess experimentally prior to commercialization (Table 8.3). If such effects occur, they are more likely to take time to show themselves (2 or more years) and be more spatially diffuse. Therefore, detecting these effects would require studies involving larger field sizes and including several crop types and native plants. Experimentation is not suggested specifically for parasitoids, because effects on predators, non-target herbivores, pollinators, species of conservation significance and other biodiversity would also need to be considered under a more holistic assessment. A monitoring system of agronomic practices for cotton in Vietnam, which

Table 8.3. Prioritization of risk hypotheses. Rank values: 1 = highest rank, 2 = intermediate rank, 3 = lowest rank. Consequence product is the multiplication product of the preceding rank numbers within potential consequence or logical relationships. The final product is the product of relative likelihood, consequence product and assessability. Ranks were determined by expert evaluation. Potential reversibility of the consequences varies from (1) irreversible to (3) readily reversed. Spatial scale of consequence varies from (1) extensive to (3) local. Potential magnitude varies from (1) severe to (3) minor. Assessability varies from (1) can be assessed in the laboratory or greenhouse (2) field experiments, or (3) post-commercialization. Final products in bold are the priority risk hypotheses. In this case, the priority is correlated strongly with relative likelihood.

Parasitoid taxon	Risk hypothesis	Relative likelihood	Relative potential consequence				Assessability	Final product
			Potential reversibility	Potential spatial scale	Potential magnitude	Consequence product		
Apanteles sp.	RH1	2	3	2.5	2–3	15–22.5	1	30–45
Apanteles sp.	RH2	1	3	2.5	2–3	15–22.5	1	**15–22.5**
Apanteles sp.	RH3	2	3	2.5	2–3	15–22.5	1	30–45
Apanteles sp.	RH4	1	3	2.5	2–3	15–22.5	1	**15–22.5**
Apanteles sp.	RH5	1	3	2.5	2–3	15–22.5	1	**15–22.5**
Apanteles sp.	RH6	2	3	2.5	2–3	15–22.5	1	30–45
Apanteles sp.	RH7	1	3	2.5	2–3	15–22.5	3	45–67.5
Apanteles sp.	RH8	2.5	1.5	2	2.5	7.5	3	56.25
Apanteles sp.	RH9	3	3	3	3	27	3	243

quantifies area of relay cropping, intercropping, amount of fertilizer and insecticide use, as well as any other changes in crop management, could be implemented instead (see monitoring below). The cropping and crop management pattern may also change due to policy-related measures to increase biodiversity in agroecosystems. If there are significant changes in cotton management, then additional research could be initiated to determine if there are any significant adverse effects on parasitoid effectiveness as biocontrol agents from such changes.

Because Vietnam has no significant likelihood of cotton gene flow outside cotton fields (Chapter 11, this volume), it is not necessary to investigate RH9 in Vietnam

8.7. Experimental Protocols to Test Risk Hypotheses

To estimate the potential risks outlined above, perhaps two pre-release experiments should be conducted prior to the commercial introduction of Bt cotton into Vietnam. Depending on the results from these experiments, additional experiments may be needed. In addition, after commercialization, it may be useful to monitor the responses of parasitoids to the introduction of Bt cotton. Experiment 1 tests Risk Hypotheses 1, 2 and 3, using adult parasitoids.

Experiment 1. Bi-trophic effects of Bt cotton on the parasitoid *Apanteles* sp., assessed by feeding of adults on nectar and pollen of Bt cotton or aphid honeydew

These experiments can be conducted in the laboratory or under GMO containment greenhouse conditions, using potted test plants (cut flowers may not produce typical nectar or pollen, so living plants are more reliable). The objectives are: (i) to confirm whether nectar, pollen and honeydew contain Bt toxin; (ii) to determine whether adult *Apanteles* sp. feed on this nectar, pollen or honeydew and acquire Bt toxin; and (iii) to quantify any effects of Bt nectar, pollen or honeydew on adult longevity and fecundity.

The experimental design should be completely randomized (CRD) with two factors (six treatments in total) as:

1. Three types of food: (i) pollen (cotton flower at blooming without petals, to eliminate nectar), (ii) nectar (cotton flower at blooming without stamens, to eliminate pollen) and (iii) aphid honeydew (actively growing aphid colonies on non-flowering cotton plants).

2. Two sources of food: (i) Bt cotton and (ii) a suitable control cotton variety (isogenic, conventional).

3. Samples of the three food types should be taken from actively growing plants and analysed immediately for the presence of Bt toxin and metabolites. Lateral flow sticks may be used if sufficiently large samples can be collected. Otherwise, quantitative ELISA would be necessary. A positive control must be used; one possibility is a small, standardized piece of Bt cotton leaf tissue. A negative control (distilled water) must also be used. Any food type that does not test positive can be eliminated from the following experiments. Liu *et al.* (2005b) found that adult parasitoids contained detectable amounts of Cry1Ac toxin after feeding on honey solution in which Cry1Ac protein was dissolved, confirming that if Bt is found in nectar, it will be detectable in the parasitoid adult.

Prior to experimentation, the experimental materials must be prepared as in the steps below:

1. It is necessary to identify and rear the most important species of *Apanteles* in the proposed cultivation area and use this one species for all of the experiments.
2. Suitable mass rearing techniques for *H. armigera* host and the parasitoid must be set up under greenhouse and laboratory conditions. A simple method for mass rearing of *H. armigera* in the tropical conditions of the Philippines was described by Ocampo *et al.* (2000). Mass rearing conditions for an *Apanteles* species was reported in Nealis and Fraser (1988).
3. Growing plants: two cotton varieties with and without Bt transgenes (preferably isogenic, apart from the introduced insecticidal gene) will be planted in separate pots over a period of time and grown under uniform conditions to produce uniformly blooming flowers for daily collection during the period of experimentation.
4. Some of these plants must be inoculated with *A. gossypii*, or conversely, *A. gossypii* must be removed from some of the plants.

At least ten replications are needed for each treatment. A replicate equates to one insect cage for each treatment, containing 5–1 newly emerged mated female adult parasitoids. Replicates can be set up on different days; however, whenever a replicate is set up, all six treatments must be set up on the same day. Adult ages should be similar among all of the treatments.

Potted plants should be replaced with a freshly flowering plant each day for at least 5 consecutive days because a cotton flower blooms for only 1 day, withering in the evening. After the 5th day, two adults (one male, one female) will be collected from each replicate and tested for the presence of Bt toxin, using either lateral flow strips or quantitative ELISA. A positive and negative control must be used. The positive control can be the Bt food type and the negative control can be the non-Bt food type.

The remaining adult parasitoids will be removed from the cages and fed water or sugar water until death. For treatments testing positive for the acquisition of Bt toxin, insects will be observed every day, to record the following biological parameters:

1. Mortality of the adults recorded at 6 and 12 h and daily after treatment, calculated as percentage of control mortality.
2. Longevity of the male and female adults (number of days) by recording the day of death for each parasitoid.

3. Egg load at death should be recorded by dissecting the dead female to count the number of eggs or the degree of ovary development. Leg length or wing length should also be measured, as these may correlate with egg load. The female parasitoids in this replicated set should be dissected at 5 of 8 days after treating for the number of eggs or the level of egg development.

4. Fecundity of the parasitoid should be recorded over the reproductive life of the adult by including another set of replications in parallel for each treatment. After the 5th day, 20 *H. armigera* second instars will be given to the females from each treatment replicate for a 24 h period. These will be replaced by a new batch of host larvae every 24 h for 5 days. Exposed larvae will be reared individually to assess successful *Apanteles* parasitism. The number and date of emerging parasitoid adults will be recorded for each individual host larva until parasitoid emergence is complete.

Data analysis will depend on each parameter and can be analysed using *t*-tests or GLM for egg load and fecundity, and parametric or Kaplan–Meier survival statistics for mortality and longevity (Southwood and Henderson, 2000).

Experiment 2. Tri-trophic effects of *Spodoptera litura* (or *S. exigua*) fed on Bt cotton to its parasitoid *Apanteles* sp.

Experiment 2 tests RH5, with two consecutive experiments. First, it is necessary to know the efficacy (lethal and sublethal effects) of Bt cotton on herbivorous insects (including both target and non-target species) feeding on the crop. These herbivores could affect their parasitoids indirectly through tritrophic interactions in the local food web. Therefore, two consecutive experiments are needed.

Pre-test with host
Pre-test for the effects of Bt cotton to the mortality and biology of *S. litura* (or *S. exigua*) in order to obtain an indication of the possible adverse effects on the host that might affect the parasitoid (i.e. host-mediated effects) (see Chapter 6, this volume, for more details).

S. litura will be mass reared on isogenic cotton leaves (control) in the laboratory. The moths needed for laying eggs will be reared in small insect screen cages of 50 × 50 cm diameter and 80 cm height, and the larvae will be reared in test (Petri) dishes or cups (plastic, of 10 cm diameter × 6 cm height). Newly emerged moths from this rearing will be collected to test for the effect of Bt cotton on their biological characteristics. *Spodoptera* spp. are target pests for some Bt cotton events such as Cry1F and VIP3A (see Chapter 4, this volume), so for those events, a different species should be used.

The experiment could be conducted under laboratory or greenhouse conditions, using a completely randomized design (CRD). There are only two treatments: (i) Bt cotton leaves and (ii) normal (isogenic control) cotton leaves.

From the mass-reared insects, one pair (male and female) of newly emerged adult moths will be collected, mated and the female allowed to lay eggs on cotton plants in individual insect cages. Sugar solutions will be provided as food for the adults and egg masses removed daily and incubated separately for hatching. Egg masses from 5–10 pairs should be used. From each pair, around 20 newly hatched larvae should be given each treatment (fed either control or Bt cotton) and reared individually in test dishes or cups through larval development, pupation and adult emergence of the first generation.

Pairs of emerging first generation adult moths will be selected for mating and laying eggs. To avoid inbreeding depression, *S. litura* individuals should be selected from different families. Around 10 pairs should be established from the Bt cotton families and 10 more pairs established from the control cotton families. The number of eggs laid should be counted daily until death. One hundred eggs will be collected from each pair to measure hatchability. Ten larvae will be selected from each pair and reared individually on the same diet as their parents to measure survival rate of larvae, development time, time of pupation, time of emergence, sex ratio and adult longevity.

Experiment with host and parasitoid

Test for the effect of Bt transgene on the parasitoid *Apanteles* sp. of *S. litura* (or *S. exigua*) feeding on Bt cotton.

Assumption: it is expected that the results of the pre-test experiment will indicate that *S. litura* (a pest which is probably not well controlled by the Cry1Ac toxin) shows some level of sublethal effects on Bt cotton but the insects will still develop to adults and survive to the second generation, with symptoms revealed as smaller larval size and slower development (Chapter 6, this volume).

Mass rearing of insects: the host *S. litura* will be mass reared on Bt cotton and non-Bt control (near isogenic) cotton to produce third-instar larvae for testing the Bt effect on the host herbivore and the parasitoid. *Apanteles* sp. will be mass reared on the normal *S. litura* host fed on control cotton. These experiments will be performed under standardized greenhouse or laboratory condition (28–30°C, 80% humidity and 14:16 h light:dark). The experimental design should be a CRD with two treatments of (i) Bt cotton and (ii) non-Bt (control) cotton, with ten replications per treatment.

In each small insect cage (individual replicate), ten third-instar larvae of the *S. litura* host will be introduced for feeding on the leaves of Bt cotton or non-Bt cotton. At the same time, two gravid females of *Apanteles* sp. will be introduced for laying eggs on the third-instar larvae. It is a good idea to observe the parasitoids' oviposition and record how many larvae are parasitized. After 3 days, the host larvae will be removed and reared individually in small cages until emergence of the adult wasp. These wasps will be used for laying eggs on the third-instar larvae, as described above, and reared for adult emergence of the second generation.

Fitness parameters will be recorded for the first and second generations of the parasitoid as follows: (i) proportion of adult wasps emerged; (ii) date of emergence (gives developmental times); (iii) weight and size of the adults; (iv) sex ratio; (v) fecundity; and (vi) per cent parasitism of host larvae (successful parasitoid emergence plus host death).

8.8. Monitoring Effects of Farming Practice Changes After Cultivation of Bt Cotton (RH8)

The objective of monitoring is to follow up the possible changes in the *H. armigera* and *Spodoptera* spp. pest populations, crop damage and particularly parasitism of the pests when Bt cotton is grown as part of an integrated pest management (IPM) programme. In addition, commercial use of Bt cotton may result in changes to cropping systems and farming practices, which could themselves affect biological control by parasitoids adversely. Associated with these changes, other pest species may escape parasitoid attack and become problematic. Their incidence and effect may be observed and measured as changes in their population density and crop damage caused.

Monitoring for parasitoid changes will be conducted under field conditions once a year, for 3 consecutive years, to record any changes in pest incidence and in pesticide use. Parasitoids are expected to respond quickly, so after 3 years, monitoring can be reduced to every other year. The location should be in an area of intensive cultivation of Bt cotton where there is variation in cropping practices, e.g. intercropping of cucumber or watermelon with maize in the dry season, followed by rice in the rainy season. These adjacent cotton-based cropping patterns will be recorded because they may give some effects on the insect pests of the experiment, such as the shifting of the host plants of *H. armigera*, density fluctuation or outbreak of secondary pests. In Vietnam, this area is likely to be in the southern coastal lowlands, where farmers practise various cropping patterns. The two main cropping seasons are (i) rain-fed cotton from July/August to November/December and (ii) irrigated cotton from December/January to April/May (Chapter 2, this volume). The irrigated cotton grown in the dry season usually gives higher yields and this season could be chosen initially for testing Bt cotton.

Observations should be set up to evaluate: (i) Bt cotton with IPM; (ii) Bt cotton with farmers' conventional practices for pest control; (iii) non-Bt (conventional, isogenic) cotton variety with IPM; and (iv) conventional, isogenic cotton with farmers' usual practices. At least four fields of each type should be observed. If all four fields are not available, observations will be conducted minimally on Bt cotton and non-Bt cotton and the pest control practices on each field will be recorded. Cultural practices (planting density, fertilizers, weed control, etc.) will be recorded. In IPM fields, pest populations and damage caused will be monitored and appropriate insecticide applications will be used when these indicators reach the action threshold. In the fourth treatment, the pest management methods will depend on the decisions of each farmer (e.g. biocontrol, IPM, insecticides).

The following measurements should be recorded for comparison of treatments (see Chapter 6, this volume, for more detailed protocols for monitoring pests and viral disease):

1. Monitor population development of *H. armigera*, *Spodoptera* spp. and other non-target lepidopteran pests: record the density of each insect life cycle stage per unit of area (e.g. $1\,\mathrm{m}^2$) every week for the whole cropping season (estimated 120 days).

2. Record damage of *H. armigera* and other pests on the crop as per cent of leaf damage and bored bolls.

3. Record percentage parasitism of *H. armigera* eggs, larvae or pupae by *Apanteles* and other parasitoid species by placing sentinels from laboratory colonies in the field or by collecting eggs, larvae and pupae from the field and rearing in the laboratory until parasitoids hatch.

4. Record population build-up and fluctuation of sucking pests (e.g. leafhoppers, thrips, aphids, whiteflies, mites) for the whole cropping season.

5. Record virus incidence as (i) percentage of plants damaged and (ii) disease index (severity of the disease).

For the 2nd and 3rd years, the observations will be repeated in the same area and using the same protocol as described above, to allow comparison between years. Collected data will be tabulated and analysed for comparing between treatments and through consecutive years, by using *t*-tests, GLM with Tukey's HSD, or contingency table analysis (Southwood and Henderson, 2000).

8.9. Sequence and Scaling of Proposed Experimentation

Suitable GMO containment facilities must be developed and made available. Laboratory and greenhouse experiments (1 and 2) can be started as soon as suitable adapted Bt cotton (with isogenic/near isogenic control lines) is available for testing in Vietnam. Depending on the results of these experiments, additional experiments may be needed. Field monitoring may be followed when the pre-release tests give results indicating a likely adverse effect. The total duration should be 3–5 years, depending on the need for additional testing of any adverse effects of Bt cotton on selected species and regional biodiversity (non-target herbivores, predators, parasitoids, flower visitors – see Chapters 6–10, this volume).

8.10. Conclusions and Future Developments

The experiments on the parasitoid *Apanteles* sp. are presented in detail because this is a common parasitoid of the most severe pests of cotton in Vietnam. Furthermore, this genus is relatively easy to mass rear and release for experimentation, using techniques developed previously for non-transgenic cotton (Nealis and Fraser, 1988; Uçkan and Gülel, 2000).

Large knowledge gaps in the ecological characteristics of parasitoids attacking pests of cotton in Vietnam have limited the identification of significant potential risks from Bt cotton. However, the diverse cropping patterns of cotton in different regions of Vietnam (see Chapter 2, this volume) provide fertile ground for observations on the knock-on effects of Bt cotton to parasitoids, which need to be developed further before additional experiments can be planned.

Experiments for other prioritized parasitoids can be designed for implementation after obtaining and analysing the results of these highest priority experiments. Next highest priority for development of experimentation would be assigned to the parasitoid *Aphelinus* sp. attacking the cotton aphid, *A. gossypii*, because changes in Bt cotton and reduced pesticide use for controlling lepidopteran pests could cause an upsurge of sucking pests and thus increase the transmission of virus diseases by such vectors.

Acknowledgements

We would like to thank Dr Phạm Văn Lâm, Institute of Plant Protection, Ha Noi, Vietnam, for checking the list of parasitoids on cotton pests in Vietnam. Dr Nick Birch, SCRI, is funded by the Scottish Executive.

References

Adamczyk, J.J. Jr and Meredith, W.R. Jr (2004) Genetic basis for variability of Cry1Ac expression among commercial transgenic *Bacillus thuringiensis* (*Bt*) cotton cultivars in United States. *Journal of Cotton Science* 8, 17–23.

Andow, D.A. and Hilbeck, A. (2004) Science-based risk assessment for non-target effects of transgenic crops. *Bioscience* 54, 637–649.

Baur, M.E. and Boethel, D.J. (2003) Effect of Bt-cotton expressing Cry1A(c) on the survival and fecundity of two hymenopteran parasitoids (Braconidae, Encyrtidae) in the laboratory. *Biological Control* 26(3), 325–332.

Bernal, C.C., Aguda, R.M. and Cohen, M.B. (2002) Effect of rice lines transformed with *Bacillus thuringiensis* toxin genes on the brown planthopper and its predator *Cyrtorhinus lividipennis*. *Entomologia Experimentalis et Applicata* 102(1), 21–28.

Birch, A.N.E. and Wheatley, R.E. (2005) GM pest-resistant crops: assessing environmental impacts on non-target organisms. In: Hester, R.E. and Harrison, R.M. (eds) *Sustainable Agriculture. Issues in Environmental Science and Technology 21*. The Royal Society of Chemistry, Cambridge, UK, pp. 31–57.

Birch, A.N.E., Wheatley, R.E., Anyango, B., Arpaia, S., Capalbo, D., Getu Degaga, E., Fontes, E.M.G., Kalama, P., Lelmen, E., Lövei, G.L., Melo, I.S., Muyekho, F., Ngi-Song, A., Ochieno, D., Ogwang, J., Pitelli, R., Schuler, T., Setamou. M., Sithnantham, S., Smith, J., Nguyen Van Son, Songa, J., Sujii, E., Tan, T.Q., Wan, F.-H. and Hilbeck, A. (2004) Biodiversity and non-target impacts: a case study of Bt maize in Kenya. In: Hilbeck, A. and Andow, D.A. (eds) *Environmental Risk Assessment of Genetically Modified Organisms, Volume 1: A Case Study of Bt Maize in Kenya*. CAB International, Wallingford, UK, pp. 117–185.

Greenplate, J.T. (1997) Response to reports of early damage in 1996 commercial Bt transgenic cotton (Bollgard) planting. *Society of Invertebrate Pathology Newsletter* 29, 15–18.

Greenplate, J.T. (1999) Quantification of *Bacillus thuringiensis* insect control protein Cry1Ac over time in Bollgard cotton fruit and terminals. *Journal of Economic Entomology* 92, 1377–1383.

Groot, A. and Dicke, M. (2002) Insect-resistant transgenic plants in a multi-trophic context. *Plant Journal* 31, 387–406.

Hilbeck, A. and Schmidt, J.E.U. (2006) Another view on Bt proteins – how specific are they and what else might they do? *Biopesticides International* 2, 1–50.

Jervis, M.A., Kidd, N.A.C. and Watson, M.A. (1992) A review of methods for determining dietary range in adult parasitoids. *Entomophaga* 57, 365–374.

Jervis, M.A., Kidd, N.A.C., Fitton, M.G., Huddleston, T. and Dawah, H.A. (1993) Flower visiting by hymenopterous parasitoids. *Journal of Natural History* 27, 67–105.

Leius, K. (1960) Attractiveness of different foods and flowers to the adults of some hymenopterous parasites. *Canadian Entomologist* 92, 369–376.

Liu, X.X., Zhang, Q.-W., Zhao, J.-Z., Li, J.-C., Xu, B.-L. and Ma, X.-M. (2005a) Effects of Bt transgenic cotton lines on the cotton bollworm parasitoid *Microplitis mediator* in the laboratory. *Biological Control* 35(2), 134–141.

Liu, X.X., Zhang, Q., Zhao, J.-Z., Cai, Q., Xu, H. and Li, J. (2005b) Effects of the Cry1Ac toxin of *Bacillus thuringiensis* on *Microplitis mediator*, a parasitoid of the cotton bollworm, *Helicoverpa armigera*. *Entomologia Experimentalis et Applicata* 114, 205–213.

Liu, X.X., Sun, C.-G. and Zhang, Q.-W. (2005c) Effects of transgenic Cry1A + CpTI cotton and Cry1Ac toxin on the parasitoid, *Campoketis chlorideae* (Hymenoptera: Ichneumonidae). *Insect Science* 12(2), 101–107.

Lövei, G.L. and Arpaia, S. (2005) The impact of transgenic plants on natural enemies: a critical review of laboratory studies. *Entomologia Experimentalis et Applicata* 114, 1–14.

Nealis, V.G. and Fraser, S. (1988) Rate of development, reproduction and mass rearing of *Apanteles fumiferanae* Vier. (Hymenoptera: Braconidae) under controlled conditions. *Canadian Entomologist* 120(3), 197–204.

Nguyen, I.T. and Nguyen, S.T. (1982) The use of *Trichogramma* in Vietnam. *Zachista Rastenii* 52. Cited in: Waterhouse (1998) *Biological Control of Insect Pests: Southeast Asian Prospects*. ACIAR Books Online, Australia. www.aciar.gov.au/publication/MN051 (accessed 12 September 2007).

Nguyen Van Cam and Viet, H.T. (1996) Influence of some factors on the formulation process and application of NPV-Ha for controlling cotton bollworm (*Helicoverpa armigera*) on tobacco. In: Nguyen Van Cam and Pham Van Lam (eds) *Selected Scientific Reports on Biological Control of Pests and Weeds (1990–1995)*. Agricultural Publishing House, Hanoi, pp. 24–33 (in Vietnamese).

Nguyen Van Huynh and Sen, L.T. (2003) *Textbook of Agricultural Entomology*. Can Tho University, Can Tho, Vietnam (in Vietnamese).

Ocampo, V.R., Tabur, M.T., Fulgencio, R.D., Lim, D.C. and Orozco, F.D. (2000) Diet development and mass rearing of the corn earworm, *Helicoverpa armigera* (Hubner) (Lepidoptera: Noctuidae) in the Philippines. *Philippine Agricultural Sciences* 83, 282–291.

Pallini, A., Silvie, P., Monnerat, R.G., Ramalho, F. de S., Songa, J.M. and Birch, A.N.E. (2006). Non-target and biodiversity impacts on parasitoids. In: Hilbeck, A., Andow, D.A. and Fontes, E.M.G. (eds) *Environmental Risk Assessment of Genetically Modified Organisms, Volume 2: A Case Study of Bt Cotton in Brazil*. CAB International, Wallingford, UK, pp. 200–224.

Pham Van Lam (1996) Contributions to the study on fauna of hymenopterous parasitoids in Vietnam. In: *Selected Scientific Reports on Biological Control of Pests and Weeds (1990–1993)*, Volume 1. Agricultural Publishing House, Hanoi, pp. 95–103 (in Vietnamese).

Pham Van Lam (2003) *The List of Natural Enemies of Major Pests on Crop Plants in Vietnam*. Agricultural Publishing House, Hanoi, (in Vietnamese).

Poppy, G.M. and Sutherland, J.P. (2004) Can biological control benefit from genetically-modified crops? Tritrophic interactions on insect-resistant transgenic plants. *Physiological Entomology* 29, 257–268.

Powell, W. (1986) Enhancing parasitoid activity in crops. In: Waage, J. and Greathead, D. (eds) *Insect Parasitoids*. Academic Press, New York, pp. 319–340.

Price, P.W. (1997) *Insect Ecology*, 3rd edn. John Wiley & Sons, Inc., USA.

Ramirez-Romero, R., Bernal, J.S., Chaufaux, J. and Kaiser, L. (2007) Impact assessment of Bt-maize on a moth parasitoid, *Cotesia marginiventris* (Hymenoptera: Braconidae), via host exposure to purified Cry1Ab protein or Bt-plants. *Crop Protection* 26(7), 953–962.

Röse, U.S.R., Lewis, J. and Tumlinson, J.H. (2006) Extrafloral nectar from cotton (*Gossypium hirsutum*) as a food source for parasitic wasps. *Functional Ecology* 20(1), 67–74.

Schuler, T.H., Denholm, I., Jouanin, L., Clark, A.J. and Poppy, G.M. (2001) Population-scale laboratory studies of the effect of transgenic plants on non-target insects. *Molecular Ecology* 10, 1845–1853.

Schuler, T.H., Denholm, I., Clark, S.J., Stewart, C.N. and Poppy, G.M. (2004) Effects of Bt plants on the development and survival of the parasitoid *Cotesia plutellae* (Hymenoptera: Braconidae) in susceptible and Bt-resistant larvae of the diamondback moth, *Plutella xylostella* (Lepidoptera: Plutellidae). *Journal of Insect Physiology* 50, 435–443.

Sekhon, B.S. and Varma, G.C. (1983) Parasitoids of *Pectinophora gossypiella* [Lep.: Gelechiidae] and *Earias* spp. [Lep.: Noctuidae] in the Punjab. *Entomophaga* 28, 45–54.

Southwood, T.R.E. and Henderson, P.A. (2000) *Ecological Methods*, 3rd edn. Blackwell Publishing, Oxford, UK.

Sullivan, D.J. and Völkl, W. (1999) Hyperparasitism: multitrophic ecology and behavior. *Annual Review of Entomology* 44, 291–315.

Takada, H. (2002) Parasitoids (Hymenoptera: Braconidae, Aphidiinae; Aphelinidae) of four principal pest aphids (Homoptera: Aphididae) on greenhouse vegetable crops in Japan. *Applied Entomology and Zoology* 37(2), 237–249.

Takada, H. and Tokumaru, S. (1996) Observations on oviposition and host-feeding behavior of *Aphelinus gossypii* Timberlake (Hymenoptera: Aphelinidae). *Applied Entomology and Zoology* 31(2), 263–270.

Uçkan, F. and Gülel, A. (2000) Effects of host species on some biological characteristics of *Apanteles galleriae* Wilkinson (Hym.; Braconidae). *Turkish Journal of Zoology* 24 (Ek), 105–113.

Waage, J. and Greathead, D. (1986) *Insect Parasitoids*. Academic Press, London, UK.

Yan, F., Bengtsson, M., Anderson, P., Ansebo, L., Xu, C. and Witzgall, P. (2004) Antennal response of cotton bollworm (*Helicoverpa armigera*) to volatiles in transgenic Bt cotton. *Journal of Applied Entomology* 128(5), 354.

Zhang, G.-F., Wan, F.-H., Lövei, G.L., Liu, W.X. and Guo, J.-Y. (2006) Transmission of Bt toxin to the predator *Propylea japonica* (Coleoptera: Coccinellidae) through its aphid prey feeding on transgenic Bt cotton. *Environmental Entomology* 35(1), 143–150.

9 Potential Effects of Transgenic Cotton on Flower Visitors in Vietnam

Agriculture depends crucially on ecological services (Tilman *et al.*, 2002) and one of the important biological services is pollination (Daily, 1999). Although agriculture has detrimental effects on biological diversity, agricultural habitats (and specifically flowers and floral resources within them) also support biodiversity significantly, especially in heavily cultivated areas. Cotton flowers are open for 1 day only from dawn to sunset, but the crop produces flowers over a 2-month period. Cotton flowers produce copious nectar and pollen and are accessible to many organisms because they are large and open (Oosterhuis and Jernstedt, 1999). Cotton flower visitors include many species of flies, wasps, butterflies, beetles, thrips and bees (Kevan and Baker, 1983).

The non-target effects of Bt cotton on flower visitors may be positive or negative. For example, pollinators or rare lepidoptera populations can be enhanced because of reduced insecticide use on Bt cotton (Chapter 1, this volume). Pollinator diversity in natural areas might also be enhanced by decreasing populations of a dominant floral competitor. In risk assessment, we assess the potentially adverse effects, not the beneficial effects. Thus, it is important to consider carefully whether an increase or a decrease in the population size of a particular floral visitor would result in a beneficial or adverse effect on the environment. The importance of the diversity and abundance of pollinator species is highlighted by the Convention on Biological Diversity (CBD, 1996) and the FAO International Pollinator Initiative.[1]

Possible adverse effects associated with flower visitors include (Figs. 9.1 and 9.2):

- Reduced pollination of cotton, resulting in lower fibre yield and seed production. Although cotton has a high rate of self-pollination, cross-pollination increases production (McGregor, 1976). However, cotton production in Vietnam is on

[1] http://www.fao.org/biodiversity/pollinat_en.asp.

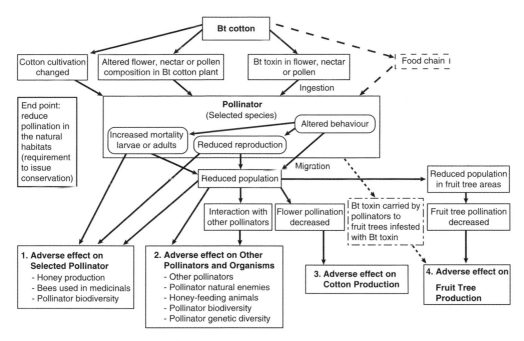

Fig. 9.1. Diagram of possible risk hypotheses associated with four adverse effects pathways (assessment end points) for pollinators visiting Bt cotton flowers. Arrows represent causal connections. Large 'pollinator' box includes the possible responses (measurement end points) of a pollinator (entity) to Bt cotton (stressor). Arrows leading into the 'pollinator' box may affect any of the measurement end points in the box. Dashed lines indicate pathways that are not discussed in detail in this chapter. Dotted lines are a potential pathway discussed in the text. The 'pollinator' box is developed in more detail in Fig. 9.3.

small areas (Chapter 2, this volume) and the yield generally is not limited by pollination.

- Reduced pollination of other crops nearby, such as fruit crops, coffee or beans, resulting in lower yields in those crops.
- Reduced pollination in natural areas, causing lower seed set and other effects on plants and animals that eat seeds.
- Reduced honeybee numbers, causing reduced production of bee products. Vietnam is the second largest honey exporter worldwide and many thousands of farmers keep *Apis mellifera* or *A. cerana* colonies, or harvest bee products from wild bee colonies (APISERVICES, 1997).
- Reduced wild bee population size, resulting in adverse effects on rare or valued species that rely on bee products such as honey.
- Reduced genetic diversity in wild bee populations. This could result from large, rapid reductions in bee population size.
- Reduced populations of species of conservation value, such as butterflies, that feed on nectar and pollen.
- Reduced populations of flower-visiting predators and parasitoids, causing reduced biological control of pests. This issue is also evaluated in Chapters 7 and 8 (this volume).

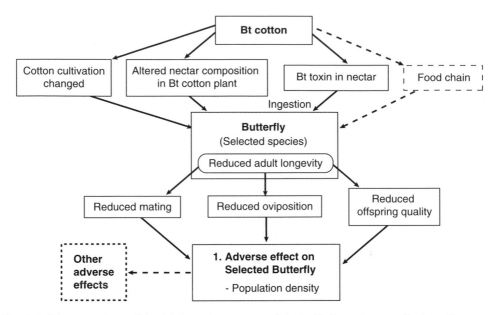

Fig. 9.2. Diagram of possible risk hypotheses associated with the adverse effects pathways (assessment end points) for butterflies visiting Bt cotton flowers. Arrows represent causal connections. Large 'butterfly' box includes the possible response (measurement end point) of a butterfly (entity) to Bt cotton (stressor). Any of the arrows leading into the 'butterfly' box may affect the measurement end points in the box. Dashed lines indicate pathways that are not discussed in detail in this chapter.

To assess the potential effects of Bt cotton on non-target flower visitors in cotton ecosystems in Vietnam we used an environmental risk assessment model originating from Arpaia *et al.* (2006). This chapter concentrates on the first three parts of the process: (i) list non-target flower visitors and identify species for further evaluation; (ii) identify potential exposure pathways and potential adverse effect pathways; (iii) formulate risk hypotheses based on the identified exposure pathways and adverse effect pathways and design experiments testing the risk hypotheses. In addition, we present some possible experimental designs and discuss the analysis and interpretation of field biodiversity studies.

In this chapter, we consider all of the adverse effects listed above by considering all flower visitors together. We then identify high priority species to evaluate: (i) the potential reduction in pollination value; (ii) the potential reduction in the economic value of bee products; (iii) the potential adverse effects stemming from reduction in wild bee populations, including effects on other species and loss of genetic diversity; and (iv) the potential reduction in species with conservation value or biological control value.

9.1. Selection of Flower-visiting Species for Further Evaluation

So far, 28 species of common flower visitors that feed on the floral resources of cotton have been identified in Vietnam (Table 9.1: based on records at the

Table 9.1. Selection Matrix for flower-visiting species associated with cotton in Vietnam, showing scores for six association criteria and seven significance criteria. High rank = 1; medium rank = 2; low rank = 3; gap of knowledge = ? Life stage that eats/comes into contact with plant tissues: A = adult; L = larva; S = social species. Plant tissue consumed: F1 = petals; F2 = anthers; F3 = stamens; N = nectar; P = pollen. Association with Crop criteria: A1 = geographic distribution; A2 = habitat specialization: degree of association with cotton habitat; A3 = prevalence: proportion of suitable habitat occupied; A4 = abundance on cotton crop; A5 = phenology: degree of overlap of herbivore life history with cotton; A6 = phenology: proportion of cotton growing season when present (what parts of the cotton season is the species present: E = early, M = middle, L = late; 1 = all, 2 = E–M; 3 = M or ML). Functional Significance criteria: PC = significance as a pollinator of cotton; PO = pollinator of other crops; PN = pollinator of natural areas; FN = food for natural enemies; DV = disease vector; SD = seed disperser; BC = biological control agent. MA = mean of Association with Crop criteria; MS = mean of Functional Significance criteria. Sum is the sum of the two means and rank is the final priority ranking used in species selection. The highest priority species ranks are in bold.

Species or species group	Order and family	Life stage	Plant tissues consumed	Association with crop							Functional significance								Summary	
				A1	A2	A3	A4	A5	A6	MA	PC	PO	PN	FN	DV	SD	BC	MS	Sum	Rank
Pamara mathias Fabr.	Lepidoptera: Hesperidae	A	N	3	2	2	2	3	3	2.50	3	3	3	3	3	3	3	3.00	5.50	11
Phalanta sp.	Lepidoptera: Nymphalidae	A	N	2	2	2	2	3	3	2.33	3	3	3	3	3	3	3	3.00	5.33	10
Precis atlites L.	Lepidoptera: Nymphalidae	A	N	2	2	2	2	3	3	2.33	3	3	3	3	3	3	3	3.00	5.33	10
Papilio polytes L.	Lepidoptera: Papilionidae	A	N	2	3	2	3	3	3	2.67	3	3	3	3	3	3	3	3.00	5.67	12
Terias sp.	Lepidoptera: Pieridae	A	N	2	3	2	2	3	3	2.50	3	3	3	3	3	3	3	3.00	5.50	11
Cletus punctiger Dallas	Hemiptera: Coreidae	A	N	2	2	2	2	2	2	2.00	3	3	3	3	3	3	2	2.86	4.86	9
Lathyrophthamus sp.	Diptera: Calliphoridae	A	N	2	2	2	2	2	2	2.00	2	2	2	3	3	3	1	2.29	4.29	5
Splaerophoria cylindria Say.	Diptera: Calliphoridae	A	N	2	2	2	2	2	2	2.00	2	2	2	3	3	3	2	2.43	4.43	7
Epistophe cinotella Zetterstedt	Diptera: Calliphoridae	A	N	2	2	2	2	2	2	2.00	2	2	2	3	3	3	2	2.43	4.43	7

Continued

Table 9.1. Continued

Species or species group	Order and family	Life stage	Plant tissues consumed	Association with crop							Functional significance								Summary	
				A1	A2	A3	A4	A5	A6	MA	PC	PO	PN	FN	DV	SD	BC	MS	Sum	Rank
Triceratopyga sp.	Diptera: Calliphoridae	A	N	2	2	2	2	2	2	2.00	2	2	2	3	3	3	2	2.43	4.43	7
Chrysomyia megacephala Fabr.	Diptera: Calliphoridae	A	N	2	2	2	2	2	2	2.00	2	2	2	3	3	3	2	2.43	4.43	7
Sarcophaga peregrina Robineau Adesvoidy	Diptera: Sarcophagidae	A	N	2	2	2	2	2	2	2.00	2	2	2	3	3	3	2	2.43	4.43	7
Sarcophaga meramira Meigen	Diptera: Sarcophagidae	A	N	1	2	2	2	2	2	1.83	2	2	2	3	3	3	2	2.43	4.26	4
Didea fasciata Macquart	Diptera: Syrphidae	A	N	1	2	2	2	2	2	1.83	2	2	2	3	3	3	1	2.29	4.12	3
Syrphus spp.	Diptera: Syrphidae	A	N	1	2	2	2	2	2	1.83	2	2	2	3	3	3	1	2.29	4.12	3
Camponotus japonicus Mayer	Hymenoptera: Formicidae	A, S	F2, P, N	2	2	2	2	3	2	2.17	2	2	2	3	3	3	2	2.43	4.60	8
Micraspis discolor Fabr.	Coleoptera: Coccinellidae	A	P, N	2	2	2	2	2	2	2.00	2	2	2	3	3	3	1	2.29	4.29	5

Species	Order: Family																			
Menochilus sexmaculatus Fabr.	Coleoptera: Coccinellidae	A	P, N	2	2	2	1	2	2	1.83	2	2	2	3	3	3	1	2.29	4.12	3
Xanthopimpla flavolineata Cameron	Hymenoptera: Ichneumonidae	A	N	2	2	2	2	2	2	2.00	2	2	2	3	3	3	1	2.29	4.29	5
Brachymeria sp.	Hymenoptera: Chalcididae	A	N	2	2	2	2	2	2	2.00	2	2	2	3	3	3	1	2.29	4.29	5
Amplex amoena Stal	Hymenoptera: Sphecidae	A	N	2	2	2	2	2	2	2.00	2	2	2	3	3	3	1	2.29	4.29	5
Xylocopa verticalis Lepeletier	Hymenoptera: Xylocopidae	A	P, N	1	2	2	2	2	2	1.83	2	2	2	3	3	3	3	2.57	4.40	6
Megachile sp.	Hymenoptera: Megachilidae	A, L, S	P, N	1	2	2	2	2	2	1.83	2	2	2	3	3	3	2	2.29	4.12	3
Bombus spp.	Hymenoptera: Apidae	A, L, S	P, N	1	2	1	2	2	2	1.67	1	1	1	2	3	3	2	1.86	3.52	2
Apis cerana Fabr.	Hymenoptera: Apidae	A, L, S	P, N	1	1	1	1	2	2	1.33	1	1	1	2	3	3	1	1.71	3.05	1
Apis mellifera L.	Hymenoptera: Apidae	A, L, S	P, N	1	1	1	1	2	2	1.33	1	1	1	2	3	3	1	1.71	3.05	1

Nhaho Research Institute for Cotton and Agricultural Development, the Southern Fruit Tree Research Institute and the Bee Research and Development Centre). Many of these are natural enemies, but some are nectar-feeding herbivores of potential conservation value and four are primary pollinators (several more species can also function as pollinators). It is likely that a thorough survey of cotton would reveal many more common flower visitors. For example, in Brazil, a systematic survey of cotton flowers found 153 flower-visiting insect species, about half of them bees, whereas only three species of bees had been mentioned previously in the scientific literature as possible cotton pollinators (Carmen Pires, Brasilia, June 2007, personal communication).

For choosing the most relevant non-target flower visitors from this list, a matrix was used to estimate association with the cotton crop (Table 9.1, Association with Crop, criteria A1–A6) and functional significance (Table 9.1, Functional Significance, 7 criteria). These criteria were ranked from 1 to 3 for each species (rank 1 = high to rank 3 = low). The ranking was based on information from experts in Vietnam and other cotton producing countries. In the absence of information for a particular species, a '?' was used to indicate a knowledge gap. Ideally, the ranking should be carried out separately for each region of Vietnam where cotton is grown (Chapter 2, this volume). This could not be done for flower-visiting insects because there is little information to allow separate regional rankings. Similarly, it would have been useful to consider separately the rainy and dry seasons, but the information was insufficient to do this.

All criteria for association with cotton were ranked for most species. However, the two phenology criteria were considered of little importance for comparing flower-visiting species because all species were present as adults during the cotton flowering period. We consider that 'Geographic distribution', 'Habitat specialization', 'Prevalence' and 'Abundance' were sufficient criteria for ranking high priority flower-visiting species according to their association with the crop. Interestingly, there was less uncertainty for the significance criteria than for the association criteria.

The criteria were combined in two ways. When criteria were weighted equally, the mean for 'Association with crop', mean for 'Functional significance' and the additive score of the two means are displayed in Table 9.1. The seven taxa with the highest priority were *A. mellifera*, *A. cerana*, *Bombus* spp., *Megachile* sp., *Didea fasciata*, *Syrphus* spp. and *Menochilus sexmaculatus*. Alternatively, the pollination services criteria were weighted much higher than the other significance criteria. Specifically, we made the combined 'Functional significance' rank equal to the highest rank of any of the criteria for significance as pollinator (Table 9.1, criteria: significance as a pollinator of cotton (PC), of other crops (PO) and in natural areas (PN); and significance as food for natural enemies (FN)). This alternative method ranked the same four hymenopteran pollinator species listed above (with an even higher priority) and did not indicate any additional pollinator species as important. Thus, there may be little need to weight the pollination service criterion more than other criteria. In addition, some lepidoptera species of potential conservation significance were identified using the unweighted method but were not identified by the weighted method. The highest ranked lepidopteran species were *Precis atlites* and *Phalanta* sp. This finding is another reason to use the unweighted method for combining ranks.

From the Selection Matrix, the following species were selected to evaluate identified adverse effects: (i) potential reduction in pollination value – *A. mellifera*, *A. cerana*, *Bombus* spp., *Megachile* sp.; (ii) potential reduction in economic value of bee products – *A. mellifera* and *A. cerana*; (iii) potential adverse effects caused by the reduction in wild bee populations – *A. cerana*; and (iv) potential reduction in species with conservation or biological control value – *D. fasciata*, *P. atlites* and *Phalanta* sp.

All four prioritized hymenopteran pollinators (*Megachile* sp., *Bombus* spp., *A. cerana* and *A. mellifera*) were selected for further evaluation because, in addition to cotton, they pollinate other crops and plants in natural areas. In addition, *A. mellifera* and *A. cerana* are economically important species in Vietnam. Vietnam has more than 1000 beekeepers who make a living from keeping *A. mellifera* and many thousands of farmers keep *A. mellifera* or *A. cerana* as a secondary income source (APISERVICES, 1997). *A. cerana* is also an important species in natural areas, so potential harm to *A. cerana* and species that depend on it should be assessed. The hover fly *D. fasciata* was selected because it was the highest ranked flower-visiting biological control agent and the two lepidopteran species, *P. atlites* and *Phalanta* sp., were selected because they were the highest ranked species of potential conservation concern.

9.2. Potential Exposure Pathways of Flower Visitors

Exposure of flower visitors to stressors associated with Bt cotton may occur directly via consumption of floral parts or products, or indirectly via the consumption of other flower visitors. Because the species that consume flower visitors are natural enemies, their indirect exposure will be assessed elsewhere (Chapters 7 and 8, this volume). Direct consumption can include eating flower tissue, pollen and/or nectar. Most flower tissues such as petals, anthers, stamens and ovule tissue of all the currently available Bt cottons contain variable amounts of Bt proteins. The Cry1Ac, Cry1Fa and VIP3A Bt proteins (though not Cry2Ab) are expressed in cotton pollen (Chapter 4, this volume). No expression of a Bt protein has been detected in floral nectar and the total protein content of floral nectar is very low. Presumably, expression is similar in extra-floral nectaries, but this remains to be confirmed.

Consequently, all pollen-feeders and flower-feeders will be exposed to Bt proteins for any of the currently available Bt cotton events, but nectar-feeders will not be exposed to Bt proteins, unless pollen contaminates the nectar and is imbibed. Some butterflies may increase their exposure by soaking or churning pollen in nectar until it partly dissolves and then imbibing the fluid (Roulston and Cane, 2000).

It cannot yet be determined if any effects could occur via other alterations in flower tissue, nectar or pollen related to the transgene and there are no data to indicate if floral tissue, nectar and pollen are modified in any other way that could affect floral visitors. For example, Bt cotton may have an altered volatile profile (e.g. Yan *et al.*, 2004), which could affect its attractiveness to pollinators and thereby alter exposure.

9.3. Potential Adverse Effects Pathways

Apis cerana, Apis mellifera, Bombus spp. [Hymenoptera: Apidae]

The Apidae are important pollinators of many crops, as well as wild plants. South-east Asia is one of the centres of diversity for the genus *Apis*, which includes honeybees, so there may be significant genetic and taxonomic diversity to consider. These bee species are eusocial and both *Apis* spp. produce honey, propolis, beeswax, royal jelly and other medicinal products used by humans.

Five adverse effects pathways should be considered for these apid species (Fig. 9.1).

*   Reduced cotton pollination, resulting in lower production (quantity or quality). This pathway will not be addressed further because cotton can self-pollinate at high rates and, in the production systems of Vietnam, cotton yield is unlikely to be pollinator-limited.
*   Reduced production of nearby fruit trees or other crops that require pollination (all three species).
*   Reductions in the production of bee products. This adverse effect pathway is not relevant for *Bombus* spp. because they do not produce any harvestable product.
*   Reductions in wild populations of *A. cerana*, which could harm populations of rare animals that depend on it for food (e.g. possibly civet cats, dragonflies, lizards, frogs, birds).
*   Adverse effects on the genetic or taxonomic diversity of *A. cerana* (Smith *et al.*, 2005).

A possible concern is that the bees act as vectors of Bt toxin, causing contamination of nearby fruit crops (dotted lines, Fig. 9.1). We consider this unlikely because of the small quantity of pollen and other materials moved from flower to flower and the long time between pollination and fruit maturation, which gives more time for the Bt toxin to break down (the toxin is UV-light sensitive). However, if necessary, this possibility could be tested relatively easily.

In addition, changes in cotton cultivation practices associated with the use of Bt cotton may also affect these bee species negatively (Fig. 9.1). Cultivation of Bt cotton might change the cropping systems, with subsequent changes in the distribution, abundance or quality of nectar and pollen resources. The feasibility and importance of this possibility requires specific information about the new Bt cotton production practices in Vietnam and their effects on floral resources.

Megachile sp. [Hymenoptera: Megachilidae], leafcutter bees

Leafcutter bees are solitary bees that build cells for their brood, often lining the cells with cut parts of leaves. The bees gather pollen and nectar to feed their brood. In the temperate zone, *M. rotundata* is an important pollinator of lucerne and, in Vietnam, *Megachile* sp. is a pollinator of cotton, as well as other crops, including pigeon pea, lily, shallot, water hyacinth (*Anthidium*), *Acacia* spp. (*A. nilotica* and *A. tortilis*) and plants in natural areas (Lê Thị Thu Hồng, Chau Thanh, January 2007, personal communication). Consequently, if their populations are reduced by their association with Bt cotton, reduced yield in other crops and/or reduced reproductive potential or fitness of some plants in natural areas might result. These bees do not produce any commercial products (e.g. honey), South-east Asia is not a centre for their biological diversity and they are not strong competitors of the eusocial apid species. Thus, the main adverse effect pathway for *Megachile* sp. is:

- Reduced pollination of other crops and/or plant species in natural areas. This requires that these other crops and plants in natural areas would become pollinator limited should *Megachile* sp. populations decrease.

Didea fasciata [Diptera: Syrphidae]

Syrphid flies are important predators of aphids during their immature (larval) stages and feed on floral nectar as adults. Females also feed on pollen to obtain proteins necessary for egg maturation. Thus, the main adverse effect pathway would involve a reduction in adult longevity due to nectar and/or pollen feeding on Bt cotton, which results in reduced mating success, oviposition or offspring quality. This, in turn, could result in reduced biological control of aphids and increased crop damage from this pest. Chapter 7 (this volume) addresses arthropod predators in cotton and therefore these species will not be considered further in this chapter. However, this species was not identified as an important biological control agent for further assessment (Chapter 7, this volume).

Precis atlites, Phalanta sp. [Lepidoptera: Nymphalidae]

The main adverse effect pathway associated with adult butterflies on Bt cotton is:

- Reduction of the population, resulting in a loss of conservation and/or aesthetic and cultural value.

The grey pansy *P. atlites* and the leopard butterfly *Phalanta* sp. occur throughout South and South-east Asia and are common cotton flower visitors in Vietnam (Table 9.1). Many species and subspecies of *Phalanta* occur in tropical environments. Neither is listed as threatened on the IUCN Red List.[2] Both species may be indicators of possible impact on protected species, but are themselves presently not of conservation concern. Therefore a change in their population density in cotton growing areas would probably not be considered

[2] http://www.iucnredlist.org/.

adverse. Further literature studies and field research may identify butterfly species associated with cotton habitats (species that live in or near cotton fields and may feed on cotton or another associated plant) that are actually threatened or endangered. If such species were identified, suitable indicator species from the same genus or family could be selected as substitutes for these rare species, as it is not appropriate to sacrifice rare species for testing.

9.4. Formulate Risk Hypotheses

A risk hypothesis is a chain of causal links, starting from the Bt cotton and ending with a characteristic (also known as an assessment end point) of the selected flower visitor that is connected closely to an undesirable adverse effect (US EPA, 1998). Figure 9.1 outlines the risk hypotheses for pollinators and Fig. 9.2 outlines the main risk hypothesis for non-pollinator flower visitors, such as butterflies. Figure 9.3 provides additional detail. Considering

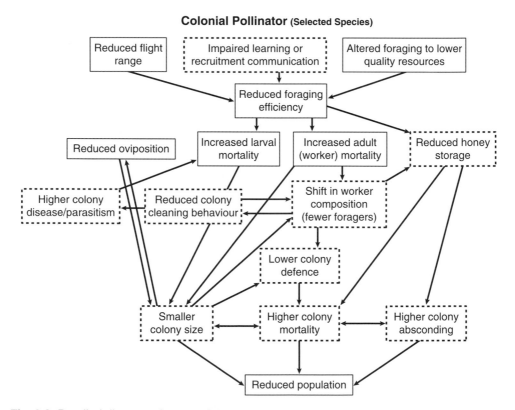

Fig. 9.3. Detailed diagram of a part of the possible risk hypotheses in Fig. 9.1 for pollinators visiting Bt cotton flowers. Dashed boxes are for eusocial pollinators only and solid boxes are for either solitary or eusocial pollinators. Any of the arrows leading into the 'pollinator' box in Fig. 9.1 could affect any of the boxes in this figure directly. Arrows are simple representations of the causal connections within the pollinator entity.

the exposure pathways and adverse effects pathways in Figures 9.1, 9.2 and 9.3, five generic risk hypotheses were formulated to describe the ways in which Bt cotton might lead to an adverse effect via pollinators and flower visitors. The following section lists the risk hypotheses and evaluates their relevance to each species associated with the adverse effect pathway in the previous section.

Each risk hypothesis involves a reduction in population density of the flower-visitor species. In each risk hypothesis, this could be caused by any one of these identified pathways:

1. Direct effects of the Bt toxin in pollen and/or nectar and/or decrease in the quality of cotton nectar and pollen in Bt cotton.
2. Changes in the attractiveness of Bt cotton flowers for flower visitors, due either to changes in volatile patterns and/or changes in the quality (and therefore attractiveness) of cotton nectar and pollen in Bt cotton.
3. Alterations in landscape structure as a consequence of growing Bt cotton.

In the subsequent text, these three pathways are not addressed in detail, although the reader should assume that any of the three is possible.

Risk hypothesis 1. Bt cotton causes a reduction in bee population density and/or colony quality, resulting in reduced production of honey and other bce products

Apis mellifera
Honey production by *A. mellifera* in the central highlands is about 4 million kg/year. Pure honey comes mainly from rubber and coffee plantations and a smaller amount (~12 t) from longan, eucalyptus, cassava, cashew and a few other crops (APISERVICES, 1997). Honey production is very important in Vietnam, but relatively little of it originates from cotton. Moreover, insecticide applications on cotton affect *A. mellifera* adversely and beekeepers avoid this by keeping their colonies away from the crop. Thus, even if Bt cotton affected *A. mellifera* adversely, the effect on honey production would be small, so this risk hypothesis is not important. However, it is important to determine if transgene products can be detected in *A. mellifera* honey from colonies used to pollinate Bt cotton, as this could affect the export markets for honey from Vietnam.

Apis cerana
Unlike *A. mellifera*, *A. cerana* is a feral/wild species in the central highlands. The relative importance of various nectar and pollen sources for this species is less well known than for *A. mellifera* (Tong *et al.*, 2005). Thus, it is possible that Bt cotton might reduce colony size or vigour of *A. cerana*, resulting in reduced honey production and harvest, but the impact on human-harvested honey and other products is likely to be less than that for *A. mellifera* because it is used less commonly by commercial beekeepers.

Risk hypothesis 2. Bt cotton causes a reduction in bee population density, resulting in reduced genetic diversity of the species

Apis mellifera

This is an unlikely risk hypothesis for *A. mellifera* because most of the colonies in the central highlands are kept by beekeepers, who can replenish their stock periodically.

Apis cerana, Bombus spp., Megachile spp.

It is possible that genetic diversity in the feral and wild populations of these species could be reduced. However, the reductions in population density would have to be very large and rapid to have an influence on genetic diversity. Genetic diversity of *A. cerana*, *Bombus* spp. and *Megachile* sp. potentially could be affected adversely.

Risk hypothesis 3. Bt cotton causes a reduction in bee population density, resulting in adverse effects on species associated with the bees

Apis mellifera

This hypothesis is unlikely for *A. mellifera* because colonies are maintained primarily by beekeepers and wild animals associated with these beehives usually are regarded as pests.

Bombus spp. and Megachile spp.

The community associations of *Bombus* spp. and *Megachile* sp. are poorly known in the central highlands, so it is not possible to assess this risk hypothesis for these species. This uncertainty could be addressed with appropriate post-commercialization monitoring.

Apis cerana

More is known about wild *A. cerana*. Several species use *A. cerana* honey in the wild, including civets (*Paradoxurus hermaphroditus* (Pallas)[3] and *Viverra zibetha* (L.)[4] [Carnivora: Viverridae]), and these protected species could be affected adversely if honey becomes scarce. Other predators that eat wild bees include dragonflies, lizards, frogs and birds.

Risk hypothesis 4. Bt cotton causes a reduction in bee forager quality or density or feral bee colony density, resulting in reduced pollination and yield of nearby crops

This hypothesis requires that these crops become pollinator-limited as a result of the reduction in the pollinator population in cotton.

[3] The Asian Palm civet *Paradoxurus hermaphroditus* is on the CITES list of species banned from international trading: http://www.cites.org/eng/resources/species.html.
[4] The Oriental or large Indian civet *Viverra zibetha* is on the IUCN Red List of endangered species: http://www.iucnredlist.org/search/details.php/41709/dist.

Apis mellifera

In the diverse cropping systems of the central highlands, several crops occur near cotton and require pollination by A. mellifera, including grapes, mango, jujube and melons. Bees from single colonies visit both cotton and these other crops, so it is possible that pollination of other crops could be affected via the direct effects of Bt toxin or associated changes in the quality of cotton nectar and pollen in Bt cotton.

Apis cerana

For A. cerana, it is necessary to determine which crops near cotton rely on this species for pollination (Crane, 1991). Bees from single colonies visit both cotton and other crops so, if A. cerana is an important pollinator of these other crops, it is possible that their pollination could be affected.

Bombus spp.

Flowering apricot trees are highly attractive to Bombus spp. in the spring and this crop would be the main one possibly affected by Bt cotton. Apricot probably does not bloom when cotton blooms.

Megachile spp.

In the case of Megachile sp., pollination of pigeon pea and shallot could be affected by Bt cotton. These crops probably do not bloom when cotton blooms.

Risk hypothesis 5. Bt cotton causes a reduction in population density of important flower visitors other than bees

This hypothesis can be considered further if butterfly species of conservation concern are identified for cotton and suitable indicator species can also be identified in order to carry out experiments. The main adverse effect pathway is that adult butterfly feeding on cotton nectar results in reduced adult longevity, with a consequent decrease in mating, oviposition or offspring quality (Fig. 9.2). Because nectar has little Bt protein, adult butterflies are likely to be exposed to Bt through pollen only. It is not known if activated Bt toxins in transgenic pollen can affect adult butterflies adversely, but this is unlikely, as most butterflies do not digest the pollen (see above). More subtle effects on adults are also possible, but these are even less plausible. For example, egg size typically declines with adult age and adult feeding can increase egg size in later adult life, resulting in healthier offspring with higher survival rates. Feeding on Bt cotton nectar may result in poorer quality offspring from older adults, where egg size declines further when feeding on Bt nectar. Effects on larvae could be examined using methods from Chapter 6 (this volume).

Summary of risk hypotheses for potential investigation

Nine possible combinations of risk hypotheses (RH) and species associated with Bt cotton in Vietnam were identified for four bee taxa: A. mellifera (RH4),

A. cerana (RH1, RH2, RH3, RH4), *Bombus* spp. (RH1, RH4) and *Megachile* sp. (RH2, RH4) (Table 9.2). No risk hypotheses were identified for the flower visitors that were not bees.

These risk hypotheses can be prioritized logically for each taxon. A logical prioritization can be based on the required size of an effect on the bees for the hypothesis to be possible. Some of these require a much larger effect on the bee species than others. For example, RH2, RH3 and RH4 require a larger effect on *A. cerana* than RH1, and RH2 requires a larger effect on *Bombus* spp. and *Megachile* sp. than RH4. If the smaller effect does not occur, then the larger effect is less likely to occur. Based on this logic, the following should be the initial focus for risk assessment: *A. cerana* – RH1; *Bombus* spp. and *Megachile* sp. – RH4.

In addition, some hypotheses are *a priori* more likely than others, some have greater potential consequences (irreversibility, spatial scale and magnitude) and some are assessed more readily prior to commercial release than others. The hypotheses that are more likely, have more serious potential consequences and are assessed more easily could be the focus of initial investigation. In general, RH1 (reduced bee products) is more likely than RH4 (reduced pollination of other crops) because the effect is immediate to the cause and both are more likely than RH2 (reduced genetic diversity) or RH3 (effects on associated species), which require much greater and more rapidly occurring effects (Table 9.2). The consequences of RH2 (reduced genetic diversity) would be irreversible, while the consequences of RH4 (reduced pollination of other crops) can be mitigated. Mitigation of the effects of RH4 for *A. mellifera* may be quite easy, because it is possible to bring in additional *A. mellifera* colonies. Finally, some of the hypotheses will be difficult to assess prior to commercialization (Table 9.2). For example, effects on civets (RH3) or the hypotheses based on changes in landscape structure will be difficult to assess. These hypotheses could be evaluated via post-commercialization monitoring. Overall, these considerations suggest that RH1 (reduced bee products) for *A. cerana* and RH4 (reduced pollination of other crops) for *Bombus* spp. and for *Megachile* sp. should be prioritized for pre-commercialization risk assessment.

9.5. Possible Experiments

RH1 for *A. cerana* and RH4 for *Bombus* spp. are two of the highest priority risk hypotheses (Table 9.2), so some experiments are suggested below to address these two risk hypotheses.

RH 1. Bt cotton causes a reduction in bee population density or colony quality of *A. cerana*, resulting in reduced production of honey

A laboratory method to measure the effects on eusocial bee larvae of Cry toxin either purified *in vitro* (Brodsgaard *et al.*, 2003) or in pollen (Hanley *et al.*, 2003) can serve as the first experiment. The pollen study might be easier to perform and could be more realistic. Laboratory studies to date have demon-

Table 9.2. Prioritization of risk hypotheses. Rank values: 1 = highest rank, 2 = intermediate rank, 3 = lowest rank. Consequence product is the multiplication product of the three preceding rank numbers. The final product is the product of relative likelihood, consequence product and assessable. Ranks were determined by expert evaluation. Potential reversibility of the consequences varies from irreversible (1) to readily reversed (3). Spatial scale of consequence varies from extensive (1) to local (3). Potential magnitude varies from severe (1) to minor (3). Assessable varies from can be assessed in the laboratory or greenhouse (1), field experiments (2), or post-commercialization (3) (some RH have a range of values because some aspects can be tested in the laboratory, such as direct toxicity, while others may need to be evaluated post-commercialization, such as changes in cropping systems). The final product of the priority risk hypotheses is in bold.

High priority pollinator	Risk hypothesis	Relative likelihood	Relative potential consequence					
			Potential reversibility	Potential spatial scale	Potential magnitude	Consequence product	Assessable	Final product
A. mellifera	RH4	2	3	3	2	18	1	36
A. cerana	RH1	1	2	2	1.5	6	1 or 3	**6**
A. cerana	RH2	3	1	2	2	4	2.5	30
A. cerana	RH3	3	1–2	2	1.5	3–6	3	27–54
A. cerana	RH4	2	3	3	2	18	1 or 3	36–108
Bombus spp.	RH2	3	1	2	2	4	2.5	30
Bombus spp.	RH4	2	2–3	3	2	12–18	1 or 3	**> 24**
Megachile sp.	RH2	3	1	2	2	4	2.5	30
Megachile sp.	RH4	2	2–3	3	2	12–18	1 or 3	**> 24**

strated small or no effects on *A. mellifera* larvae, pupae and adults from Cry toxins (reviewed in Malone and Pham-Delègue, 2001; Liu *et al.*, 2005). However, cowpea trypsin inhibitor (CpTI) has affected honeybees (Picard-Nizou *et al.*, 1997) and soybean trypsin inhibitor proved toxic to larval and adult bumblebees (Malone *et al.*, 2000; Brodsgaard *et al.*, 2003).

If the above-mentioned experiment were to give inconclusive results, more extensive (and expensive) studies on whole bee colonies might be done. Bt cotton and its parental or isogenic line (non-transgenic cotton) should be planted separately in ten greenhouses (600–1000 m²). Cotton can be planted at different times to extend the availability of flowering cotton. Two colonies of *A. cerana* (or *A. mellifera*) should be released during cotton flowering in one greenhouse. Comb occupation, reproduction rate, growth rate, adult size, adult lifespan and age structure can be measured for each colony (Tong *et al.*, 2005). Comb occupation is the proportion of the comb being used for brood production and honey storage and is an indicator of colony health. Reproduction can be measured using a transparent film with grids (one grid contains 100 cells) to facilitate counting eggs. The grid is placed on the comb surface for counting newly laid eggs and can be removed after counting. The newly laid eggs should be checked and recorded daily. Growth rate can be measured by checking development in 12–20 cells every 1 or 3 days, when developmental stage (egg, larvae, pupae) should be recorded. Adult size can be measured on newly emerged adults and adult lifespan can be measured by marking these individuals in the colony and following them until they die. The age structure of the colony can be estimated from these measurements. At least five replicate samples should be taken from each colony. If resources are available, behaviour can also be measured, such as aggression (Pearce *et al.*, 2001), cleaning (Bozic and Valentincic, 1995) and foraging behaviour (Picard-Nizou *et al.*, 1995), learning capacity (Picard-Nizou *et al.*, 1997) and honey accumulation.

RH 4. Bt cotton causes a reduction in bumblebee forager quality, density or bumblebee colony density, resulting in reduced pollination and yield of nearby apricot

While it is possible to rear and test the effects of Bt toxin on bumblebees, it may be quicker and less costly to test a later link in the risk hypothesis (Fig. 9.1). For example, it may be easier than colony rearing (some *Bombus* sp. are difficult to rear) to determine if a reduction in bumblebee numbers results in a reduction in apricot pollination and fruit production. If fruit production in apricot is not affected by a reduction in bumblebees, then the risk hypothesis is rejected; there is no risk. As an experimental treatment, bumblebees should be excluded from apricot blossoms, while allowing other pollinators free access to the same blossoms. If this is not possible, then it may be necessary to quantify the number of visits to receptive apricot blossoms by bumblebees and all other potential pollinators. A subsequent pollination study with controlled pollinations could then be conducted with four experimental treatments: no pollinators (negative control), only bumblebees at their typical visitation rate, only other pollinators (no bumblebees) at their typical visitation rate and both bumblebees and other pollinators together (positive control).

If this study is inconclusive, another link in the risk hypothesis should be tested: does a reduction in bumblebees in cotton during the summer result in a reduction in bumblebees in apricot during the next spring? This probably can be tested in the present cotton landscapes, as cotton receives a considerable amount of insecticides that are toxic to most bees. For this experiment, find apricot trees near two different non-Bt cotton fields, one that receives insecticides which kills nearly all bumblebees that visit and one that receives no insecticides or insecticides that are not toxic to bumblebees. Measure the visitation rate by bumblebees to the apricot trees. If this test does not falsify the risk hypothesis, toxicity tests or field experiments with Bt cotton could be conducted. A similar sequence of studies could be conducted for *Megachile* sp.involving either pigeon pea or shallot as the pollinated crop.

Field experiments with Bt cotton

If the laboratory or greenhouse experiments or the pollination experiments do not falsify the risk hypotheses, then field experiments conducted with the transgenic crop at the pre-commercial stage may be needed. This experiment could have four treatments: (i) conventionally managed cotton with normal pesticide treatments (conventional plot); (ii) cotton under integrated pest management regime (may be the same as (i)); (iii) Bt cotton with no pesticides; and (iv) Bt cotton with pesticides as needed. The field experiment should be replicated validly (preferably 4–5 replicates), taking into consideration the logistical and land constraints. Field size should be large enough to represent actual production fields. Visual observation of plants and flowers is the main sampling method to be used. Starting shortly after flowering commences, a visual inspection of flowers for a fixed period of time (e.g. 10 min) at several locations/plot should be done every 7 days during good weather (sunny, little wind) until the end of the flowering period. The locations should be distributed regularly within the field, avoiding the field edges and margins. Sampling across evident gradients (e.g. slope or soil moisture gradient) in the plots should be considered. During each census, the observer collects all insects inside and under the flowers separately, including insects that land on or in the flowers. Insects under the flowers may be feeding on the nectaries, while insects in the flower may be feeding primarily on the pollen. All insects should be collected and taken to the laboratory for identification. Many flowers can be observed simultaneously. The starting position for sampling should be changed every time. The duration of each sample should be recorded. The unit of observation is the number of insects per 10 min.

Diversity comparisons

Diversity is an important component of ecological communities and effects on biodiversity are a routine part of many risk assessment requirements. However, this almost always requires field-scale experiments, and often for long time periods, before effects on diversity are detectable. In this chapter, we present only a general scheme for diversity evaluation, but not detailed sampling methods. Field methods for arthropod monitoring are well documented in the literature (e.g. Southwood and Henderson, 2000).

To compare the diversity of the different assemblages, the system suggested by Southwood and Henderson (2000) might be used. This starts with an evaluation of the rank-abundance curves. The *x*-intercept shows species richness in the assemblage and the slope of the curves is related to diversity: a steeper curve indicates lower diversity than a less steep one.

After this, a scalable diversity method could be employed. There are several publicly available software packages to do this. One way of generating diversity profiles that can be compared is to use a generalized Rényi diversity function (Tóthmérész, 1995). The Rényi diversity, $HR(a)$, is defined as:

$$HR(a) = \frac{1}{1-a} \ln \sum_{i=1}^{S} p_i^a \,,$$

where p_i is the relative abundance of the *i*th species, S is the total number of species in the sample, a is a scale parameter with no direct biological meaning, and $a \geq 0$, $a \neq 1$. This function is sensitive to the rare species for small values of the scale parameter (close to zero), whereas it is sensitive to the abundant species for larger values of the scale parameter. Diversity profiles were calculated by the DivOrd package (Tóthmérész, 1993) or programmes in R (Tóthmérész, 2005). Four special scale parameter values merit extra consideration:

1. $a = 0$. At this point, the Rényi diversity function is equal to the logarithm of the species richness of the assemblage; $HR(0) = \log S$. In this case, the method is extremely sensitive to the contribution of the rare species to the diversity of the assemblage.
2. $a \to 1$ ($a \neq 1$, see above). The function gives the value of the Shannon diversity index, $HR(a \to 1) = -\sum p_i \ln p_i$. In this case, the diversity is sensitive to the rare species, although not so extremely as for $a = 0$.
3. $a = 2$. The function is related to the quadratic or Simpson diversity, $HR(2) = \ln (1/\sum p_i^2)$. In this case, the method is more sensitive to the frequent species than to the rare ones.
4. $a \to +\infty$ (a is very large). The function converges to the relative abundance of the most common species. This is the logarithm of the reciprocal value of the Berger–Parker or dominance index (Southwood and Henderson, 2000).

When comparing two Rényi diversity profiles, there are three possible cases (Fig. 9.4): (i) when the two profiles do not cross, the upper profile represents a more diverse assemblage under any common measure of diversity; (ii) when the curves cross once, one assemblage is more diverse when rare species are weighted more heavily (low a) and the other assemblage is more diverse when common species are weighted more heavily (high a); and (iii) the curves may cross twice. This indicates a complex relationship. Assemblage A is more diverse when the rare or common species are weighted more, but for interme-

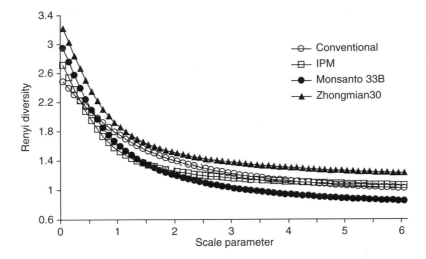

Fig. 9.4. Rényi diversity profiles for spider assemblages in transgenic Bt and non-transgenic cotton fields at Nan-Pi Research Station, Hebei Province, north central China, in 1998. The relative positions of the four profiles did not change when the scale parameter $a > 6$, so the curves were truncated at $a = 6$ on the x-axis. Reproduced with permission from Liu *et al.* (2004).

diate weights, assemblage B is more diverse. This can occur when assemblage B has a higher evenness and lower species richness than assemblage A. Such a situation may indicate that assemblage B is under environmental stress (such as a pesticide effect which results in lower densities of the common species, inflating the evenness component of the assemblage).

Rényi diversity profiles of the different treatments can be constructed using available free software (Tóthmérész, 1993, 2005; Tóthmérész and Magura, 2005) and compared, thus giving a well-defined description of the diversity relationships among the assemblages/treatments to be compared. This comparison can be informative even when descriptive, but an approximate statistical test could be a non-parametric comparison of the Rényi-index values at selected, fixed points.

9.6. Monitoring

RH1, RH3 and aspects of the other risk hypotheses that relate to landscape change cannot be assessed easily prior to commercialization. All of these can be evaluated by monitoring landscape structure, pollinator densities and diversity after commercialization. Landscape structure would be useful to monitor for many reasons and can be done by surveys of farmers or by remote sensing (see also Chapter 8, this volume).

Pollinator densities and diversity can be monitored by selecting a few crops and native plants and a few localities as monitoring stations where pollinator visitation is measured every year or alternate years (more frequently in the beginning). The parameters (species richness, number of rare species, visitation rate, abundance, diversity profile), methods (direct observation, sweep sampling) and analysis tools have been mentioned above. As pollinator diversity is threatened worldwide (Allen-Wardell *et al.*, 1998), this monitoring system may be helpful for ensuring the long term sustainability of pollinator communities in Vietnam.

References

Allen-Wardell, G., Bernhardt, P., Bitner, R., Burquez, A., Buchmann, S., Cane, J., Cox, P.A., Dalton, V., Feinsinger, P., Ingram, M., Inouye, D., Jones, C.E., Kennedy, K., Kevan, P., Koopowitz, H., Medellin, R., Medellin-Morales, S., Nabhan, G.P., Pavlik, B., Tepedino, V., Torchio, P. and Walker, S. (1998) The potential consequences of pollinator declines on the conservation of biodiversity and stability of crop yields. *Conservation Biology* 12, 8–17.
APISERVICES (1997) L'Apiculture au Viêt-Nam en 1995. Etude de Faisabilité. http://www.beekeeping.com/articles/us/vietnam/vietnam_us.htm (accessed 28 October 2007).
Arpaia, S., Fonseca, V.L.I., Pires, C.S. and Silveira, F.A. (2006) Non-target and biodiversity impacts on pollinators and flower-visiting insects. In: Hilbeck, A., Andow, D.A. and Fontes, E.M.G. (eds) *Environmental Risk Assessment of Genetically Modified Organisms, Volume 2: Methodologies for Assessing Bt Cotton in Brazil*. CAB International, Wallingford, UK, pp. 155–174.
Bozic, J. and Valentincic, T. (1995) Quantitative analysis of social grooming behavior of the honeybee *Apis mellifera carnica. Apidologie* 26, 141–147.
Brodsgaard, H.F., Brodsgaard, C.J., Hansen, H. and Lövei, G.L. (2003) Environmental risk assessment of transgene products using honeybee (*Apis mellifera*) larvae. *Apidologie* 34, 139–145.
CBD (1996) Agricultural Diversity: International Initiative for the Conservation and Sustainable Use of Pollinators. Secretariat of the Convention on Biological Diversity, Montreal, Canada. www.cbd.int/programmes/areas/agro/pollinators.aspx (accessed 30 October 2007).
Crane, E. (1991) Apis species of tropical Asia as pollinators, and some rearing methods for them. *Acta Horticulturae* 288, 29–48.
Daily, G.R. (1999) Developing a scientific basis for managing earth's life support systems. *Conservation Ecology* 3(2), 14. http://www.ecologyandsociety.org/vol3/iss2/art14/ (accessed 24 November 2007).
Hanley, A.V., Zachary, Y.H. and Walter, L.P. (2003) Effects of dietary transgenic Bt corn pollen on larvae of *Apis mellifera* and *Galleria mellonella. Journal of Apicultural Research* 42, 77–81.
Kevan, P.G. and Baker, H.G. (1983) Insects as flower visitors and pollinators. *Annual Review of Entomology* 28, 407–453.
Liu, W.-X., Wan, F.-H., Guo, J.-Y. and Lövei, G.L. (2004) Spiders and their seasonal dynamics in transgenic Bt- vs. conventionally managed cotton fields in north-central China. In: Samu, F. and Szinetár, Cs. (eds) *European Arachnology 2002. Proceedings of the 20th European Colloquium of Arachnology*. Plant Protection Institute and Berzsenyi College, Budapest, pp. 337–342.
Liu, B., Xu, C.R., Yan, F.M. and Gong, R.H. (2005) The impacts of the pollen of insect-resistant transgenic cotton on honeybees. *Biodiversity and Conservation* 14(14), 3487–3496.
McGregor, S.E. (1976) *Insect Pollination of Cultivated Crop Plants*. http://gears.tucson.ars.ag.gov/book/ (accessed 26 October 2007).

Malone, L.A. and Pham-Delègue, M.H. (2001) Effects of transgene products on honeybees (*Apis mellifera*) and bumblebees (*Bombus* sp.). *Apidologie* 32, 287–304.

Malone, L.A., Burgess, E.P.J., Stefanovic, D. and Gatehouse, H.S. (2000) Effects of four protease inhibitors on the survival of worker bumblebees, *Bombus terrestris* L. *Apidologie* 31, 25–38.

Oosterhuis, D.M. and Jernstedt, J. (1999) Morphology and anatomy of the cotton plant. In: Smith, C.W. and Cothren, J.T. (eds) *Cotton: Origin, History, Technology and Production*. John Wiley & Sons, New York, pp. 175–206.

Pearce, A.N., Huang, Z.Y. and Breed, M.D. (2001) Juvenile hormone and aggression in honey bees. *Journal of Insect Physiology* 47, 1243–1247.

Picard-Nizou, A.L., Phamdelègue, M.H., Kerguelen, V., Marilleau, R., Olsen, L., Grison, R., Toppan, A. and Masson, C. (1995) Foraging behavior of honeybees (*Apis mellifera* L.) on transgenic oilseed rape (*Brassica napus* L. var *oleifera*). *Transgenic Research* 4, 270–276.

Picard-Nizou, A.L., Grison, R., Olsen, L., Pioche, C., Arnold, G. and Pham-Delègue, M.H. (1997) Impact of proteins used in plant genetic engineering: toxicity and behavioral study in the honeybee. *Journal of Economic Entomology* 90, 1710–1716.

Roulston, T.H. and Cane, J.H. (2000) Pollen nutritional content and digestibility for animals. *Plant Systematics and Evolution* 222, 187–209.

Smith, D.R., Warrit, N., Otis, G.W., Thai, P.H. and Dinh Quyet Tam (2005) A scientific note on high variation in the non-coding mitochondrial sequences of *Apis cerana* from South East Asia. *Journal of Apicultural Research* 44(4), 197–198.

Southwood, T.R.E. and Henderson, P.A. (2000) *Ecological Methods*, 3rd edn. Blackwell Publishing, Oxford, UK.

Tilman, D., Cassman, K.G., Matson, P.A., Naylor, R. and Polasky, S. (2002) Agricultural sustainability and intensive production practices. *Nature* 418, 671–677.

Tong, X.C., Boot, W.J. and Sommeijer, M.J. (2005) Production of reproductives in the honeybee species *Apis cerana* in northern Vietnam. *Journal of Apicultural Research* 44(2), 41–48.

Tóthmérész, B. (1993) DivOrd 1.50: A program for diversity ordering. *Tiscia* 27, 33–44.

Tóthmérész, B. (1995) Comparison of different methods of diversity ordering. *Journal of Vegetation Science* 6, 283–290.

Tóthmérész, B. (2005) Diversity characterisations in R. In: Lovei, G.L. and Toft, S. (eds) *European Carabidology 2003*. Danish Institute of Agricultural Sciences Report Series No. 114, pp. 333–344.

Tóthmérész, B. and Magura, T. (2005) Scalable diversity and scalable diversity characterizations. In: Lovei, G.L. and Toft, S. (eds) *European Carabidology 2003*. Danish Institute of Agricultural Sciences Report Series No. 114, pp. 353–368.

US EPA (1998) *Guidelines for Ecological Risk Assessment*. United States Environmental Protection Agency, Risk Assessment Forum, Washington, DC, EPA/630/R095/002F.

Yan, F., Bengtsson, M., Anderson, P., Ansebo, L., Xu, C. and Witzgall, P. (2004) Antennal response of cotton bollworm (*Helicoverpa armigera*) to volatiles in transgenic *Bt* cotton. *Journal of Applied Entomology* 128(5), 354.

10 Potential Effects of Transgenic Cotton on Soil Ecosystem Processes in Vietnam

Pham Van Toan, Hoang Ngoc Binh, Beatrice Anyango, Claudia Zwahlen, Barbara Manachini, David A. Andow and Ron E. Wheatley

Soils support many processes that are vital to the continued functioning of the biosphere. They support plant growth by providing a matrix for root development and the provision of water and nutrients. Many vital ecosystem processes occur in soils, including recovery of nutrients from the residues of previous crops and other life forms, the suppression of plant pathogens, carbon sequestration and the bioremediation of various wastes. Soils contain possibly the most biologically diverse ecosystems, containing many thousands of different species of bacteria, protozoa, fungi, micro- and macrofauna, the numbers and activities of which are both temporally and spatially very variable. The functional dynamics of these processes are determined by complex interactions between the physical, chemical and biological factors in soils. Plant inputs are a major driver of activity in soil ecosystems by providing both nitrogen and energy from fixed carbon. During plant growth, these inputs come from roots and root exudates, root debris and cellular remains and plant residues that fall on the soil surface. After a plant dies, root and above-ground residues are the inputs.

The root rhizosphere is a particularly active location, where root exudates stimulate the growth and activity of various soil microorganisms. The highest numbers of bacteria, fungi, protozoa and microfauna associated with living plants are found in the rhizosphere, increasing in abundance and number of species from seedling to maturity of the plant (Ronn et al., 1996) when numbers and activity are at the maximum. As the plant senesces, the microbial population size reverts to that in non-rhizosphere soil. Rhizosphere microorganisms are influenced by plant species, plant age, root exudate and soil type, moisture, temperature and pH. The constituents of the exudates influence rhizosphere development and, similarly, these rhizosphere organisms affect the host plant's growth and development.

Bacillus thuringiensis is an aerobic spore-forming bacterium that produces crystal proteins during sporulation. Although *B. thuringiensis* has been found in soils all over the world (Martin and Travers, 1989), they do not survive or grow well (Saleh *et al.*, 1970). Little toxin is found in soils, even when *B. thuringiensis* is used as an insecticidal spray. Transgenic insecticidal plants such as Bt cotton produce and release relatively large amounts of transgenic proteins (Morse *et al.*, 2006), such as the active Bt toxins, and microbial processes have been shown to be particularly responsive to protein substrates (Wheatley *et al.*, 1997, 2001). Some Bt cotton varieties may release Bt proteins actively via root exudates (Gupta *et al.*, 2002; Rui *et al.*, 2005), while others may not (Saxena *et al.*, 2004). Proteins from root exudates and dead plant material may also persist in the soil, particularly in soils with high clay content (Venkateswerlu and Stotzky, 1992; Stotzky, 2004), so it is possible that changes in the composition of root exudates from a transgenic plant may alter rhizosphere communities significantly (Saxena *et al.*, 1999; Saxena and Stotzky, 2000; Brusetti *et al.*, 2004). Similarly, transgenic plant residues may affect non-target species and soil ecosystem processes (Zwahlen *et al.*, 2003a,b). As both the soil microbial communities and associated meso- and macroorganisms contribute to the maintenance of soil fertility and plant production, effects on them should be assessed.

This chapter concentrates on the three major cotton growing areas in Vietnam: the coastal lowlands region, the central highlands (eastern and western Truong Son Mountain Range) and the south-eastern region (Chapter 2, this volume). Soils in these three regions are very different, so it will be necessary to assess the effects of transgenic cotton on typical soils from all three regions. The soils in the south-eastern region are luvisols, andosols and acrisols, on which high yields of cotton can be achieved. According to Le (1998a), the soils are well aggregated, have good water absorption, are slightly acidic with an average pH of 5.0 and an organic content between 3 and 5%. The soils are rich in total phosphorus, poor in soluble phosphorus and very poor in soluble potassium, but have a high cation exchange capacity.

In the central highlands, the soils are mainly luvisols, rhodic ferrasols and haplic acrisols. The luvisols are fertile and well drained, with a pH of 5.8, and are high in humic substrates, exchangeable Ca^{2+} and Mg^{2+}, soluble phosphorus and potassium. However, the topsoil is thin and mixed with stones. The acrisols tend to be infertile, well drained with sand and stone, and again the topsoil is shallow. The soil is slightly acidic, poor in soluble phosphorus, potassium and exchangeable Ca^{2+} and Mg^{2+}. In contrast, the ferrasols have a deep cultivation layer, with good structure, high organic matter (2–5%) and are rich in total phosphorus and potassium, but poor in soluble phosphorus (Tran, 1995).

The soils in the coastal lowlands region are mainly delta soils, consolidated occasionally by grey light soils. The soil is acidic, pH below 4.5, organic content 1%, with low total but high soluble phosphorus (Le, 1998b). In the dry coastal region of central Vietnam, the new alluvial deposits are good for growing cotton (Le, 1998a). The red soil in central Vietnam is not acidic, but is low in organic matter content (< 1.3%), poor in total phosphorus (0.03% P_2O_5) and rich in exchangeable cations.

10.1. Assessment Strategy

The vast complexity of soil ecosystems means that it is both impractical and unreliable to study such systems based on species lists. Instead, the focus should be on ecosystem processes. Numerically, there are far fewer ecosystem processes than species, but it is still impractical to measure all of the processes. Choices of reliable end points that can be assessed have to be made. In this chapter, we consider soil ecosystem processes and prioritize these for case-specific testing. This prioritization is based first on possible exposure and the significance of the process in cotton ecosystems in Vietnam, and then on the potential exposure pathways and adverse effect scenarios. Testable risk hypotheses are formulated and experiments to test these hypotheses in both the laboratory and field described.

A study commissioned by the UK Environment Ministry (DEFRA, 2004) described three different approaches to assessing the effects of GMOs on soils. The first two are based on an approach proposed by Bruinsma *et al.* (2003). If specific microbial indicators can be identified that might be affected in the crop–soil ecosystem, then these microbial species can be used as assessment entities in laboratory experiments (first approach). If not, then general testing of the microbial community would be carried out in the laboratory, using profiling methods (second approach). A third approach stems from work on soil quality indicators (Allan *et al.*, 1995), based on the identification of soil processes/services. All three methods are intricately related and can be used in combination. However, DEFRA (2004) point out that the first two methods may be unable to assess impacts related to poor degradation of a xenobiotic chemical. The first two methods have the advantage of being laboratory-based and can therefore be used early in the risk assessment process, while the third approach is fundamentally a field-based approach that would be best implemented later during small-scale field testing. The third method has the advantage of being tied most closely to assessment end points of concern, while the first two methods require extensive extrapolation and many assumptions. In addition, the first two methods require the development of the laboratory infrastructure to conduct the tests, while the third method requires lower capital investment. Hence, for Vietnam, we have relied primarily on the third approach, with some consideration of the other approaches.

10.2. Selection of Soil Processes

Soil ecosystems are driven by the types and amounts of carbon-containing compounds entering soils, providing both energy and nutrients (Wheatley *et al.*, 1990, 1991, 2001). These inputs are from plant residues and organic compounds released by the roots of growing plants. The constituents of these inputs vary according to the type of plant cultivated, its physiology, structure and the stage of growth.

The main soil ecosystem processes were identified and listed (Table 10.1) and ranked to identify suitable assessment end points (1 is a high rank and 3 a

Table 10.1. Ranking of soil processes and microbial communities for Bt cotton in Vietnam. High rank = 1, medium rank = 2, low rank = 3. A = link to above-ground plant residues that fall on to soil or are incorporated into soil; B = link to root residues in soil or incorporated into soil; C = link to root exudates; D = link to root cells; E = important plant nutrients or products (E_1) and significance (E_2) for the functioning of the cropping system; F = significance as a soil health indicator.

Ecosystem process	Indicator organisms	Potential exposure					Potential significance				Summary	
		A	B	C	D	Mean	E_1	E_2	F	Mean	Sum	Rank
Biomass decomposition	Macrofauna, fungi, bacteria	1	1	3	3	2	C	1	1	1	3	2
Cellulose and lignin breakdown	Fungi, bacteria	1	1	3	3	2	C	1	1	1	3	2
Ammonification	Bacteria	3	2	2	2	2.25	N	1	2	1.5	3.75	3
Nitrification	Bacteria	3	3	2	3	2.75	N	1	1	1	3.75	3
Denitrification	Bacteria	3	3	2	2	2.5	N	1	3	2	4.5	3
Nitrogen fixation	Bacteria (free living, not nodulated)	3	1	1	2	1.75	N	2	2	2	3.75	3
Phosphorus and micronutrient uptake	Bacteria, fungi and mycorrhiza	3	1	1	3	2	P and micronutrients	1	1	1	3	2
Horizontal gene transfer	Bacteria	3	2	3	1	2.25	–	3	3	3	5.25	3
Plant pathogens	Viruses, bacteria, fungi, nematodes, insects	1	1	2	1	1.25	–	1	3	2	3.25	3
Soil particle aggregation	Microorganisms	1	1	3	3	2	C, N, micronutrients	1	1	1	3	2
Water-holding capacity	Microorganisms	2	1	3	3	2.25	C, N, micronutrients	1	1	1	3.25	2
Microbial communities												
Plant/microbe interactions	Microorganisms	1	1	1	1	1	C, N, micronutrients	1	2	1.5	2.5	1
Biodiversity	Microorganisms, fungi	1	1	1	1	1	C, N, micronutrients	2	1	1.5	2.5	1

low rank). Processes were ranked in relation to their association with the main possible exposure pathways to transgene proteins from transgenic cotton and their significance in the crop ecosystem and as indicators of soil health. Rankings were based on published and expert knowledge.

The highest ranked soil processes were biomass decomposition, cellulose and lignin breakdown, phosphorus and micronutrient uptake, soil particle aggregation and water-holding capacity (Table 10.1). Horizontal gene transfer between Bt cotton and microbial populations appeared to be unlikely and so was ranked low. Because Bt cotton is likely to provide Bt proteins to the living root rhizosphere and to the soil as crop residues, biomass decomposition and cellulose and lignin breakdown may be the first processes affected. In the cotton rhizosphere, phosphorus dynamics, which depend on the mycorrhizal associations in the rhizosphere, might also be affected similarly. Changes in biomass decomposition may affect soil aggregate formation and water-holding capacity (Carter, 2002), although the mechanisms by which this might occur are not clear. In addition, the microbial community measures (plant/microbe interactions and biodiversity) were highly ranked. Interactions between microorganisms occur in soils, resulting in either stimulation or inhibition of plant growth. Such interactions can be examined by adding to soil the specific compounds in the plant exudates affecting specific soil processes. Carbon inputs can also affect microbial function in more subtle ways, as they have a role in microbial interactions and signalling. Temporally variable interactions between portions of the general microbial population and specific groups of microorganisms, such as the autotrophic nitrifiers, in arable soils in response to additions of carbon and nitrogen have been reported (Wheatley et al., 2001). These effects may be transient, but will be determined by inputs from the plant presently growing in the system, such as root exudates, and inputs from the previous crop.

Horizontal gene transfer

De Vries and Wackernagel (2004) state that intraspecific and interspecific horizontal gene transfers, e.g. conjugation and transduction, are part of the lifestyle of prokaryotes and have shaped microbial genomes throughout evolution. However, in their review of many components of the soil ecosystem, they conclude that although transfers between plants and bacteria are possible, each of the many steps involved in the release of intact DNA from the plant to integration into the prokaryotic genome has such a low probability that successful transfer events will be extremely rare. For natural transformations to occur in soils, free DNA and competent bacteria have to be in close proximity (Smalla et al., 2001), which is more likely close to roots where Bt toxin-containing residues are actively being degraded. Marker genes from some transformed plants have been detected in soil (Nielsen et al., 2000; Widmer et al., 2001). However, there are no reports of the transformation of plant DNA to indigenous soil microorganisms. In ideal conditions in the laboratory, horizontal gene flow from genetically modified sugarbeets to *Acinetobacter* sp. was detected at very low rates (Gebhard and Smalla, 1999; Nielsen and van Elsas, 2001). In fact, horizontal transfer from the transgenic plant is much less likely to occur than from the organism, in this case *B. thuringiensis*, from which the gene was derived originally. As the soils in Vietnam are different from those used in these studies, the possibility of horizontal gene transfer could be considered, but is ranked as a low priority.

10.3. Identifying Potential Exposure Pathways in Soil

Bt proteins are present, as active toxins, in most cells in the Bt-transformed plant. There are various pathways by which Bt plant material and Bt proteins enter soils (Table 10.2). During the growing season, Bt proteins can enter via root exudates, leachates from damaged roots and sloughed off root debris, e.g. root cap cells and root hairs, or via above-ground plant parts, e.g. leaves, squares, bolls, or pollen that fall to the ground. These sources will increase as the plants grow larger. At the end of the growing season, Bt proteins enter the soil via plant residues that remain in the field after harvest, the amounts of which depend on crop management practices.

The fate of both the Bt protein-containing plant materials and the free Bt proteins in soil ecosystems depends on the balance between input and degradation rates. Laboratory studies with Bt cotton plant material showed that the toxins were still detectable in soils and retained insecticidal activity when the experiments were terminated after 28 days (Palm *et al.*, 1994; Donegan *et al.*, 1995) or 120 days (Sims and Ream, 1997). However, free Bt proteins can be both degraded rapidly by microbial degradation and leached from soils in groundwater and runoff (Palm *et al.*, 1996). Degradation rates vary with soil pH, temperature and water availability, and soil structure, texture and the organic matter content (Donegan *et al.*, 1995). Free Bt proteins are bound rapidly to clay particles and organic compounds, such as humic acids, where they retain their insecticidal activity (Venkateswerlu and Stotzky, 1992; Crecchio and Stotzky, 1998). Generally, protein degradation will be slower in clayey soils than sandy soils and slower in large pieces of residue than small ones. Bt protein in shredded and ploughed-in cotton residues was not detectable in soil samples 3 months after Bt cotton harvest (Head *et al.*, 2002).

Although the possibility that Bt proteins in soils might be absorbed by crop plants following Bt cotton has not been investigated, uptake from soils in which Bt maize had been grown previously has been studied. None of the test plants, maize, carrot, radish and turnip, absorbed any Bt proteins from the different soil types (Saxena and Stotzky, 2001a). Indirect exposure via movement of Bt protein may occur with the harvest of cotton (Table 10.2). Harvested material is removed from the fields and by-products are disposed of near cotton gins. These by-products may be incorporated into the soil, where they could have effects on soil ecosystem processes. This possibility should be considered to determine if it could lead to an adverse effect. Although exposure could occur as a consequence of gene flow to relatives (Table 10.2), this is unlikely to be significant as the only relatives are other cotton cultivars and volunteers (Chapter 11, this volume).

Table 10.2. Potential exposure assessment for soil processes associated with Bt cotton. Assessment is based primarily on the literature on Cry1Ac cotton and '?' indicates uncertainty. As several of the other Bt cottons use a similar promoter, their expression is expected to be similar, although this should be confirmed.

Exposure type	Assessment
Direct exposure	
Exposure via living roots	
Do roots contain transgene products/metabolites?	Yes
How long do the products or metabolites persist?	Growing season
Exposure via root secretions and senescent root tissue	
Do roots secrete transgene products or metabolites?	Possibly
Does senescent root tissue contain transgene products or metabolites?	Yes
How long do the products or metabolites persist?	? < Growing season
Exposure to plant parts that fall on the soil during growing season (pollen, flowers, residue)	
Do any of the plant parts that fall on the soil contain transgene products or metabolites?	Yes
Do transgene products or metabolites leach from these plant parts?	?
How long do the products or metabolites persist	? < Growing season
Exposure to plant residue in soil (leaf, stem, stalk, root)	
Do any plant residues contain transgene products or metabolites?	Yes
Do transgene products or metabolites leach from these residues?	Yes?
How long do the products or metabolites persist in residue?	200–300 days
How long do the products or metabolites persist in soil?	200–300 days?
Indirect exposure	
Exposure to transgene products or metabolites after they have moved	
Are there any transport processes that would move plant part, transgene products or metabolites? Specify the process and address the following questions.	Yes, ploughing, cultivation and harvest
Where are the plant parts, transgene products or metabolites moved to?	Incorporated into soil, moved to harvest areas
How long do the products or metabolites persist?	200–300 days in soil?; unknown for harvest areas
Exposure after gene flow	
Can exposure via gene flow occur	
What are the likely recipients of gene flow (plants and recipients of horizontal gene flow)?	Other cotton cultivars and volunteers
Where are they located in the landscape (what habitat, landscape element, location relative to crop)?	Cotton near Bt cotton fields
Do these recipient organisms express the transgene?	Probably yes

In summary, the probability that all of the selected ecosystem processes are exposed to Bt protein is high. The largest exposures are likely to be via roots or root exudates associated with living plants or via plant residues. However, if Bt cotton is not replanted, exposure is likely to be transient, lasting no more than 1 or possibly 2 years.

10.4. Identifying Potential Adverse Effects Pathways

Identifying potential adverse effects is critical for establishing meaningful assessment end points and relating measurements of ecosystem processes and microbial communities to those end points. Table 10.3 provides potential adverse effects pathways starting with a change in the ecosystem process or microbial community and ending in adverse effects. These adverse effects could be linked directly to the hypothesized change, or may be a consequence of other changes. In addition, we have provided an estimated time frame in which an adverse effect might appear.

Table 10.3. Potential adverse effects assessment for soil processes associated with Bt cotton in Vietnam.

Ecosystem or community process	Potential change	Adverse effect	Time frame	Consequent adverse effects
Biomass decomposition	Lower SOM	Reduction in soil fertility	Several years	Yield loss; increase in soil erosion
	Higher SOM	Increased soil-borne disease	Several years	Yield loss
Cellulose and lignin breakdown	Faster	Changes in soil fertility	Several years	Unknown: possible, but unspecified
	Slower	Changes in soil fertility	Several years	Unknown: possible, but unspecified
Phosphorus & micronutrient uptake	Slower uptake	Nutrient stress	Several years	Reduced soil fertility, yield loss
	Faster uptake	Nutrient imbalance	Several years	Yield loss
Soil particle aggregation	Lower	Reduced water infiltration	Several years	Increase in soil erosion; increased drought, yield loss
	Higher	None	–	–
Water-holding capacity	Lower	Increased droughtiness	Several years	Increase in soil erosion, yield loss
	Higher	None	–	–
Plant/microbe interactions	Changed	Unspecified	Unknown	Unknown: possible, but unspecified
Biodiversity	Lower	Unspecified	Unknown	Unknown: possible, but unspecified
	Higher	None	–	–

Changes in ecosystem processes may lead to specific adverse effects (Table 10.3). For example, reduced soluble organic matter (SOM) may reduce fertility, reduced phosphate (P) uptake may cause nutrient stress and lower water-holding capacity may cause increased droughtiness. In addition, some of these effects may have additional consequences, leading to other potential adverse effects, such as increased erosion, yield loss and reduced soil fertility. Each process is hypothesized to lead to some specific subset of the possible adverse effects and no process is hypothesized to lead to all of the possible adverse effects. In addition, some of the hypothesized changes in the processes could lead to beneficial effects (e.g. increased SOM may lead to improved fertility, better soil structure and decreased erosion) but, because a risk assessment focuses on the potential adverse effects, these are not addressed here. Notably, cellulose and lignin breakdown is associated less clearly with a specific adverse effect; this lack of specificity is problematic and an adverse effect should be specified fully before this ecosystem process is used for risk assessment. The other ecosystem processes are suitably structured conceptually, making it possible to use them for assessing ecological risks.

Plant–microbe interactions and microbial biodiversity are not associated with any specific adverse effect (Table 10.3). Changes in microbial community structure have been used frequently as general indicators of stress and recent developments in molecular approaches have provided methods for the direct assessment of the degree of structural diversity in microbial communities. Previous investigations of whole microbial community DNA by denaturing gradient gel electrophoresis (DGGE) have shown changes in the whole soil community profile under different cultivated crops. So, it can be hypothesized that the soil-microbial communities will be affected when a Bt-transformed crop is cultivated. This effect may be similar to other varieties of the same crop, or it may be different from these. The DGGE method, however, is not quantitative and changes are not permanent, so it will be necessary to conduct extensive baseline sampling to establish levels of natural variation and enable the result to be interpreted. Petras and Casida (1985) reported increases in bacterial, fungal and actinomyces populations after soils were treated with commercial preparations of *B. thuringiensis* subsp. *kurstaki* and suggested that the proteins were used as growth substrates. Sun *et al.* (2007) found the addition of Bt cotton tissue to laboratory microcosms stimulated some soil enzyme activities. For transgenic Bt maize, several authors suggested that, if some bacterial communities had the potential to degrade Cry1Ab, this extra protein in the environment might cause proliferation in certain microorganism populations, which might then lead to a faster decomposition of Bt versus non-Bt maize (Blackwood and Buyer, 2004; Baumgarte and Tebbe, 2005). Muchaonyerwa *et al.* (2005) examined the response of microbes to additions of *B. t.* subsp. *kurstaki*, *B. t.* subsp. *israelensis* and *B. t.* subsp. *tenebrionis* in three Zimbabwean soil types (vertisoil, oxisol and alfisol), but found no effects on bacterial or fungal populations. Several other studies found no or minor effects of Cry1Ab Bt plants on soil microbial communities (Donegan *et al.*, 1995; Saxena and Stotzky, 2001b; Blackwood and Buyer, 2004; Baumgarte and Tebbe, 2005; Griffiths *et al.*, 2005; Icoz *et al.*, 2008). What implications, if any, that changes in microbial interactions or diversity may have on ecosystem processes would need to be determined. Thus, although these methods are being proposed for use in GMO ERA (Bruinsma *et al.*, 2003), it may be more useful in Vietnam to focus ERA on concrete end points of important ecosystem processes.

10.5. Risk Hypotheses and Analysis Plan

Selecting the processes and community indicators with highest potential exposure and with concrete assessment end points, we identify biomass decomposition, cellulose and lignin breakdown, phosphorus uptake, soil particle aggregation and water-holding capacity as the five processes and indicators most critical for further assessment. Changes in soil particle aggregation and water-holding capacity are likely to be consequences of other changes in the microbial communities and soil management practices, and could be monitored over time, so we describe simple risk hypotheses for the first three ecosystem processes.

Risk hypothesis 1

Decomposition rates of Bt cotton plant residues are faster than those for conventional cotton plant residues. If decomposition is more rapid, then soil organic matter content may drop, decreasing soil fertility, soil aggregate structure and water-holding capacity. This could result in reduced yield quantity and quality or higher input requirements, such as fertilizer or water.

Risk hypothesis 2

Symbiosis with mutualistic mycorrhizal fungi is reduced in Bt cotton, resulting in lower phosphorus uptake. A reduction in mutualism may decrease root development. This could lead to increased susceptibility to drought, lower nutrient absorption and reduced plant growth, which could result in reduced yield quantity and quality or higher input requirements, such as fertilizer or water.

Recent work on the biological basis of soil aggregate stability has shown that mycorrhizal fungi can play an important role via the secretion of a glycoprotein, glomalin (Purin and Rillig, 2007). Individual species of arbuscular mycorrhizal fungi differ in the production of this compound and, correspondingly, in their ability to stabilize soil aggregates. As a result, Bt effects on mycorrhizal fungal density or composition could also result in changes in soil aggregate stability.

Analysis plan

Both risk hypotheses are speculative and, if possible, should be tested with rapid, relatively inexpensive methods. The first hypothesis requires that Bt cotton decomposes faster than non-Bt cotton, so this requirement could be the target for the initial experiments. Chen *et al*. (2004) found that some Bt cotton cultivars had higher leaf tissue nitrogen concentration. If, at the same time, lignin and C content stay the same, this might suggest that plant residues will decompose more quickly. The second hypothesis requires that mutualistic symbioses are reduced in Bt cotton. This is more difficult to test directly, but could be initiated by examining the overall benefits to the plant from mycorrhizal symbioses. Not all of the mycorrhizal fungal species forming symbioses will be mutualistic (Bever, 2002). If either of these requirements are verified, the consequent effects (the ultimate assessment end points) may take a few years to appear (Table 10.3), if they appear at all. Thus, experimental evaluation of these consequences would take several years and require considerable resources to maintain and repeat the experiments at suitably large spatial scales. An alternative to such long-term, expensive experiments is to monitor key soil characteristics, as discussed in the following section.

10.6. Protocols for Assessing or Monitoring Risk

Risk hypothesis 1

Testable scientific prediction: decomposition rates of Bt cotton plant residues are faster than those for conventional cotton plant residues.

Litterbag experiment with plant residues
Grow Bt and non-Bt cotton either in a contained facility or, preferably, in a small field plot. It is critical that plant residues are as similar to field conditions as possible. Plants should be harvested using normal practices and the residues collected immediately after harvest. Residues should be air-dried and sorted into stem, root and leaf tissue, but not ground. The relative biomass of root, stem and leaf litter should be measured. A representative subsample of each residue tissue (root, stem, litter) should be dried at 60°C to constant weight and stored for further analyses (see below). Weighed quantities of litter should be placed in litterbags. A representative mixture of root, stem and leaf litter can be placed in each bag, or separate bags for each type of litter can be used. Litterbags should be placed horizontally in soils (10 cm depth) and on the soil surface (covered with a thin layer of soil or residue, < 1 cm). They should be constructed of inert plastic material (such as greenhouse shade cloth), with a *c*.5 × 5 mm mesh on top and a *c*.1 × 3 mm mesh on the bottom (other materials of similar mesh can be used, as long as they will not decompose rapidly in the soil). The finer bottom mesh reduces litter loss from the bag and the coarse upper mesh allows soil macroorganism ingress and egress to and from the bag. Field locations for the bags should be in the main cotton growing region and main cotton soil types in Vietnam. If the hypothesis is confirmed, the experiment could be repeated in other cotton growing regions of Vietnam. All bags should be paired with one Bt and one non-Bt bag next to each other and between 30–50 replicate pairs should be placed in the field. Bags can be labelled using aluminum tags that are

stapled on each litterbag. For additional methodological details, see House and Stinner (1987) and Bradford *et al.* (2002).

Several parameters can be measured; however, it will be essential to collect data necessary to estimate decomposition rates. These data are starting litter weight, C/N ratio, lignin, cellulose and mineral ash of starting litter, ending litter weight, C/N ratio, lignin, cellulose and mineral ash of all ending litterbags (to correct for influx of soil). In addition, optional data can be collected on respiration rate, microbial community structure and macrobiota. For sampling certain soil invertebrates, such as collembolans, mites and myriapods, litterbags can be placed in Berlese, Tullgren, or the more efficient MacFadyen extractor (MacFadyen, 1961, 1962) for approximately 1 week. Invertebrates can be caught in vials filled with isopropanol. After the extraction procedure, the isopropanol in the vials should be exchanged with 70% (v/v) ethanol. Depending on the soil organism group of interest, extraction methods other than the one described above might be more efficient.

Risk hypothesis 2

Testable scientific prediction: Bt cotton will grow more slowly than conventional cotton plants in the presence of mycorrhizal fungi typically found in cotton fields.

Greenhouse experiment with living plants
Cotton plants and cotton soils are collected to create root and soil inocula of the mycorrhizal communities, as in Bever (1994). Although this method does not eliminate all soil pathogens completely, it minimizes their effect and is much simpler than using sterile pure fungal cultures (Bever, 2002). Once the inocula are established, they can be used to inoculate potting soils with the background microbial communities either with or without mycorrhizal inocula. Surface sterilized Bt and non-Bt cottonseeds are then planted into each soil. After standardizing for germination, plant growth rates are measured. Each replicate of the four pots of each treatment should be grown together, so that the growth enhancement effect of the mycorrhizal fungi can be calculated more easily. About 10–20 replicates are probably needed. Growth enhancement can be tested to determine if it is the same or different between Bt and non-Bt plants.

Monitoring soil quality

If either of the two scientific predictions above is confirmed, then it will be important to implement a simple, formal soil monitoring programme. If both are falsified, then a formal soil-monitoring programme might not be essential, although it still would be beneficial. If the first prediction is verified, then monitoring of soil organic matter, soil aggregate structure (bulk density and porosity) and water-holding capacity should be implemented. These should be

measured in a cotton field prior to the planting of Bt cotton and measured
every 1–3 years afterwards (more frequently at the beginning). Sufficient samples should be taken with an Uhlen sampler from representative fields to
enable detection of meaningful changes in organic matter, aggregate structure
and water-holding capacity. If the second prediction is verified, then Bt cotton
growth should be monitored for signs of phosphorus, nutrient or drought
stress that would not normally occur in conventional cotton. If a possible
adverse effect is detected during monitoring, then experiments should be initiated to confirm that Bt cotton is causing the effect and possible mitigation
responses considered.

References

Allan, D.L., Adriano, D.C., Bezdicek, D.F., Cline, R.G., Coleman, D.C., Doran, J.W., Haberern,
    J., Harris, R.F., Juo, A.S.R., Mausbach, M.J., Peterson, G.A., Schuman, G.E., Singer, M.J.
    and Karlen, D.J. (1995) Soil quality: a conceptual definition. *Soil Science Society of
    America Agronomy News* June 7.
Baumgarte, S. and Tebbe, C.C. (2005) Field studies on the environmental fate of the Cry1Ab Bt-
    toxin produced by transgenic maize (MON810) and its effect on eubacterial communities in
    the maize rhizosphere. *Molecular Ecology* 14, 2539–2551.
Bever, J.D. (1994) Feedback between plants and their soil communities in an old field commu-
    nity. *Ecology* 75, 1965–1977.
Bever, J.D. (2002) Negative feedback within a mutualism: host-specific growth of mycorrhizal
    fungi reduces plant benefit. *Proceedings of the Royal Society* 269, 2595–2601.
Blackwood, C.B. and Buyer, J.S. (2004) Soil microbial communities associated with Bt and non-
    Bt corn in three soils. *Journal of Environmental Quality* 33, 832–836.
Bradford, M.A., Tordoff, G.M., Eggers, T., Jones, T.H. and Newington, J.E. (2002) Microbiota,
    fauna, and mesh size interactions in litter decomposition. *Oikos* 99, 317–323.
Bruinsma, M., Kowalchuk, G.A. and Veen, J.A. van (2003) Effects of genetically modified plants
    on microbial communities and processes in soil. *Biology and Fertility of Soils* 37,
    329–337.
Brusetti, L., Francia, P., Bertolini, C., Pagliuca, A., Borin, S., Sorlini, C., Abruzzese, A., Sacchi,
    G., Viti, C. and Giovannetti, L. (2004) Bacterial rhizosphere community of transgenic Bt176
    maize and its non-transgenic counterpart. *Plant Soil* 266, 11–21.
Carter, M.R. (2002) Soil quality for sustainable land management. *Agronomy Journal* 94,
    38–47.
Chen, D., Ye, G., Yang, C., Chen, Y. and Wu, Y. (2004) Effect after introducing *Bacillus thur-
    ingiensis* gene on nitrogen metabolism in cotton. *Field Crops Research* 87, 235–244.
Crecchio, C. and Stotzky, G. (1998) Insecticidal activity and biodegradation of the toxin from
    Bacillus thuringiensis subsp. *kurstaki* bound to humic acids from soil. *Soil Biology and
    Biochemistry* 30, 463–470.
De Vries, J. and Wackernagel, W. (2004) Microbial horizontal gene transfer and the DNA release
    from transgenic crop plants. *Plant and Soil* 266, 91–104.
DEFRA (2004) *Mechanisms for Investigating Changes in Soil Ecology Due to GMO Releases*.
    Ref EPG 1/5/214. Department of Environment, Food and Rural Affairs, UK Government.
    http://www.defra.gov.uk/environment/gm/research/epg-1-5-214.htm (last accessed 27
    November 2007).
Donegan, K., Palm, C., Fieland, V., Porteous, L., Ganio, L., Schaller, D., Bucao, L. and Seidler,
    R.J. (1995) Changes in levels, species and DNA fingerprints of soil microorganisms

associated with cotton expressing the *Bacillus thuringiensis* var. *kurstaki* endotoxin. *Applied Soil Ecology* 2, 111–124.

Gebhard, F. and Smalla, K. (1999) Monitoring field releases of genetically modified sugar beets for persistence of transgenic plant DNA and horizontal gene transfer. *FEMS Microbiology Ecology* 28, 261–272.

Griffiths, B.S., Caul, S., Thompson, J., Birch, A.N.E., Scrimgeour, C., Andersen, M.N., Cortet, J., Messéan, A., Sausse, C., Lacroix, B. and Krogh, P.H. (2005) A comparison of soil microbial community structure, protozoa and nematodes in field plots of conventional and genetically modified maize expressing the *Bacillus thuringiensis* Cry1Ab toxin. *Plant and Soil* 275, 135–146.

Gupta, V.V.S.R., Robert, G.N., Neate, S.M., McClure, S.G., Crisp, P. and Watson, G. (2002) Impact of Bt-cotton on biological processes in Australian soils. In: Akhurst, R.K., Beard, C.E. and Hughes, P. (eds) *Biotechnology of Bacillus thuringiensis and its Environmental Impact. Proceedings of the 4th Pacific Rim Conference*. CSIRO, Canberra, pp. 191–194.

Head, G.P., Surber, J.B., Watson, J.A., Martin, J.W. and Duan, J.J. (2002) No detection of Cry1Ac protein in soil after multiple years of transgenic Bt cotton (Bollgard) use. *Environmental Entomology* 31(1), 30–36.

House, G.J. and Stinner, R.E. (1987) Decomposition of plant residues in no-tillage agroecosystems: influence of litterbag mesh size and soil arthropods. *Pedobiologia* 30, 351–360.

Icoz, I., Saxena, D., Andow, D.A., Zwahlen, C. and Stotzky, G. (2008) Microbial populations and enzyme activities in soil *in situ* under transgenic corn expressing cry proteins from *Bacillus thuringiensis*. *Journal of Environmental Quality* 37, 647–662.

Le Cong Nong (1998a) Nutritional need of cotton trees and cultivational techniques for highbreed cotton. In: *Cotton Production Technology for High Productivity*. Agriculture Publishing House, Hanoi, pp. 151–163 (in Vietnamese)

Le Xuan Dinh (1998b) Natural condition in some cotton areas. In: Nguyen Huu Binh (ed.) *Cotton Production Technology for High Productivity*. Agriculture Publishing House, Hanoi, pp. 11–38 (in Vietnamese).

MacFadyen, A. (1961) Improved funnel-type extractors for soil arthropods. *Journal of Animal Ecology* 30, 171–184.

MacFadyen, A. (1962) Soil arthropod sampling. *Advances in Ecological Research* 1, 1–34.

Martin, P.A.W., and Travers, S. (1989) Worldwide abundance and distribution of *Bacillus thuringiensis* isolates. *Applied and Environmental Microbiology* 55, 2437–2442.

Morse, S., Bennett, R. and Ismael, Y. (2006) Environmental impact of genetically modified cotton in South Africa. *Agriculture Ecosystems and Environment* 117, 277–289.

Muchaonyerwa, P., Waladde, S., Nyamugafata, P., Mpepereki, S. and Ristori, G.G. (2005) Persistence and impact on microorganisms of *Bacillus thuringiensis* proteins in some Zimbabwean soils. *Plant and Soil* 266, 41–46.

Nielsen, K.M. and van Elsas, J.D. (2001) Stimulatory effects of compounds present in the rhizosphere on natural transformation of *Acinetobacter* sp BD413 in soil. *Soil Biology and Biochemistry* 33, 345–357.

Nielsen, K.M., van Elsas, J.D. and Smalla, K. (2000) Transformation of *Acinetobacter* sp. strain BD413 (pFG4 Delta nptII) with transgenic plant DNA in soil microcosms and effects of kanamycin on selection of transformants. *Applied and Environmental Microbiology* 66, 1237–1242.

Palm, C.J., Donegan, K., Harris, D. and Seidler, R.J. (1994) Quantification in soil of *Bacillus thuringiensis* var. *kurstaki* delta-endotoxin from transgenic plants. *Molecular Ecology* 3, 145–151.

Palm, C.J., Schaller, D.L., Donegan, K.K. and Seidler, R.J. (1996) Persistence in soil of transgenic plant produced *Bacillus thuringiensis* var *kurstaki* delta-endotoxin. *Canadian Journal of Microbiology* 42, 1258–1262.

Petras, S.F. and Casida, L.E. (1985) Survival of *Bacillus thuringiensis* spores in soil. *Applied and Environmental Microbiology* 50, 1496–1501.

Purin, S. and Rillig, M.C. (2007) The arbuscular mycorrhizal fungal protein glomalin: limitations, progress, and a new hypothesis for its function. *Pedobiologia* 51, 123–130.

Ronn, R., Griffiths, B.S., Ekelund, F. and Christensen, S. (1996) Spatial distribution and successional pattern of microbial activity and micro-faunal populations on decomposing barley roots. *Journal of Applied Ecology* 33, 662–672.

Rui, Y.-K., Yi, G.-X., Zhao, J., Wang, B.-M., Li, Z.-H., Zhai, Z.-X., He, Z.-P. and Li, Q.X. (2005) Changes of Bt toxin in the rhizosphere of transgenic Bt cotton and its influence on soil functional bacteria. *World Journal of Microbiology and Biotechnology* 21, 1279–1284.

Saleh, S.M., Harris, R.F. and Allen, O.N. (1970) Recovery of *B. thuringiensis* from field soils. *Journal of Invertebrate Pathology* 15, 55–59.

Saxena, D. and Stotzky, G. (2000) Insecticidal toxin from *Bacillus thuringiensis* is released from roots of transgenic Bt corn *in vitro* and *in situ*. *FEMS Microbiology Ecology* 33, 35–39.

Saxena, D. and Stotzky, G. (2001a) Bt toxin uptake from soil by plants. *Nature Biotechnology* 19, 199.

Saxena, D. and Stotzky, G. (2001b) *Bacillus thuringiensis* (Bt) toxin released from root exudates and biomass of Bt corn has no apparent effect on earthworms, nematodes, protozoa, bacteria, and fungi in soil. *Soil Biology and Biochemistry* 33, 1225–1230.

Saxena D., Flores, S. and Stotzky, G. (1999) Insecticidal toxin in root exudates from Bt corn. *Nature* 402, 480.

Saxena D., Stewart, C.N., Altosaar, I., Shu, Q. and Stotzky, G. (2004) Larvicidal Cry proteins from *Bacillus thuringiensis* are released in root exudates of transgenic *B. thuringiensis* corn, potato, and rice but not *B. thuringiensis* canola, cotton, and tobacco. *Plant Physiology and Biochemistry* 42, 383–387.

Sims, S.R. and Ream, J.E. (1997) Soil inactivation of the *Bacillus thuringiensis* subsp. *kurstaki* CryIIA insecticidal protein within transgenic cotton tissue: laboratory microcosm and field studies. *Journal of Agricultural and Food Chemistry* 45, 1502–1505.

Smalla, K., Wieland, G., Buchner, A., Zock, A., Parzy, J., Kaiser, S., Roskot, N., Heuer, H. and Berg, G. (2001) Bulk and rhizosphere soil bacterial communities studied by denaturing gradient gel electrophoresis: plant-dependent enrichment and seasonal shifts revealed. *Applied and Environmental Microbiology* 67, 4742–4751.

Stotzky, G. (2004) Persistence and biological activity in soil of the insecticidal proteins from *Bacillus thuringiensis*, especially from transgenic plants. *Plant and Soil* 266, 77–89.

Sun, C.X., Chen, L.J., Wu, Z.J., Zhou, L.K. and Shimizu, H. (2007) Soil persistence of *Bacillus thuringiensis* (Bt) toxin from transgenic Bt cotton tissues and its effect on soil enzyme activities. *Biology and Fertility of Soils* 43, 617–620.

Tran An Phong (1995) Evaluation of land use status from point of view of ecological and sustainable development. Project 02–09, Agriculture Publishing House, Hanoi (in Vietnamese).

Venkateswerlu, G. and Stotzky, G. (1992) Binding of the protoxin and toxin proteins of *Bacillus thuringiensis* subsp. *kurstaki* on clay minerals. *Current Microbiology* 25, 225–233.

Wheatley, R.E., Ritz, K. and Griffiths, B. (1990) Microbial biomass and mineral N transformations in soil planted with barley, ryegrass, pea or turnip. *Plant and Soil* 127, 157–167.

Wheatley, R.E., Griffiths, B.S. and Ritz, K. (1991) Variations in the rates of nitrification and denitrification during the growth of potatoes (*Solanum tuberosum* L.) in soil with different carbon inputs and the effect of these inputs on soil nitrogen and plant yield. *Biology and Fertility of Soils* 11, 157–162.

Wheatley, R.E., Ritz, K. and Griffiths, B. (1997) Application of an augmented nitrification assay to elucidate the effects of a spring barley crop and manures on temporal variations in rates. *Biology and Fertility of Soils* 24, 378–383.

Wheatley, R.E., Ritz, K., Crabb, D. and Caul, S. (2001) Temporal variations in potential nitrification dynamics in soil related to differences in rates and types of carbon and nitrogen inputs. *Soil Biology and Biochemistry* 33, 2135–2144.

Widmer, F., Fliessbach, A., Laczko, E., Schulze-Aurich, J. and Zeyer, J. (2001) Assessing soil biological characteristics: a comparison of bulk soil community DNA-, PLFA-, and Biolog-analyses. *Soil Biology and Biochemistry* 33, 1029–1036.

Zwahlen, C., Hilbeck, A., Gugerli, P. and Nentwig, W. (2003a) Degradation of the Cry1Ab protein within transgenic *Bacillus thuringiensis* corn tissue in the field. *Molecular Ecology* 12, 765–775.

Zwahlen, C., Hilbeck, A., Howald, R. and Nentwig, W. (2003b) Effects of transgenic Bt corn litter on the earthworm *Lumbricus terrestris*. *Molecular Ecology* 12, 1077–1086.

11 Environmental Risks Associated with Gene Flow from Transgenic Cotton in Vietnam

JILL JOHNSTON WEST, VŨ ĐỨC QUANG, MARC GIBAND, BAO RONG LU, DAVID A. ANDOW, NGUYỄN HỮU HỔ, VƯƠNG THỊ PHẤN AND LÊ QUANG QUYỀN

Gene flow contributes naturally to the evolution of species but, in the specific case of transgenic crops, gene flow of one or more transgenes could have adverse ecological, economical or social effects. Gene flow between transgenic crops and conventional (non-transgenic) varieties or their wild relatives has been cited as one of the central ecological risks associated with the application of biotechnology to crop production (NRC, 1989), because of potential adverse effects on natural biodiversity. Assessing gene flow consequences is challenging, because it is difficult to predict the ecological effects of transgenes that are integrated into different genetic backgrounds or expressed in different ecological contexts. Indeed, plants that acquire transgenes will continue to evolve, subject to natural and artificial selection pressures in the agricultural setting and beyond. Finally, once transgenes have moved into new populations, they may be impossible to remove from the environment.

This chapter discusses the potential risks of gene flow from genetically modified insect-resistant cotton in Vietnam, using a general framework to guide gene flow risk assessment. This framework includes assessment stages beginning with the conception of the specific transgenic crops (Pre-transformation assessment) and extending until after their release into the environment (Post-release monitoring, Box 11.1). We focused on the factors that affected the likelihood of transgene flow to each type of non-transgenic cotton population present in Vietnam. We considered the variety of environmental factors found in cotton growing regions of Vietnam, including the distributions of potential recipient plant populations and the types of agricultural and social practices that might influence gene flow in each area. We evaluated the potential adverse consequences of transgene introgression and spread in recipient populations, to the degree possible. In this chapter, we assess what we anticipate to be the most serious potential adverse consequences of transgene flow from transgenic insect-resistant cotton in Vietnam.

Box 11.1. Gene flow and its consequences: a question framework to guide risk assessment

Pre-transformation event
Stage 1. Consideration of the biology of the crop to be transformed.
Stage 2. Consideration of the site where the crop is to be released.
Stage 3. Evaluation of relevant crop–weed hybrids.

Post-transformation event, pre-release
Stage 4. Potential spread of the transgene.
Stage 5. Potential environmental consequences of transgene spread.

Post-release
Stage 6. Monitoring and management of transgene flow and its consequences (not addressed in this chapter).

Stage 1. Consideration of the biology of the crop to be transformed
Before crops are transformed, there are some basic biological features such as pollination syndromes that will set the foundation for gene flow risks. Wind-pollinated species may have higher average pollen movement than species that are self-pollinated or insect-pollinated. Pollen longevity and crop height also may affect the distance that pollen travels.

1.1   What type of breeding system does the crop have (e.g. obligate outcrossing, self-pollinated, mixed)?
1.2   To what extent does the crop outcross?
1.3   What type of vectors lead to pollen movement from the crop (e.g. wind, insects)?
1.4   How far does pollen disperse from crop fields (see Arriola and Ellstrand, 1996)?

Gene flow can also occur due to the movement of seeds and vegetative propagules. Propagule mediated gene flow can occur when crop volunteers or feral crop populations arise, or through human-mediated propagule movement.

1.5   How frequently are crop offspring observed in agricultural fields?
1.6   How frequently do crop populations arise along roadsides or in disturbed areas near agricultural fields?
1.7   Can seed banks develop from crop species? (For this to occur, seeds must possess or evolve some level of seed dormancy.)
1.8   Can the crop be propagated vegetatively?
1.9   Are feral or volunteer plants capable of being perennial?

Stage 2. Consideration of site where transgenic crop is to be released
The opportunity for gene flow will vary from location to location because of variation in the geographic distribution of relatives. The presence of wild, weedy, or rare populations of related species may create more concerns for a region or country than for one with no close relatives to the crop. In most areas of the world, there will be some information available on the presence of crop wild relatives. Sources include herbaria, literature on crop germplasm, seed bank or germplasm collections and any country-specific survey records of weedy or rare species distributions.

2.1   What other members of the crops' botanical family occur within the area where the transgenic crop would be released (see Johnston *et al.*, 2004)?

Continued

Box 11.1. *Continued*

2.2 Are there any members of different genera within the family that might be able to outcross with the crop?
   a. Is the locale a centre of origin or a centre of diversity for the crop genus?
   b. Are some related species known to be sources of germplasm for crop breeders? (If no specific information is available for a given crop, refer to Harlan and de Wet (1971) for initial guidance.)
   c. Are there differences in ploidy among members of the crop genus or family that might change the likelihood of gene flow with some relatives?
2.3 Will non-transgenic versions of the same crop be grown nearby?
2.4 Is the biology in questions 1.1–1.4 known for each related species, as well as the crop?
2.5 If feral populations arise, are they capable of acting as 'genetic bridges' by bringing crop genes in contact with related populations that are otherwise geographically isolated from the crop?

Stage 3. Evaluation of relevant crop-relative hybrids
The formation of viable F_1 hybrids can be assessed using commercial, non-transgenic crop varieties. Sample crosses should be made with each type of potential gene flow recipient (Song *et al.*, 2004). For example, if the populations that might experience gene flow include a weedy variety of the crop species and two congeneric species, one cultivated and one rare, at least three types of crosses need to be evaluated (crop × weed, crop × crop and crop × wild). Replication of crosses with different populations increases the credibility of results.

3.1 Do hand crosses between the crop and potential gene flow recipient populations result in F_1 hybrids?
3.2 Do hybrids arise spontaneously in the field?
3.3 Are F_1 hybrids vigorous and fertile under field conditions?
3.4 Are backcross hybrids observed in the field?
3.5 How fit are backcross hybrids in the field? (If they do not occur in the field naturally, make some in the greenhouse and put them in the field.)

Stage 4. Potential spread of the transgene
Once greenhouse or small-scale field trials can begin, information about the likelihood of transgene spread can be gathered. A key factor driving the spread of the transgene is the fitness effect that it has on the population that has experienced gene flow. The fitness effects will differ somewhat in wild versus managed populations because farmers likely will exert additional selection on crop populations. Snow *et al.* (2003) provide an example of how to study fitness following hybridization.

4.1 What is the fitness effect of the transgene when incorporated into each type of gene flow recipient population?
4.2 In wild populations, does the transgene increase fitness?
   a. Is transgene expression in wild populations similar to that of crops?
   b. Does the fitness increase depend on a specific ecological factor (e.g. the presence of insect herbivores) that may vary over seasons and geographic regions?
   c. Is the fitness benefit likely to be observed near agricultural fields only (e.g. herbicide tolerance will not increase fitness in the absence of herbicide sprays)?

Continued

Box 11.1. *Continued*

4.3 In volunteers or crop populations, does the presence of the transgene increase yield?
    a. Will farmers be more likely to select and save seeds from transgenic individuals than non-transgenics?

Stage 5. Potential environmental consequences of transgene spread
The potential environmental consequences of transgene flow, introgression and spread generally fall into four categories: reduction in genetic diversity of the recipient population, increased weediness of the recipient population, complications for resistance evolution management and biodiversity impacts of the transgene that reach outside the agricultural setting. In addition, gene flow may affect socio-economic issues, such as contamination of non-transgenic crops.

The genetic diversity of recipient populations may be altered by rapid gene flow and introgression from transgenic crops. The change in genetic diversity generally is thought to be of special concern in two types of scenarios. First, if a rare or endangered recipient species experiences gene flow, the crop genes may change the character of the species to the point that it is no longer recognizable. Second, if landraces or other sources of germplasm for plant breeding experience gene flow, they may lose some unique genetic variation that could have been useful to breeders.

5.1 Will the crop be grown in an area that is a centre of origin or centre for diversity for its family or genus?
5.2 How much crop to-wild or crop-to-landrace gene flow is occurring already?
5.3 Are there any germplasm collections that might have already been captured and stored the genetic diversity of interest?
5.4 Are farmers growing landraces of the crop geographically isolated from the areas where the transgenic crop will be grown?
5.5 Is the transgene under strong selection, or is the transgenic crop widely grown, which would enable rapid displacement of landrace gene pools?

The problem of increased weediness requires that the recipient population increase in abundance and improve competitive ability against other plants.

5.6 Could the transgenic trait allow free-living relatives to become more abundant within their typical habitats?
5.7 Could the transgenic trait allow free-living relatives to occupy new ecological niches (e.g. due to cold or drought tolerance)?
5.8 If free-living relatives of the crop become more widespread due to the transgene, are they likely to displace competitively any native plant species in their typical habitat or by the extention of their native range?

Gene flow could complicate resistance management. Are there possible effects of transgenic insect resistance on resistance management, integrated pest management, or beneficial non-target species?

5.9 Will gene flow reduce the efficacy of the refuge?

Gene flow can generate biodiversity impacts outside the agricultural field (Letourneau *et al.*, 2003).

Continued

Box 11.1. *Continued*

5.10   How could the crop-to-crop spread of transgenes affect local agricultural production?

   5.10.1 If transgenic herbicide resistance spread to wild relatives, would herbicide efficacy be reduced?

   5.10.2 Are there other ways in which transgenic crops will alter local agricultural practices?

5.11   What is the transgene product and what are the likely effects of the transgene product on other plants, herbivores and beneficial organisms in the region (see Chapters 6–10, this volume)?

5.12   Does the transgene confer any unintentional but ecologically significant changes in the chemical composition of hybrid and backcrossed progeny that could affect plant competition, plant–insect or plant–soil interactions?

   Gene flow can have socio-economic effects.

5.13   Could the spread of any type of transgene compromise the market value of non-GM crops?

11.1. Gene Flow and Introgression

We defined gene flow as the *movement* and introgression as the *integration* (via backcrosses) of genes or alleles, which originated in one plant population, into another plant population. There are three avenues for gene flow: pollen-, seed- and vegetative organ-mediated gene flow. In the first case, gene flow occurs when pollen from one plant or population fertilizes an individual in another population. In the second case, gene flow occurs through the natural dispersal of seeds by animals, wind, water, or other means. In the third case, gene flow occurs through the natural dispersal of vegetative organs (e.g. tillers, tubers and rhizomes) of plant species by animals, wind, water, or other means. Humans also may play a role in seed and vegetative-organ dispersal, as would be the case of seeds falling on the ground during picking or harvesting, or during transportation to the processing plant. Cotton has no vegetative reproduction. Introgression occurs when the offspring of gene flow mate with the recipient population, producing viable offspring that are incorporated into the recipient population.

   Humans may move seeds intentionally. The official release of a new variety and its adoption by farmers will result in the movement of large numbers of seeds around the country. This is not considered risky because it is a normal agricultural practice. Seed movement also may occur within a small or large geographical area through practices such as seed saving from one season to the other, or seed trading among farmers. This latter form of uncontrolled seed movement may be exacerbated in some specific cases, such as the overpricing or the market unavailability of varieties perceived by farmers to be superior. These events are part of the seed adoption process and the rates and patterns of these phenomena require sociological and economic analysis and cannot be predicted using plant biology parameters only. The intentional movement of seeds by humans was not considered in our analyses, even though it may increase the probability of gene flow.

Gene flow may result in detrimental effects on natural plant and animal biodiversity. In addition, transgene flow could have additional specific impacts on the environment (e.g. development of insect resistance in the case of insect-resistant varieties), on agronomic practices (e.g. disturbance of insect pest management strategies or threats to organic farming), or socio-economic conditions (e.g. loss of revenue due to the difficulty in segregating conventional and transgenic products). All of these effects can be considered in an assessment of gene flow and its consequences.

The likelihood of gene flow will depend largely on a number of factors, such as the geographical overlap between the donor and recipient populations, the sexual compatibility between donor and recipient plants, the nature and level of expression of the transgene, as well as the biology of the recipient population. To evaluate the likelihood of gene flow and its potential impact, it is thus important to identify the recipient plant species and populations. To this end, potential recipient populations were classified into groups consisting of wild relatives (Malvaceae) and species (*Gossypium*) of cotton, unattended populations (including feral and some dooryard), volunteers and traditional and improved varieties (see Box 11.2).

11.2. Cotton and Related Species in Vietnam

The genus *Gossypium*, to which the potential donor for gene flow belongs (Upland cotton – *G. hirsutum*), is a member of the Malvaceae family. The genus *Gossypium*, as well as other genera from the same family, such as *Abelmoschus*, *Abutilon*, *Althea*, *Cenocentrum*, *Decaschistia*, *Hibiscus*, *Kydia*, *Lavatera*, *Malva*, *Malvastrum*, *Malvaviscus*, *Pavonia*, *Sida*, *Thespesia*,

---

Box 11.2. Definitions of plant populations that may be recipients of gene flow from related transgenic crops

---

Dooryard cotton: cotton that is grown near homes for self-consumption.
Feral: unattended sexual or vegetative progeny that are derived from crop plants and have established populations.
Improved variety: commercial varieties that are the result of a formal breeding programme, which may be a protected variety.
Traditional variety: traditional varieties of cotton that are cultivated by small farmers and the seeds are saved.
Unattended: plants that largely are unattended by human cultivation. These include feral and some dooryard populations.
Volunteer: progeny from crop plants that arise in farmers' fields, either from seeds or vegetative propagules.
Wild: free-living populations not of crop origin.

Urena and Wissadula have been described in Vietnam (Hoàng, 1983; Phạm, 1999, pp. 516–532; Nguyễn and Đỗ, 2003; Đỗ and Nguyễn, 2004). In a number of areas around the world, some of these species are found in cotton fields and are considered weeds (e.g. Sida spp., see Table 2.2 in Chapter 2, this volume). Despite the fact that they belong to the same family, the great evolutionary distance, the difference in chromosome number and the differ-ence in reproductive habit make it extremely improbable that hybridization between any Gossypium species and these other genera could occur. None of these genera are considered part of the G. hirsutum gene pool. Thus, even in the absence of any known studies assessing such a possibility, it is reasonable to conclude that such hybridization is not possible. Therefore, non-Gossypium Malvaceae will not be considered further in this chapter.

Vietnam is not a centre of origin or diversity for Gossypium species and there are no truly wild species of cotton. In addition to G. hirsutum, three other Gossypium species are present in Vietnam (taxonomy follows Hutchinson et al., 1947; Hutchinson, 1950; Fryxell, 1979, 1992): G. arboreum (with three subspecies or types: ssp. neglectum, ssp. sanguineum and ssp. nan-king), G. herbaceum and G. barbadense (with two varieties: G. barbadense race barbadense and G. barbadense race brasiliensis [= G. barbadense var. acuminatum]). These four species can be found in different types of popula-tions, from unattended populations (feral and dooryard cotton) to improved varieties (Table 11.1). Roots, stems, leaves, fibres and seeds of most of the Gossypium species are used as traditional medicines in Vietnam (Phạm, 1999, pp. 530–531; Nguyễn and Đỗ, 2003; Đỗ et al., 2004, pp. 254–257).

Gossypium arboreum

G. arboreum (diploid species, AA) was the first species to be introduced into Vietnam, probably more than 2000 years ago from India (Vũ, 1962). It was still grown widely in the 18th and 19th centuries in a number of cotton growing regions of Vietnam (Tôn, 1974; Vũ, 1978;), before gradually being replaced by G. hirsutum varieties. Until the 1990s, G. arboreum varieties were still popu-lar in some regions, mainly in the northern regions and the northern coastal lowlands in Thanh Hoa and Nghe An. At the end of the 1980s, over half the cotton area in the north of Vietnam (> 10,000 ha) was under G. arboreum (Napompeth, 1987) and it is still grown by ethnic minority groups in the high-lands of north-west Vietnam.

Vietnamese reports mention local varieties, derived mainly from G. arboreum ssp. nanking and to a lesser extent from G. arboreum ssp. neglectum in north-ern and north-western Vietnam, that were selected and passed on by farmers as traditional varieties or landraces, including highland and lowland varieties (Vũ, 1962, 1971, 1978; Tôn, 1974). Highland varieties are medium- or long-season varieties and include a variety with brown fibre. They are taller and have larger bolls and more monopodial branches than the lowland varieties. Thanh Hoa variety, commonly seen in Thanh Hoa Province, is tolerant to drought and has long fibre. Nghe An variety has bolls that quickly fall to the ground after boll

Table 11.1. Occurrence of potential recipient populations (X) of main cotton relatives in Vietnam by geographic region.

Region	G. hirsutum				G. barbadense				G. arboreum			
	Unattended	Volunteer	Traditional	Improved	Unattended	Volunteer	Traditional	Improved	Unattended	Volunteer	Traditional	Improved
South-east				×				×				
Central highlands				×				×		×	×	
Central coast	×	×		×		×		×			×	×
Northern region		×		×		×					×	×

opening. Bac Ai variety is very popular in Ninh Thuan Province. Quynh Coi variety has the longest fibre (23–24 mm), few monopodial branches and is divided into types with violet or yellow branches and leaves. Improved varieties also exist and have the advantage of tolerance of high humidity and rainfall and resistance to leafhoppers *Amrasca devastans* Distant (Homoptera: Cicadellidae), though the yield is low and fibre length is short. Perennial *G. arboreum* cotton still can be found as individual plants or small groups of plants growing unattended along roadsides or as hedges. Even though this species has attracted less attention in modern breeding programmes, its long history in Vietnam makes *G. arboreum* a potentially valuable source of adaptive traits. Furthermore, subspontaneous forms of *G. arboreum* might occur in Vietnam, which would represent additional valuable germplasm. Currently, the Nhaho Research Institute for Cotton and Agricultural Development maintains only six accessions of *G. arboreum* of Vietnamese origin (Table 11.2).

Gossypium herbaceum

G. herbaceum (diploid species, AA) was introduced to Vietnam from South Africa but, unlike *G. arboreum*, has never been grown widely. It is used as a medicinal plant. It crosses readily with *G. arboreum* and may be the sister taxon to *G. arboreum* (Desai *et al.*, 2006).

Gossypium barbadense

G. barbadense race *barbadense* (allotetraploid, AADD) was introduced into Vietnam through seed exchange programmes in the 20th century. Introduced varieties included Ghiza and Pima varieties in southern Vietnam in the early 20th century and varieties such as Truong Nhung, Menufi and Tien Vot in northern Vietnam in the 1950s; cultivation was successful only in the dry season (Vũ, 1962, 1971, 1978; Tôn, 1974). The varieties that were grown

Table 11.2. Cotton accessions maintained at the Nhaho Research Institute for Cotton and Agricultural Development in 2003.

Species	Number of accessions	Origin					
		Vietnam	India	USA	China	France	Others
Gossypium hirsutum	1840	50	265	218	120	81	1106
Gossypium arboreum	53	6	45	–	–	–	2
Gossypium barbadense	67	1	10	2	–	–	44
Total	1960	57	320	220	120	81	1162

appear to be reselections of introduced material that was adapted to local conditions. At present, *G. barbadense* is not grown in Vietnam, but it is used as breeding material in improvement programmes (Table 11.1). However, small, unattended perennial populations of *G. barbadense* growing as dooryard cottons, hedges and feral populations in uncultivated fields in the central and southern provinces of Vietnam have been reported. In view of its recent introduction in Vietnam and the little effort to adapt it to local conditions, *G. barbadense* in Vietnam probably does not represent unique germplasm of great value. On the other hand, it may act as a genetic bridge to other recipient populations of *G. hirsutum* (Table 11.1) because it is found commonly in various regions of Vietnam as unattended populations and it hybridizes readily with *G. hirsutum* (Stephens, 1967).

   G. barbadense race *brasiliensis* (allotetraploid, AADD) has been reported from Vietnam and one accession was reported from Quang Nam Province in the central coastal lowlands region. This variety has never been distributed widely in Vietnam and is not used in breeding programmes, so it is unlikely to act even as a genetic bridge in Vietnam.

Gossypium hirsutum

G. hirsutum (tetraploid, AADD) varieties were introduced into Vietnam in the late 19th to early 20th century and gradually have replaced *G. arboreum* in most cotton growing regions of the country. *G. hirsutum* varieties grown until the 1970s originate mostly from Cambodian varieties. The Cambodian varieties are a distinct South and South-east Asian group within the *G. hirsutum* race *latifolium* group (Hutchinson, 1951; Fryxell, 1968). Local varieties planted widely in southern Vietnam included Khlay, Phu Yen, Batri, Dat, Ba Ria and Baghe cotton. Of these, Baghe cotton had the highest yield potential. In northern Vietnam, popular varieties included Quay Vit in Son La Province, Nam Dan, Quynh Luu, Van Chan in Lang Son Province and Trung Khanh in Cao Bang Province (Vũ, 1962, 1971, 1978; Tôn, 1974; Napompeth, 1987). These varieties are long-season, high-branching varieties with hairy leaves, good resistance to leafhoppers and low boll production.

   Modern varieties of *G. hirsutum* race *latifolium* were then introduced from a number of countries including France, the USA, China, Australia, India and the former Soviet Union, and were grown as imported varieties or as local selections thereof (Table 11.2). More recently, in the 1990s, breeding efforts were undertaken by the Nhaho Research Institute for Cotton and Agricultural Development to increase yields and expand cultivated areas. This effort culminated with the development of intraspecific hybrids that currently represent up to 95% of the area cultivated with cotton for commercial production. In addition to these improved varieties, *G. hirsutum* still can be found as unattended feral populations with a perennial habit along roadsides, as dooryard cotton, or as volunteers in uncultivated fields, mainly in the central coastal region (Table 11.1). As in the case of *G. barbadense*, *G. hirsutum* in Vietnam probably does not represent unique germplasm. Nevertheless, it must be mentioned that the collection of accessions has made it possible to preserve material that was used 100 years ago (Table 11.2).

11.3. Likelihood of Gene Flow and Introgression

For pollen-mediated gene flow from cotton to occur, two conditions must be met. First, the donor and recipient populations should be close enough to allow cross-pollination and, second, these cross-pollination events must lead to the formation of fertile hybrids. Gene flow associated with seed movement by humans could occur during transport of harvested bolls for delinting, which could affect roadside populations.

The first condition requires that transgenic *G. hirsutum* varieties are in close geographic proximity with other *Gossypium* species present in Vietnam to allow insect pollinators to cross-pollinate (Table 11.3). Transgenic Bt cotton is planned for introduction with the aim of increasing cotton productivity in Vietnam (Chapter 1, this volume). It is very likely that transgenic and improved conventional (non-transgenic) *G. hirsutum* varieties will be found in close proximity in all the cotton growing regions. Similarly, transgenic cotton will be in close contact with volunteer *G. hirsutum* and *G. barbadense* race *barbadense* populations that have been documented in the central coastal lowlands and the northern regions (Table 11.1). Contact with *G. barbadense* race *brasiliensis* is much less likely than contact with *G. barbadense* race *barbadense*, because the former is rarer.

Presently, *G. arboreum* and *G. hirsutum* have little geographic overlap. *G. arboreum* is grown mainly in the northern region of Vietnam where *G. hirsutum* is not cultivated widely (see Chapter 2, this volume). The Vietnam Cotton Company plans to introduce Bt cotton in the central highlands, where *G. arboreum* is not grown. It is possible that there are small, unattended populations of *G. arboreum* remaining in the central coastal lowlands region and in the northern coastal lowlands (Thanh Hoa and Nghe An Provinces) where *G. arboreum* traditionally was cultivated in the past, though there are no reports of such occurrence. Thus, geographic overlap of transgenic *G. hirsutum* and *G. arboreum* may be restricted to small areas in central and northern Vietnam.

Another aspect to consider is the possibility of cultivated transgenic cotton varieties being able to establish unattended populations via seed movement. Indeed, if established, these unattended populations could act as a reservoir for the transgene, thereby possibly aiding in its spread. The intensive land use in agricultural areas in Vietnam minimizes the chances of feral or volunteer populations establishing, even along roadsides (Nguyễn Hữu Hồ, Ho Chi Minh City, 2004, personal communication). However, reports of small, unattended cotton populations that could have originated from cotton crops do exist, particularly in the central coastal region (Table 11.1), which indicates that it is possible for small unattended populations to persist.

The second condition requires sexual compatibility and fertile hybrid formation between *G. hirsutum* and the recipient *Gossypium* species (Table 11.3). Only *G. barbadense* and *G. hirsutum* are tetraploid and likely to cross easily

Table 11.3. Factors of the main recipient populations that affect the likelihood of gene flow.

What factors affect the likelihood of gene flow?	Non-Bt *G. hirsutum*	*G. barbadense* race *barbadense*	*G. arboreum*
Will GM crop overlap with recipient populations?	Overlap likely in all cotton producing regions	Overlap likely in the south-east and in the central coast	Overlap possible in the northern and central regions
To what extent and over what distance does the recipient population outcross?	Outcrossing estimate is 2–10%. More insects may result in higher outcrossing; wind and plant density may also have an effect. A 'safe distance' to prevent outcrossing is 500–1000 m	Outcrossing estimate is 5–30%. More insects may result in higher outcrossing; wind and plant density may also have an effect. A 'safe distance' to prevent outcrossing is 500–1000 m	Outcrossing estimate is 5–30%. More insects may result in higher outcrossing; wind and plant density may also have an effect. A 'safe distance' to prevent outcrossing is 500–1000 m
Are volunteers or natural plants considered weeds?	No	No	No
Do hand crosses with Bt cotton result in fertile hybrids?	Yes	Yes	No
Does spontaneous hybridization occur in greenhouse or nature?	Yes	Yes	Yes
Does direction of pollination affect the success of crossing?	No	Not known?	No
Are F$_1$ hybrids with *G. hirsutum* viable and vigorous?	Yes	Yes, when they form successfully	Rarely
Do F$_1$ hybrids reproduce successfully?	Yes	Yes, but with slightly lower fitness compared to either parent	No. There is an almost zero chance of reproduction
Are backcross hybrids vigorous and reproductively fit?	Yes	Yes, advanced generation hybrids will tend to back-cross and regain a similar form to one parent or the other	No
Can hybrids have seed dormancy or form seed banks?	No	No	No

Continued

Table 11.3. *Continued*

What factors affect the likelihood of gene flow?	Non-Bt *G. hirsutum*	*G. barbadense* race *barbadense*	*G. arboreum*
Do crop plants volunteer in years following cultivation? Can they establish feral populations?	Yes, they may volunteer. Feral populations are seen infrequently and are poorly documented	There is not much active cultivation of this species, but unattended plants may persist for many years	Yes, they may volunteer. Feral populations are seen infrequently and are poorly documented
Is the transgene likely to persist in the agricultural environment for several years in feral or hybrid populations?	Yes. Introgressed Bt genes might increase yield. However, farmers may convert to Bt cotton quickly, reducing the opportunity for this type of gene flow	Yes, but less likely than for *G. hirsutum*	Unlikely
Could transgenes spread clonally?	No	No	No
Are feral or natural plants considered weeds?	No	No	No

with a transgenic *G. hirsutum* variety under field conditions (Shepherd, 1974; Wang *et al.*, 1995). Based on expertise available from cotton breeders, we understand that F₁ hybrids and backcross introgressants will have a similar fitness to the cultivated varieties if transgenic cotton crosses with other varieties of *G. hirsutum*, and a somewhat lower fitness in crosses with *G. barbadense* (Stephens, 1946, 1950, 1967).

In contrast, *G. arboreum* and *G. herbaceum* are diploid species and do not cross easily with *G. hirsutum* (Desai *et al.*, 2006). They are part of the secondary *Gossypium* gene pool (Percival *et al.*, 1999). Nevertheless, although the majority of F₁ hybrids will be sterile, a few rare, fertile hybrids may be obtained because the chromosome structure of *G. arboreum* (A genome) is similar enough to the A subgenome of *G. hirsutum* (AD genome). The fate of these rare, fertile F₁ hybrids will be determined by the likelihood of introgression to the *G. arboreum* population, which is nearly zero (Percival *et al.*, 1999). In summary, because the geographic overlap of *G. arboreum* and *G. hirsutum* in Vietnam is low and hybridization and introgression is unlikely, gene flow and introgression from *G. hirsutum* to *G. arboreum* is highly unlikely. Similarly, *G. hirsutum* is highly unlikely to hybridize and introgress into *G. herbaceum* in Vietnam.

Likelihood of gene flow – unknowns

The preceding analysis identifies several unknowns regarding the likelihood of gene flow from transgenic cotton in Vietnam. First, the presence, identity and distribution of unattended populations and traditional varieties are not well documented in Vietnam. It is likely that published surveys of cotton (*G. hirsutum* and *G. barbadense* race *barbadense*) have not recorded small, unattended populations, so a potentially important recipient of gene flow is poorly documented. Second, there are few data available on the pollination biology of cottons in Vietnam. It is necessary to extrapolate from studies in other countries to estimate outcrossing rates, pollination distances and types of pollinators likely to visit *Gossypium* species in Vietnam. Finally, the effects of intercropping on gene flow, a practice that is common in cotton production in some parts of Vietnam, are unknown. It may be possible that intercropping reduces the likelihood of gene flow through physical interference with pollinators, but too little information is available to discuss the possibility.

11.4. Likelihood of Transgene Establishment and Spread

If gene flow occurs, the rate and extent to which introgressants will establish and spread will depend on the fitness of the F_1 hybrid, the fitness imparted by the particular transgene and the geographic extent and frequency of use of transgenic cotton (Haygood *et al.*, 2003; Andow and Zwahlen, 2006). If the transgene increases the fitness of the introgressed plant, establishment and spread can be rapid. On the contrary, if the introgressed transgene results in neutral or lower fitness, establishment and spread is unlikely, unless the geographic extent of use of transgenic cotton is very large compared to the recipient population.

The fitness imparted by an introgressed insect resistance transgene in a recipient population will be determined largely by the level of pest suppression it provides to individuals in the recipient population. This, in turn, will be determined by the level of transgene expression and level of insect herbivore pressure that can be alleviated by the transgene in the recipient population. As this insect herbivore pressure increases, the fitness benefit due to the transgene will increase and higher expression levels should provide greater suppression of this pressure. The wider the outcrossing event, the less predictable transgene expression will be. Without the possibility of testing every outcrossing event, we assumed that the level of transgene expression would be similar to or lower than that in the original transgenic variety. The analysis for the different recipient populations for transgene establishment and spread is shown in Table 11.4.

Crosses between transgenic cotton (*G. hirsutum*) varieties and non-transgenic improved *G. hirsutum* varieties will result in viable offspring with similar fitness as the parental populations. The level of expression of the introgressed transgene may vary, but it is likely similar to or perhaps slightly lower than that in the donor transgenic variety. Herbivore pressure on recipient *G. hirsutum* populations with an introgressed insect resistance transgene will

Table 11.4. Likelihood of transgene spread and establishment. NA = not applicable.

What is the likelihood a Bt transgene from cotton would spread following gene flow from crops?	Hybrids with Non-Bt *G. hirsutum*	Hybrids with *G. barbadense* race *barbadense*	Hybrids with *G. arboreum*
Will introgressed transgene increase yield, making farmers more likely to select for it?	Unlikely. Seed saving is not practised much in Vietnam	No. These are unattended populations, which are not selected by farmers	N/A
Is transgene stably inherited over several generations?	Yes, highly likely	Harder to predict with wider cross	No
Are seeds or vegetative prop-agules dispersed from the crop naturally or by human activity?	Yes, by human activity and possibly by birds or small mammals	Yes, by human activity and possibly by birds or small mammals	N/A
Is the transgene likely to spread and persist in 'natural' areas beyond the agricultural setting?	No, only in fairly disturbed areas, such as field edges or roadsides	Possibly in the unattended populations	No
What are the fitness effects of the transgene in hybrids? Can they be tested empirically?	Not known. Likely to be similar to fitness effect in crop. Fitness advantage seen only if target insect pressure is present	Not known. Likely to be similar to fitness effect in crop. Fitness advantage seen only if target insect pressure is present	N/A
What ecological factors have the greatest effect on fitness components of the wild popula-tion and how will the transgene interact with these factors?	Not known	Not known	Not known
Is there a fitness cost associated with the transgene in the absence of selective pressure?	Not known	Not known	Not known

Continued

Table 11.4. *Continued*

What is the likelihood a Bt transgene from cotton would spread following gene flow from crops?	Hybrids with Non-Bt *G. hirsutum*	Hybrids with *G. barbadense* race *barbadense*	Hybrids with *G. arboreum*
Could hybrid seed have dormancy or establish seed banks?	No	No	No
Do hybrid and feral populations have mechanisms for dispersing seeds further than cultivated varieties?	No	No, may have different opportunity for dispersal, due to different environment	N/A
Could populations invade new habitats after transgene introgression?	No	No	No
Could spread of hybrid or feral populations act as a 'genetic bridge' by bringing the transgene in contact with additional compatible relatives?	Not likely	Possible, but no recipient relatives except *G. hirsutum* to receive the transgene	No
What is the likelihood a Bt transgene from cotton would spread following gene flow from crops?	Spread after introgression into *G. hirsutum* crops could happen. Farmer conversion to Bt likely to happen more quickly	It may spread if insect pressure is high	Highly unlikely

be similar to that of the transgenic cotton. It thus seems likely that recipient individuals with the insect resistance transgene would experience a significant fitness and/or yield benefit, which could result in transgene spread in the recipient population. This could be exacerbated if cotton farmers selected as seed parents individual plants in their fields with enhanced fitness or yield. In Vietnam, however, farmer selection may be unlikely because the Vietnam Cotton Company (VCC) occupies a central role in seed distribution and in seed cotton purchase, and ginning facilities are centralized.

Crosses between transgenic varieties and unattended feral, dooryard, or volunteer populations of G. *hirsutum* and G. *barbadense* race *barbadense* are expected to occur with some frequency and F_1 hybrids are expected to be vigorous and fairly fit (Stephens and Phillips, 1972). Introgressed transgene expression may be variable, but likely similar to or somewhat lower than levels in the original donor transgenic cotton. Herbivore pressure on the unattended populations may be lower than on cotton crops due to the low density of plants in unattended populations and the high diversity of plant species found in uncultivated areas. Thus, if an insect resistance transgene establishes in such populations, the plants likely will gain little fitness benefit and the likelihood of transgene spread is lower than for crop recipients. Nevertheless, these populations could act as bridges for transgene flow.

Establishment and spread – unknowns and other considerations

Some data gaps were identified. Uncertainty in the existence, identity and location of unattended *Gossypium* populations does not allow for a complete analysis (Table 11.2). Furthermore, data on the fitness effect of an insect resistance transgene on hybrids in unattended populations outside an agricultural environment are not available, nor are data on herbivory due to insect pests in such environmental settings. A fitness effect will influence the potential spread of the transgene strongly once it is established in these cotton populations.

As each possible avenue of transgene spread and establishment was evaluated, estimates of the relative effects of intentional human spread of transgenic cotton were made. Insect-resistant cotton varieties are expected to result in improved economic returns and will be adopted quickly, if and when they are introduced into Vietnam. This rapid adoption of transgenic varieties could result in the intentional spread of the transgene that most likely would outweigh the spread of the transgene due to other means of gene flow.

11.5. Potential Ecological Consequences of Gene Flow

The potential loss of unique germplasm from gene flow could be detrimental to breeders, farmers and society as a whole. In the case of Vietnam, the tetraploid cottons (G. *hirsutum* and G. *barbadense*) could hybridize with cultivated transgenic varieties. G. *barbadense* has a short history in Vietnam, having been introduced into the country in the mid-20th century. It is no longer cultivated, but unattended populations have been described as dooryard and feral populations. Due to its recent introduction, and thus to the small possibility of it having accumulated a significant level of genetic variability, G. *barbadense* populations present in Vietnam are highly unlikely to represent unique germplasm that would be at risk from gene flow.

Similarly, G. *hirsutum* has been present in Vietnam for a fairly short period of time. It was grown as introduced varieties or as reselections thereof. It was only fairly recently that locally bred varieties and hybrids were released. Furthermore, this cotton has been grown in a small area and a small range of environments in Vietnam, which limits the pressures that could generate and maintain unique genetic variation. It is thus also unlikely that unique genetic variability is present within this species in Vietnam. The collection and preservation of material adapted to local conditions has been undertaken. However, it still might be useful to conduct more surveys to verify that unattended populations of *G. hirsutum* do not have unique genetic variation.

Although *G. arboreum* is likely to harbour substantial unique genetic variation for traits such as tolerance to abiotic conditions or disease organisms in Vietnam, loss of this unique genetic diversity is unlikely. While some traditional varieties of *G. arboreum* are grown in areas where transgenic cotton may be grown, the species is genetically incompatible with tetraploid *Gossypium* species.

It is uncertain if *G. herbaceum* has unique genetic diversity in Vietnam. Similar to *G. arboreum*, *G. herbaceum* has been in Vietnam for a long time, but is genetically incompatible with tetraploid cotton. Thus, this species is not at risk of loss of unique genetic diversity.

Gene flow and introgression are likely only for the more common tetraploid species in Vietnam, *G. hirsutum* and *G. barbadense* race *barbadense*. Although a transgenic insect resistance trait may increase the relative fitness of recipient cultivated or unattended populations, it is unlikely to increase their weediness. The life history traits of these species, such as low seed shedding and dispersal, absence of dormancy in local environmental conditions and absence of vegetative propagation, minimize this risk. Furthermore, even if these species are protected from insect herbivore damage and they have increased relative fitness, they are unlikely to become weedier. They have not been weedy in the past, so increased relative fitness will not increase their population size. An insect resistance gene would have to enable the unattended populations to increase significantly in competitiveness and population size before weediness is increased. If the transgene under consideration had been a herbicide tolerance gene, it might become a more problematic weed in agricultural settings where that herbicide is used to control weeds because its competitiveness and population size would be enhanced by the herbicide.

In view of these considerations, we conclude that it is unlikely that gene flow from insect-resistant transgenic cotton varieties will worsen or create a weed problem for farmers in Vietnam.

Gene flow can move transgenes out of the agricultural environment, increasing the numbers and diversity of non-target organisms that come into direct or indirect contact with the transgene and its products. In the case of transgenic cotton in Vietnam, gene flow is likely to occur to any unattended *G. hirsutum* or *G. barbadense* race *barbadense* population, but these populations would each be quite small, made up of only one or a few individual plants. The small population size of these recipients minimizes the risk to non-target organisms and biodiversity outside cotton fields. Indeed, the predicted rapid adoption of insect-resistant cotton varieties in Vietnam is much more likely to affect non-target organisms than would occur via gene flow.

Another possible consequence of gene flow is its effect on insect resistance management. To slow the evolution of insect resistance, it is recommended that sufficient areas of non-transgenic cotton or alternative host plants occur nearby to serve as a refuge for susceptibility in the target insect populations (see Chapter 12, this volume). Gene flow could contaminate conventional cotton refuges, thus reducing the effectiveness of resistance management and speeding up the evolution of insect resistance. On the other hand, cotton often is cultivated in intercropping and relay cropping systems (see Chapter 2, this volume), and some of these associated crops may be effective refuges (see Chapter 12, this volume). In this case, gene flow would have no impact on resistance management.

Potential socio-economic impacts of transgene flow also should be addressed. These can be difficult to assess, but should not be overlooked. One example of such a concern is the possibility of transgene flow from transgenic cotton varieties to a neighbouring farmer's cotton that was targeted for a GMO-free market (e.g. organic cotton). Presently, Vietnam is a net importer of cotton fibre, local production is used domestically and there are no speciality markets for fibres in Vietnam. If an organic cotton market were to develop, then this could become a significant societal issue.

In addition, Vietnam exports cottonseed oil and delinted cottonseed for the crushing industry. Thus, the market for cotton by-products, such as cottonseed oil, cottonseed meal and cottonseed flour, may be affected if importing countries impose limitations or regulations on GM-containing products. Additional costs associated with the segregation of GM-containing and conventional cottonseed products could be expensive and sometimes difficult to implement fully. Compared to the value of the cotton lint, these markets are small and alternative seed products and uses may be developed to replace these export markets, if necessary.

11.6. Conclusions and Caveats

An assessment, based on a general framework (Box 11.1), of the potential impact of gene flow was carried out, taking as a case study the introduction in Vietnam of insect-resistant transgenic cotton (*G. hirsutum*) varieties. Such an analysis must take into account the crop species biology and the transgene trait in the Vietnamese environment.

Based on the available data, the overall conclusion of this particular case study is that gene flow from insect-resistant transgenic cotton has a very low likelihood of any adverse ecological impact in Vietnam. A corollary is that adverse environmental effects of gene flow most likely will be outweighed by other possible effects related to the expected rapid adoption of transgenic cotton in Vietnam.

Vietnam is not a centre of origin or diversity for *Gossypium*. There are no wild cotton species in Vietnam and none of the related Malvaceae is sexually compatible with *G. hirsutum*. The local tetraploid cotton germplasm consists of cultivars that were imported recently, and any unattended populations are derived from these recent introductions. Thus, unique tetraploid germplasm is not likely to be present in Vietnam. Diploid *Gossypium* species may harbour unique germplasm. *G. arboreum*, which was introduced into Vietnam more than 2000 years ago, probably represents valuable germplasm with specific adaptive traits that are worthwhile preserving and *G. herbaceum* may have some valuable germplasm. The existence, characteristics and distribution of such potentially unique material are poorly documented but, in any case, diploids mostly are genetically incompatible with the transgenic tetraploids and hybrid formation and introgression into the diploid populations is extremely unlikely.

These conclusions contrast sharply with those drawn about Bt cotton in Brazil (Johnston *et al.*, 2006). Brazil is a centre of origin of cotton and a number of unique landraces and sexually compatible wild species (*G. mustelinum*) are present. The diversity of situations and the resulting diversity of the conclusions of the risk assessments argue strongly for a case-by-case approach when analysing the potential impacts of the introduction of transgenic crops. Such a case-by-case study should take into account not only the crop species, but also the nature of the transgene, as well as the environment in which the transgenic crop is to be introduced.

References

Andow, D.A and Zwahlen, C. (2006) Assessing environmental risks of transgenic plants. *Ecology Letters* 9, 196–214.

Arriola, P.E. and Ellstrand, N.C. (1996) Crop-to-weed gene flow in the genus *Sorghum* (Poaceae): spontaneous interspecific hybridization between Johnsongrass, *Sorghum halepense*, and crop sorghum, *S. bicolor. American Journal of Botany* 83(9), 1153–1160.

Desai, A., Chee, P.W., Rong, J.K., May, O.L. and Paterson, A.H. (2006) Chromosome structural changes in diploid and tetraploid A genomes of *Gossypium. Genome* 49, 336–345.

Đỗ Huy Bích, Đặng Quang Chung, Bùi Xuân Chương, Nguyễn Thượng Dong, Đỗ Chung Đàm, Phạm Văn Hiến, Vũ Ngọc Lộ, Phạm Duy Mai, Phạm Kim Mạn, Đoan Thị Nhu, Nguyễn Tập and Tran Toan (2004). *Cây Thuốc Và Động Vật Làm Thuốc ở Việt Nam*. (The medicinal plants and the animals used as medicines in Vietnam). Publishing House Khoa Hoc Ky Thuật, Hanoi (in Vietnamese).

Đỗ Thị Xuyến and Nguyễn Khắc Khôi (2004) Một số kết quả nghiên cứu chi Bông – *Gossypium* L. (họ Bông – *Malvaceae* Juss.) ở Việt Nam. (Some research results on genus *Gossipium* L. (family *Malvaceae* Juss.) in Vietnam). *Tạp Chí Sinh Học (Journal of Biology)* 4(26), 61–63 (in Vietnamese).

Fryxell, P.A. (1968) The typification and application of the Linnaean binomials in *Gossypium. Brittonia* 20(4), 378–386.

Fryxell, P.A. (1979) *The Natural History of the Cotton Tribe*. Texas A&M University Press, College Station, Texas.

Fryxell, P.A. (1992) A revised taxonomic interpretation of *Gossypium* L. (Malvaceae). *Rheedea* 2(2), 108–165.

Harlan, J.R. and de Wet, J.M.J. (1971) Toward a rational classification of cultivated plants. *Taxon* 20, 509–517.

Haygood, R., Ives, A.R. and Andow, D.A. (2003) Consequences of recurrent gene flow from crops to wild relatives. *Proceedings of the Royal Society of London, Series B* 270, 1879–1886.

Hoàng Đức Phương (1983) *Cây Bông* (Cotton). Publishing House Nông Nghiệp, Hanoi (in Vietnamese).

Hutchinson, J.B. (1950) A note on some geographical races of Asiatic cottons. *Empire Cotton Growing Review* 27, 123–127.

Hutchinson, J.B. (1951) Intra-specific differentiation in *Gossypium hirsutum*. *Heredity* 5(2), 161–193.

Hutchinson, J.B., Silow, R.A. and Stephens, S.G. (1947) *The Evolution of Gossypium and the Differentiation of the Cultivated Cottons*. Oxford University Press, Oxford, UK.

Johnston, J., Blancas, L. and Borem, A. (2004) Gene flow and its consequences: a case study of Bt maize in Kenya. In: Hilbeck, A. and Andow, D.A. (eds) *Environmental Risk Assessment of Genetically Modified Organisms, Volume 1: A Case Study of Bt Maize in Kenya*. CAB International, Wallingford, UK, pp. 187–208.

Johnston, J.A., Mallory-Smith, C., Brubaker, C.L., Gandara, F., Aragão, F.J.L., Barroso, P.A.V.,Vu Duc Quang, Carvalho, L.P. de, Kageyama, P., Ciampi, A.Y., Fuzatto, M., Cirino, V. and Freire, E. (2006) Assessing gene flow from Bt cotton in Brazil and its possible consequences. In: Hilbeck, A., Andow, D.A. and Fontes, E.M.G. (eds) *Environmental Risk Assessment of Genetically Modified Organisms, Volume 2: Methodologies for Assessing Bt Cotton in Brazil*. CAB International, Wallingford, UK, pp. 261–299.

Letourneau, D.K., Robinson, G.S. and Hagen, J.A. (2003) Bt crops: predicting effects of escaped transgenes on the fitness of wild plants and their herbivores. *Environmental Biosafety Research* 2(4), 219–246.

Napompeth, B. (1987) FAO Consultancy Report on Cotton Integrated Pest Management in Vietnam: Second Mission Report (2 April to 28 May 1987). Project VIE/84/001 – Cotton Improvement and Extension in Vietnam. UNDP/FAO, Hanoi.

Nguyễn Tiến Bân and Đỗ Thị Xuyến (2003) 97. Malvaceae Juss. – Họ Bông (Bụp). (Nr. 97. Family Malvaceae Juss). In: Nguyễn Tiến Bân (ed.) *Danh Lục Các Loài Thực Vật Việt Nam*. ('Checklist of plant species of Vietnam'). Vol. II. Publishing House Nông Nghiệp, Hanoi, pp. 556–569.

NRC (National Research Council) (1989) *Field Testing Genetically Modified Organisms: Framework for Decision*. National Academy Press, Washington, DC.

Percival, A.E., Wendel, J.F. and Stewart, J.M. (1999) Taxonomy and germplasm resources. In: Smith, C.W. and Cothren, J.T. (eds) *Cotton: Origin, History, Technology, and Production*. Wiley & Sons, New York, pp.33–63.

Phạm Hoàng Hộ (1999) *Cây cỏ Việt Nam* (An illustrated flora of Vietnam). Vol. 1. Publishing House Trẻ, Ho Chi Minh City, Vietnam (in Vietnamese).

Shepherd, R.L. (1974) Breeding root-knot resistant *Gossypium hirsutum* L. using a resistant wild *G. barbadense* L. *Crop Science* 14, 687–691.

Snow, A.A., Pilson, D., Rieseberg, L.H., Paulsen, M.J., Pleskac, N., Reagon, M.R., Wolf, D.E. and Selbo, S.M. (2003) A Bt transgene reduces herbivory and enhances fecundity in wild sunflowers. *Ecological Applications* 13, 279–286.

Song, Z.P., Lu, B.R., Wang, B. and Chen, J.K. (2004) Fitness estimation through performance comparison of F-1 hybrids with their parental species *Oryza rufipogon* and *O. sativa*. *Annals of Botany* 93, 311–316.

Stephens, S.G. (1946) The genetics of 'corky' I. The New World alleles and their possible role as an interspecific isolating mechanism. *Journal of Genetics* 47, 150–161.

Stephens, S.G. (1950) The genetics of 'corky' II. Further studies on its genetic basis in relation to the general problem of interspecific isolating mechanisms. *Journal of Genetics* 50, 9–20.

Stephens, S.G. (1967) Evolution under domestication of the New World cottons (*Gossypium* spp.). *Ciência e Cultura* 19, 118–134.

Stephens, S.G. and Phillips, L.L. (1972) The history and geographical distribution of a polymorphic system in New World cottons. *Biotropica* 4, 49–60.

Tôn Thất Trình (1974) *Improving Cotton Production Sector in Vietnam*. Saigon Publisher, Ho Chi Minh City, Vietnam (in Vietnamese).

Vũ Công Hậu (1962) *Cotton in Vietnam.* Agriculture Publishing House, Hanoi (in Vietnamese)

Vũ Công Hậu (1971) *Development of Cotton Planting in Vietnam and Cotton Varieties.* Scientific and Technical Publishing House, Hanoi (in Vietnamese).

Vũ Công Hậu (1978) *Cotton Production Techniques.* Agriculture Publishing House, Hồ Chí Minh City, Vietnam (in Vietnamese).

Wang, G.-L., Dong, J.-M. and Paterson, A.H. (1995) The distribution of *Gossypium hirsutum* chromatin in *G. barbadense* germplasm: molecular analysis of introgressive plant breeding. *Theoretical and Applied Genetics* 91, 1153–1161.

12 Resistance Risk Assessment and Management for Bt Cotton in Vietnam

GARY P. FITT, DAVID A. ANDOW, NGUYỄN HỮU HUÂN, MIKE CAPRIO, CELSO OMOTO, NGUYỄN THƠ, NGUYỄN HỒNG SƠN AND BÙI CÁCH TUYẾN

This chapter addresses the risk that insect pests associated with *Bacillus thuringiensis* (Bt) cotton may evolve resistance to Bt proteins deployed in transgenic cotton varieties in Vietnam. This risk assessment follows a framework developed through the GMO ERA Project, applied initially as a series of concepts and questions to issues of resistance risk for Bt maize in Kenya (Fitt *et al.*, 2004), then integrated more formally into a risk analysis process for the case of Bt cotton in Brazil (Fitt *et al.*, 2006).

Vietnam is considering the introduction of Bt cotton varieties in order to stabilize rainfed production and increase dry season cotton production and yields and to reduce insecticide use. To achieve this, Vietnam may introduce Bt varieties which have been commercialized elsewhere, such as Bollgard II cotton expressing Cry1Ac and Cry2Ab transgenes, owned by Monsanto; or WideStrike cotton with Cry1F and Cry1Ac transgenes, owned by Dow AgroSciences; or Chinese Bt varieties with a fused Cry1Ac/Ab transgene, owned by the Chinese Academy of Agricultural Sciences and licensed to Chinese biotechnology companies. Bt cotton expressing the VIP3A trait owned by Syngenta, but not yet commercialized, may also be considered in early introductions. Vietnam is also developing the capability to use some of these transgene combinations to transform cotton varieties bred and adapted for Vietnamese conditions.

This chapter establishes a series of informational needs that are essential to completing an assessment of resistance risk and the development of a practical resistance management plan for the deployment of Bt cotton. It deals with the specific cases of the introduction of either Bollgard II cotton or a VIP3A cotton to the cropping system of Vietnam. We concentrate on a comprehensive assessment of the pest/plant system and ecological attributes of the pests that help to define the risk of resistance and indicate possible resistance management approaches. Additional research during field-testing may be needed to address key assumptions and develop an effective, workable and acceptable resistance management plan and to establish details of the monitoring and response system.

The evolution of resistance is a real risk. Experience with insecticides and basic consideration of evolutionary theory implies that if the Bt crop is used extensively without any targeted or informed intervention, resistance could be an inevitable consequence. If the outcome of deploying Bt cotton varieties is the development of Bt-resistant pests, it is possible that Vietnam's long-standing commitment to integrated pest management (IPM) approaches could be compromised by renewed application of disruptive pesticides. It is noteworthy, however, that integration of Bt cotton into an IPM context could assist with the preservation of Bt technology.

The real issue with resistance management is how to delay increases in resistance (R) gene frequencies significantly so they remain below the point where field failures may occur. For any given crop, there are usually multiple pest species that require control and any given pest control tactic usually affects multiple pest species. Cotton in Vietnam has many insect pest species, so it is important to assess which species are at risk of resistance and then which of these is most at risk. In this chapter, we identify these most-at-risk species and then devise risk management practices that could delay the onset of resistance in these species and all of the others. As a benchmark, we use the pragmatic goal of seeking to delay resistance evolution for at least 15 years, although, in practice, much longer delays are likely.

To assess the relative resistance risk of a Bt crop, it is necessary to have a list of species that occur on the crop and are susceptible to the Bt proteins in use. Resistance risk can then be assessed by considering:

- the likely 'dose' of the transgenic toxin to which each species is likely to be exposed [influenced by characteristics of the transgene, interactions with plant chemistry and variety, climatic and agronomic factors]
- potential exposure of each species to the dose that may lead to selection in favour of resistance [influenced by association of the species with the crop as opposed to other host plants, generations per crop cycle, other hosts in the farming system, pest mobility and behaviour]

which together allow a determination of pest species at risk of evolving resistance to the transgene.

Dose is a measure of relative fitness of the three possible genotypes associated with resistance evolution. These genotypes are the *RR* homozygotes (with two *R* alleles), the *SS* homozygotes (with two susceptibility, *S*, alleles) and the *RS* heterozygotes (with one of each kind of allele). Dose is a measure of the relative fitness of the *RS* heterozygote relative to the difference between the *RR* and *SS* homozygotes. If the fitness of the *RS* heterozygote is similar to the *RR* homozygote, resistance is said to be dominant and resistance evolution can be extremely fast. If the fitness of the *RS* heterozygote is similar to the *SS* homozygote, resistance is said to be recessive and resistance evolution can be delayed for a long time with the appropriate management. Dose influences the rate of resistance evolution strongly and, coupled with information on potential exposure, we can assess the relative resistance risk of the various pest species and identify the species that is most likely to evolve resistance before the others – this might be the main target for pre-emptive management.

Resistance management first focuses around the biological attributes of this main target or weak link species. We then confirm that the resistance management strategy constructed around the weak link species would also delay resistance evolution in the other species at risk.

While doing this, it is essential that the resistance management plan be practicable; that is, growers actually can implement it. The resistance management plan builds on the information from the risk assessment using the following three steps:

- Determination of the likely requirements for resistance management, including refuges.
- Development of the likely requirements of a potentially workable resistance management plan.
- Specification of monitoring needs and development of potential contingency responses.

Vietnam currently grows only a small area of cotton (fluctuating between 15,000 and 30,000 ha since 2000) distributed across four agroecological regions (Chapter 2, this volume). The area of cotton fluctuates from year to year as farmers choose to plant or not, based on the relative price of cotton compared to other annual crops, especially rice (Chapter 2, this volume). Currently, 90% of production is rainfed, being sown in July during the rainy season (April–November) and harvested early in the dry season (December–March). Irrigated production occurs during the dry season with sowing in November–January, but currently is limited to around 1500 ha in the coastal lowlands region, mainly in the southern provinces of Ninh Thuan and Binh Thuan, and in the central coastal lowlands from Thua Thien Hue to Phu Yen. Current plans from the Vietnam Cotton Company and the Vietnamese government are to increase *irrigated production* substantially (perhaps tenfold) while maintaining rainfed production at present or slightly higher levels.

Farmers use mainly neonicotinoid pesticides and the insect growth regulator buprofezin on rainy season cotton to control sucking pests and the insect growth regulator lufenuron, abamectin and spinosad to control Lepidoptera (Chapter 2, this volume). Seed treatments of imidacloprid are used for preventative management of aphids and cotton leafhoppers. On dry season cotton, farmers use pyrethroid insecticides against *Helicoverpa armigera* and other Lepidoptera, as well as the neonicotinoids against sucking pests. In the areas where cotton blue disease (CBD) is prevalent, farmers spray against aphids from early in the season.

The two production systems (rainfed and irrigated) have very different levels of productivity and pest problems. Insect pests in the rainfed system are managed through well-validated IPM approaches, which have been implemented very effectively and largely limit the need for insecticide sprays. The rainfed IPM system relies on the use of pesticide seed treatments [largely the neonicotinoid imidacloprid] to provide an extended period of early season control of aphids, leafhoppers and thrips (although, due to the prevalence of CBD, farmers have now started to spray neonicotinoids early in the season against the aphid vectors); on some host-plant resistance characters for leafhoppers; and on conservation of the high densities of beneficials which typify the system and usually provide sound management of lepidopteran pests. Beneficial insects are generated by the diversity of other crops grown in association with cotton, particularly maize crops, which largely are unsprayed. The effectiveness of IPM in this rainfed cotton system is sustained by the high crop diversity, small field size and understanding of the production system by the farmers.

Irrigated, dry season production is a potentially favourable option because of increased yield and enhanced fibre quality but, in contrast to the rainfed crop, it experiences significantly higher pest densities and requires considerably more pesticide input to protect the higher yield potential. As a result three to four pesticide applications are required for sucking pests and another three to four for *Helicoverpa* and other Lepidoptera. One of the challenges of pest management during the dry season in the coastal lowlands region is to suppress extensive infestations of early season aphids and leafhoppers and to deal with more persistent and significant infestations of both *H. armigera* and *Spodoptera exigua*. There is some indication that applications of pesticides to control early season sucking pests and the CBD transmitted by aphids may exacerbate later problems with *Helicoverpa* by disrupting populations of beneficial insects.

The highly divergent character and phenology of these two cotton production systems provide an opportunity to consider deployment of Bt cotton in one system only where the need and potential pay-off is greatest. Because ginning is centralized and there is little seed saving, it may be possible operationally to use Bt cotton during the dry season only. Hence, we focus here on the scenario where use of Bt cotton is restricted to the irrigated, dry season component and assume that rainfed production continues with conventional varieties, supported by effective and improved IPM practices. This restriction of use to one component of cotton production is, in itself, the first element of a resistance management strategy for Bt cotton in Vietnam.

12.1. Potential Adverse Consequences of Resistance

The main potential adverse consequences of resistance are control failures, yield loss and economic hardship when the pest is otherwise difficult to control; increased use of pest management tactics, such as insecticides, that have significant adverse effects on human health and the environment, and reduced management options for growers, which can increase production costs.

In the dry season irrigated situation, we might expect Bt cotton to remove the need for up to four insecticide applications for *Lepidoptera* – an approximately 50% reduction over current usage. It may thus be possible to introduce a more coordinated IPM approach to this system once the disruptive effects of insecticides have been reduced, which may introduce further options for dealing with some of the sucking pests. Resistance evolution would jeopardize these potential advances and may also constrain the expansion of irrigated production. Whether resistance evolution would have adverse consequences for the much larger rainfed cotton industry is uncertain. Given that Bt sprays are little used in rainfed production, there may be few consequences of Bt resistance for the rainfed system, provided the IPM systems continue to be applied and are supported by extension services.

The information presented here and in Chapter 2 reflects the complexity and dynamic nature of current Vietnamese cropping systems and serves to highlight the additional challenges that may be faced during a proposed major expansion of irrigated cotton production. There are numerous gaps in knowledge and specific expertise to support potential deployment of Bt cotton and also limited quantitative understanding of pest ecology in these systems. An additional constraint for any future management strategy for Bt cotton is the small-scale nature of the production system where individual farmers control only a fraction of a hectare and where the capacity to change practices or to coordinate practices over large areas may be limited.

Despite these uncertainties, it is possible to conduct a resistance risk assessment for various Bt transgenic cottons and we believe it is likewise possible to propose an initial resistance management plan based on some realistic future scenarios.

12.2. Resistance Risk Assessment

Definition of resistance

Resistance is caused by genes that reduce susceptibility to a toxin and is a trait of an *individual*. However, it will often happen that resistance is not yet known in a target species at the pre-release stage of development of the transgenic crop. Thus, it is important to define resistance operationally, so that resistance can be looked for in advance. This definition, by necessity, will be modified as information becomes available about the expression and inheritance of resistance. It is discussed in the section on dose and efficacy, below.

In addition, resistance occurs in a field population when there are enough resistant individuals to cause economic damage to the target crop. Hence, it is also necessary that we have an operational definition of *control failure from resistance*; this will be a characteristic of a *population* and should be implemented easily and unambiguously. An operational definition of control failure from resistance is necessary so that we know what we want to avoid during resistance management and we know when we should admit failure and move on. Operationally, a control failure from resistance occurs when the pest causes significant economic damage to the crop. There are several alternative ways to implement this concept. For example, a control failure could be defined as occurring when the pest causes detectable economic damage to the crop, when the pest causes economic damage that is similar to that caused by susceptible insects on a non-resistant crop variety, or when the economic damage is considered unacceptable to the grower.

Identification of pest species at risk

Identification of key pest species that could evolve resistance to Bt cotton involves first identifying the key target pests in each of the major geographic regions and then evaluating the resistance history of each species. In some cases, identification of the key target species can be difficult because the transgenic crop has not been tested against all relevant species. While there is considerable information about the key lepidopteran pests which are known to be susceptible to Bt proteins that may be used in Vietnam and hence can be considered target species, there is almost no quantitative information from Vietnam on the efficacy of these potential Bt cotton varieties to control them in Vietnamese cropping systems.

Relatively few species are regarded as significant pests of both rainfed and irrigated production. These pests are the Lepidoptera – *H. armigera*, *S. exigua*, *Pectinophora gossypiella* – and the sucking pests – aphids (*Aphis gossypii*), leafhopper (*Amrasca devastans*), thrips (*Thrips palmi* and *Scirtothrips dorsalis*). Occasionally, a number of other sucking bugs and mites are also present.

Seven Lepidoptera can be identified as potential targets of Bt cotton in Vietnam (Table 12.1). Based on evidence of relative susceptibility to a range of Bt proteins (Cry1Ac, Cry2Ab, Cry1F, VIP3A) from studies elsewhere, we identify the key target species as *H. armigera*, *S. exigua* and *P. gossypiellu*. The latter species is highly susceptible to Cry1A and Cry2A toxins in particular, while the other two species vary in susceptibility to different proteins. All seven of these lepidopteran species occur across all cotton production regions in Vietnam, although only *H. armigera* causes serious damage in all regions (Table 12.2). *S. exigua* is most prevalent in the coastal lowlands, while *P. gossypiella* is mainly in the central highlands and the northern region, where all production is rainfed.

A prior history of resistance evolution to conventional pesticides can also provide considerable insight into the risks associated with transgenic insecticidal crops. Species with a history of repeated resistance evolution should be prioritized in any risk assessment, since their population ecology, host relationships, genetic structure and behaviour may predispose them to respond rapidly to selection pressure from a Bt crop. Only very limited information is available to assess the past history of resistance in lepidopteran pests of cotton in Vietnam, because there have been only sporadic monitoring programmes in place. Current knowledge is based largely on field failures and anecdotal information, plus results from some monitoring activity. Table 12.2 summarizes the general history of resistance and indicates that *H. armigera* and *S. exigua* have developed resistance to all classes of pesticides available, whereas the remaining five lepidopteran species have no known examples of pesticide resistance. While Bt sprays are used occasionally against *H. armigera* in irrigated cotton (Chapter 2, this volume), there is no indication of Bt resistance from past usage. Thus, of the seven potential target lepidopteran pests, *H. armigera*, *S. exigua* are likely to be at greater risk of resistance evolution than the others. However, we add *P. gossypiella* to this list due to its host specificity on cotton.

Table 12.1. Probable susceptibility of the main lepidopteran pest species in Vietnam to Cry transgenes. × = susceptible, ××× = highly susceptible.

Pest species	Cry1Ac/Ab Chinese Bt cotton	Cry1F and Cry1Ac WideStrike™	VIP3A	Cry1Ac and Cry2Ab Bollgard II™	References
Helicoverpa armigera	××	×	××	×××	Wan *et al.*, 2005; Wu and Guo, 2005; Llewellyn *et al.*, 2007; Luo *et al.*, 2007
Anomis flava	×××	×××	×××	×××	Cui and Xia, 2000
Pectinophora gossypiella	×××	×××	×××	×××	Tabashnik *et al.*, 2002; Haile *et al.*, 2004
Earias vittella	××	×	×	×××	Kranthi *et al.*, 2004
Spodoptera exigua	No data	××	××?	××	Yu *et al.*, 1997; Chitkowski *et al.*, 2003; Cloud *et al.*, 2004; Haile *et al.*, 2004
Spodoptera litura	No data	××	××?	××	Selvapandiyan *et al.*, 2001; Guo *et al.*, 2003; Zhang *et al.*, 2006
Sylepta derogata	×××	×××	×××	×××	Huang and Liu, 2005

Potential exposure of target pests to Bt cotton

Association with Bt cotton

The association of the target species with Bt cotton is the *maximum period of overlap* of the species on the target crop in terms of area, spatial distribution and seasonal availability of the crop. Overlap can be evaluated on the basis of presence/absence and general knowledge about the species. More precise, quantitative evaluations may become necessary to develop realistic resistance management plans (see next section).

The three main species differ markedly in host range and association with cotton. Of the three, *P. gossypiella* has the tightest association with cotton being a specialist on *Gossypium* species. While some other malvaceous hosts may be

Table 12.2. Regional differences in severity of key lepidopteran pest species on cotton and history of resistance to pesticides in Vietnam.

		Cotton production regions				History of resistance to pesticides in Vietnam
	Northern	Central coastal lowlands	Southern coastal lowlands	Central highlands	South-east	
Production system	Rainy season rainfed	Dry season irrigated Dec–May	Mix of rainfed and irrigated	Rainy season rainfed July–Dec	Rainy season rainfed July–Dec	
Area [ha] in 2003/04	2,200	1,450	Rainfed: 3,500 Dry: 166	10,100	3,300	
Helicoverpa armigera	High	High	High	High	High	Resistance to all classes of pesticides used (OCs, OPs, Cbms, SPs)
Spodoptera exigua	Low	High	High	Low	Low	High level resistance to all classes of pesticides used (OCs, Ops, Cbms, SPs)
Anomis flava	Low	Low	Low	Low	Low	Not been detected
Pectinophora gossypiella	High	Low	High	High	Low	Not been detected
Earias vittella	Low	Low	Low	Low	Low	Not been detected
Spodoptera litura	Low	Low	High (rainy) Low (dry)	High	High	Not been detected
Sylepta derogata	Low	Low	Low	Low	Low	Not been detected

Note: OC = organochlorine; OP = organophosphate; Cbm = carbamates; SPs = synthetic pyrethroids.

used, most of the *P. gossypiella* population is associated with cultivated cotton in Vietnam and crop hygiene between seasons is an important management tactic. *P. gossypiella* may complete several generations per year in cotton. This close association suggests that *P. gossypiella* may be exposed to more intense selection in Bt cotton than *H. armigera* or *S. exigua*. However, the lack of evidence for past resistance of *P. gossypiella* to pesticides may suggest the risk is lower. This species is also considerably more susceptible to Bt proteins (Cry1Ac, Cry2Ab and VIP3A) than other two species (Table 12.1). *H. armigera* has a wide range of

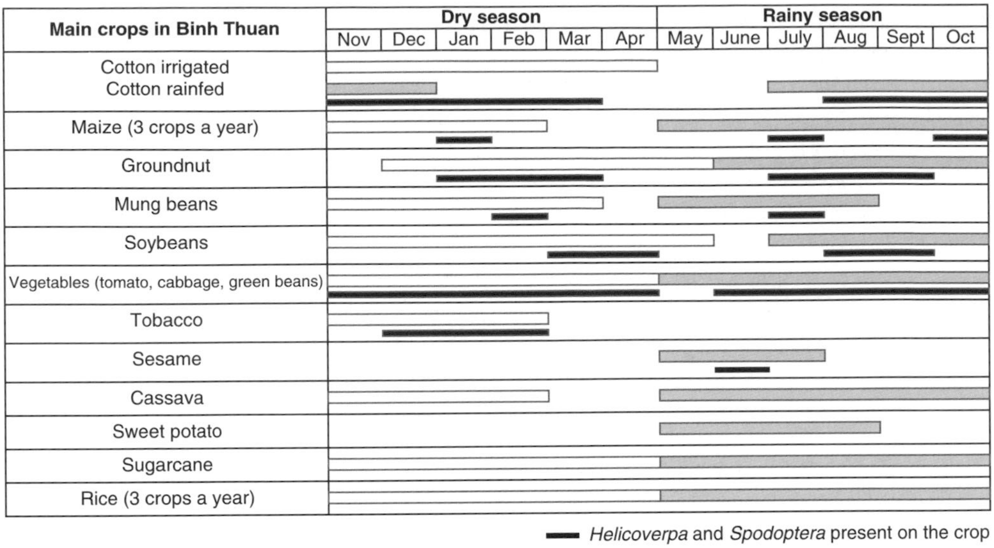

304

Main crops in Binh Thuan	Dry season						Rainy season					
	Nov	Dec	Jan	Feb	Mar	Apr	May	June	July	Aug	Sept	Oct
Cotton irrigated / Cotton rainfed												
Maize (3 crops a year)												
Groundnut												
Mung beans												
Soybeans												
Vegetables (tomato, cabbage, green beans)												
Tobacco												
Sesame												
Cassava												
Sweet potato												
Sugarcane												
Rice (3 crops a year)												

━━ *Helicoverpa* and *Spodoptera* present on the crop

Fig. 12.1. Phenology of major crops in Binh Thuan Province (central coastal lowlands region) of Vietnam. Black horizontal bars indicate the periods when *Helicoverpa* and *Spodoptera* are present on cotton, maize, groundnut, mung beans, soybeans, vegetables, tobacco and sesame. Cassava, sweet potato, sugarcane and rice are non-hosts. Light shading is the phenology during the dry season, dark shading is the phenology during the rainy season.

recorded hosts, including many crops and wild hosts, but in the coastal region where both rainfed and irrigated cotton is grown, it is possible that multiple generations will occur on cotton throughout the year. During the dry season, *H. armigera* may complete three to four generations on cotton and another two to three during the rainy season crop. However, cotton currently makes up only 5% of the crop area in this region (Binh Thuan Province). Many of the other crops (Fig. 12.1), such as maize, groundnut, soybean, mung bean and tobacco, are also suitable host plants and, in some cases (e.g. maize), are preferred more highly than cotton, so the association of *H. armigera* populations with cotton may not be close. Likewise, *S. exigua* has a wide host range incorporating not only the hosts listed above, but also a broad range of vegetable crops (Fig. 12.1), where it is likely exposed to even greater pesticide pressure than in field crops.

Association with Bt toxin on other host plants

While no other Bt crops are deployed currently in Vietnam, Bt sprays are used to some extent in vegetable crops and it is expected that Bt maize and Bt soybean will be commercialized. Hence, *S. exigua* may experience greater exposure to Bt through the vegetable crop production sector. Maize and soybean are important alternative crops for *H. armigera* and maize is a significant crop in the central coastal region of Vietnam, representing 3–5 times the area of cotton, with both rainy season and dry season crops. As a highly preferred host plant for *H. armigera*, and also a host for *S. exigua*, Bt maize would dramatically increase exposure to *Bt toxins* and could compromise the stability of a Bt cotton system severely. Soybean is not cropped as extensively as maize, so it represents a smaller risk. The risk depends largely on the exact proteins deployed. Cotton may be transformed to express VIP3A, Cry1Ac, Cry2Ab, Cry1Ac/Ab, or various combinations. The Cry1 proteins are also likely candidates in maize and soybean. It is anticipated that 50–70% of hybrid maize eventually may be planted to Bt maize expressing Cry1Ab or Cry1F. Because it shares similar binding sites to Cry1Ac, Cry1Ab maize would heighten the risk of resistance significantly through a mosaic of Bt protein exposure and selection in two host plants of *H. armigera* and *S. exigua*, which overlap extensively in time and space (Fig. 12.1). By contrast, Bt rice, which may also be considered for Vietnam, would not provide an added risk to Bt cotton since none of the target pests are common across those two crops. Overall, when the potential of these other Bt crops are considered, resistance risk of *H. armigera* is high.

Scale of adult and larval movement

Adult movement, mating and oviposition will affect exposure among plants in a field and between fields. Estimates of adult female movement should be separated into pre-mating and post-mating movement, while estimates of adult male movement should concentrate on pre-mating movement. Movement of larval stages should also be examined, since between-plant movement can compromise some management options, such as seed mixtures. If transgene expression varies among plant parts, larval movement among those plant parts also should be evaluated.

The scale of adult movement determines how much mixing and mating can occur between individuals emerging from different fields. For the purposes of relative resistance risk assessment of the target species, it is not necessary to have precise quantitative data on the species. In general, the less dispersive a species is, the greater the risk for resistance evolution (Caprio, 2001; Carrière *et al.*, 2004a). This occurs because sedentary species will be more likely to mate with individuals from the field in which they emerge and to oviposit in the same fields, which is likely to lead to greater selection pressure on that subpopulation. Hence, in assessing the resistance risk, it can suffice to rank the dispersiveness of the target species.

Vietnamese production systems are typified by small field sizes (in 2006, the 21,000 cotton farmers grew on average less than 0.7 ha of rainy season cotton each and 8204 farmers grew dry season cotton on an average of 0.35 ha each), considerable levels of intercropping and high crop diversity set in a matrix of diverse natural vegetation. There is little information specific to Vietnam on adult movement of the three target species, although *H. armigera*, *P. gossypiella* and *S. exigua* have been studied extensively elsewhere. *H. armigera* is highly mobile and capable of extensive interregional movements, although at times, populations may appear quite sedentary (Fitt, 1989; King *et al.*, 1990; Feng *et al.*, 2005). *S. exigua* is capable of extensive local and interregional movement, although it is markedly less mobile than congeners such as *S. litura* (Saito, 2000), while *P. gossypiella* is the most sedentary of the three species at risk (Tabashnik *et al.*, 1999; Carrière *et al.*, 2001, 2004a,b). For *P. gossypiella* in Arizona, it was determined that refuges should

Table 12.3. Association, mobility and fecundity of target lepidopteran species with cotton.

Species	Number of generations in cotton	Number of generations per year	Adult dispersiveness	Fecundity
H. armigera	2–4	5–7	Very high	1500–3000
P. gossypiella	2	4	Low	250–500
S. exigua	1–2	8	High	750–2000

be no further than 0.75 km away from Bt cotton fields (Carrière *et al.*, 2004a,b). Given this, and knowledge from elsewhere, it seems reasonable to rank the dispersiveness as: *P. gossypiella* < *S. exigua* < *H. armigera* (Table 12.3). However, because Vietnamese field sizes are so small, it is likely that all species will move sufficiently to leave the field where they emerge at high rates. Vietnamese researchers report that *H. armigera* moths move freely from a variety of neighbouring crops and from relay and intercrops on to cotton (Chapter 2, this volume). In the most popular cropping system, cotton is relayed after maize and intercropped with pulses and other crops, providing a continuous series of suitable host plants for *H. armigera*. These alternative intercrops can provide an important refuge for Bt-susceptible genotypes of *H. armigera*.

12.4. Dose and Efficacy

The dose of insecticidal toxin in Bt cotton is a major factor determining the level of resistance risk. Dose depends on both the concentration of the Cry toxin in the Bt plant and the genetic characteristics of the target pest. A 'high dose' is defined as one that kills a high proportion (> 95%) of heterozygous resistant genotypes similar to homozygous susceptible genotypes (Tabashnik, 1994a; Roush, 1997; Andow *et al.*, 1998; Gould, 1998). For a high dose, resistance is recessive, or nearly so. A 'low dose' is anything that is not a high dose.

Resistance management will differ for high dose versus low dose plants. Simulation models show clearly that a high dose can delay the evolution of resistance more effectively than a low dose (Roush, 1994; Alstad and Andow, 1995; Gould, 1998; Caprio, 1998a; Tabashnik *et al.*, 2003). A high dose may also allow greater options for resistance management with fewer restrictions on how non-transgenic refuges are managed (Carrière and Tabashnik, 2001; Ives and Andow, 2002; Onstad *et al.*, 2002; Storer *et al.*, 2003) and so resistance management may be implemented more readily than for low dose events. Low dose events will require larger non-transgenic refuges and/or restrictions on the management of these refuges. Indeed, in Australia, growers agreed to cap the area of single gene Bt cotton [low dose for *H. armigera*] to 30% of the total crop, in addition to the requirement for refuges (50% sprayed cotton refuge or 10% unsprayed cotton refuge) (Fitt, 2004). In the USA, it has been argued that a 50% refuge may be needed for low dose plants (Gould and Tabashnik, 1998).

Simulations have indicated that a 50% refuge would be needed for low dose plants in Brazil (Fitt *et al.*, 2006).

To evaluate the 'dose', it is essential to have insects resistant to the Bt crop that can be used to create heterozygous individuals which can be challenged with the Bt plant. However, in most cases prior to field release, resistant insects will not have been discovered. When resistance in a target species is not yet known, it is not possible to evaluate heterozygous genotypes, so it is impossible to determine if a transgenic plant is high dose or not. Instead, a temporary, provisional, operational definition of 'high dose' must be used. One such definition is· a plant is provisionally high dose if it expresses toxin at a concentration that is 25 times the LC_{99} of the target pest (Gould and Tabashnik, 1998). This operational definition has been accepted for use by the US Environmental Protection Agency (US EPA). One alternative definition is: a high dose produces at least 99.99% mortality of homozygote susceptibles relative to a non-Bt control (ILSI, 1999). Unfortunately, both these definitions link dose with efficacy, the kill rate for susceptible homozygotes.

Efficacy provides a measure of the selection pressure favouring the *R* allele in Bt crops at the beginning of the evolution of resistance, when resistance is rare. The three target pests differ in the efficacy of the Bt insecticidal proteins (Table 12.1). These evaluations are based on preliminary bioassay and field performance data derived elsewhere and could be revised as rigorous evaluations are completed in Vietnam. Cry1Ac and Cry2Ab may be present at high concentrations in some Bt cotton varieties (Tables 4.1, 4.2 and 4.3 in Chapter 4, this volume). While individually they do not achieve high efficacy against *H. armigera*, combined they can achieve high efficacy for this species. Current levels of expression of either protein result in high efficacy for *P. gossypiella*. By contrast, efficacy of these proteins is low for *S. exigua*. The toxicity of VIP3A against the target pests in Vietnam is not known completely, but evidence of its high efficacy against *H. armigera* (Llewellyn *et al.*, 2007) suggests it may have high efficacy for *H. armigera* and *P. gossypiella* at least.

Comprehensive laboratory or field information is not available for any combination of pests and specific Bt varieties in Vietnam. Based on data from elsewhere, we believe that cotton expressing both Cry1Ac and Cry2Ab represents a moderate to high dose for *H. armigera* and *S. exigua* and a high dose for *P. gossypiella*. We had insufficient information about VIP3A or any Cry1Ac/Ab cottons to evaluate their dose fully, although in the simulations described later, we assumed that VIP3A could be expressed at a high dose for both *H. armigera* and *S. exigua*.

Considering dose, efficacy, association with cotton and association with other crops, we conclude that the species most at risk for evolving resistance to Bt cotton in Vietnam is *H. armigera*.

Bioassays for estimating LC_{99}

Bioassays estimating LC_{50} or LC_{99}, or sublethal effects, are recommended to determine possible dose (Sims *et al.*, 1996). It will be most convenient to conduct the bioassays with purified toxin which is equivalent to that produced by the transgenic plant. The use of purified toxin allows experiments to evaluate the effects of toxin concentrations many times higher than that present in the transgenic plant, which may be necessary to estimate an LC_{50} or LC_{99}.

There are many ways to conduct bioassays. First, the carrier of the toxin should be selected. This can be a natural food source (plant tissue) or artificial diet. Generally, the plant tissue is treated with the toxin by surface application with a series of toxin dilutions. With an artificial diet, it can have toxin incorporated in the mixture (Gould *et al.*, 1997; Hilbeck *et al.*, 1998) or just apply on the surface. The surface treatment of artificial diet can be done by applying each dilution to the diet surface in a 128-well bioassay tray (Marçon *et al.*, 1999). This method conserves toxin and is valuable when only small amounts of toxin are available, but it is important that the surface is uniform. The method underexposes larvae that feed by boring into the diet and thus pass quickly through the treated layer (Bolin *et al.*, 1999). In all cases, neonate larvae should be used. Normally, bioassay trays are incubated for at least 7 days at 27°C, 80% RH and 24 h scotophase or photophase, after which mortality and larval biomass are measured to estimate the lethal concentration (LC) and growth inhibition (GI) (Marçon *et al.*, 1999).

Transgenic plants can also be used to create a series of toxin concentrations by diluting the tissue into an artificial diet (e.g. Olsen and Daly, 2000). This is advantageous because the toxins are in the same form as expressed in the plant, but is disadvantageous because the maximum toxin concentration that can be evaluated is less than what actually occurs in the plant (Andow and Hilbeck, 2004). Plant tissues expressing the highest concentrations of toxins will be preferable because they will allow a greater range of toxin concentrations to be tested. However, other secondary plant compounds in the plant tissues can interfere with the assay by confounding the source of mortality in the assays (Olsen and Daly, 2000). The concentration of the transgene product should also be quantified.

Use of Bt plant tissue typically will not allow estimation of an LC_{99}. However, undiluted tissue can be used in some circumstances as a discriminating concentration to separate resistant and susceptible phenotypes. Excised maize leaf tissue has proven suitable (Huang *et al.*, 2007b), although intact plants may be less suitable (Zhao *et al.*, 2002).

Need to find resistance

Because the actual dose cannot be determined until resistance alleles are recovered in natural populations, assessments of risk without this information are preliminary. We use the limited information available to estimate dose and we make the precautionary assumption that, unless there is evidence that the Bt plant expresses a high dose indicated by consistently high efficacy against a range of field colonies of the pest, then the plant expresses a low dose. Hence, it is of considerable importance to identify resistance alleles in field populations and test their inheritance in the laboratory on Bt plants. Such tests should provide definitive evidence that the Bt plant is a high or low dose plant.

For potential low dose pest species, mass selection on laboratory colonies comprising of freshly collected individuals from the field should be initiated (Akhurst *et al.*, 2003). Considerable thought is required in defining the selection pressure applied in the laboratory, i.e. whether to apply a consistent selection intensity or increasing selection intensity, as this could change the outcome of selection. Caution is also required in extrapolating from laboratory selection to the outcomes of field selection. For potential high dose target species, mass selection may be less likely to recover resistance (e.g. Bolin *et al.*, 1999; Huang *et al.*, 1999) but, in some cases, it can be successful (Gould *et al.*, 1997). For high dose species, additional methods include F_2 screens (Andow and Alstad, 1998; Genissel *et al.*, 2003; Huang *et al.*, 2007a), in-field screens (Tabashnik *et al.*, 2000, 2002; Venette *et al.*, 2000; Morin *et al.*, 2003) and any other approach that can maximize the probability of finding resistance.

If resistance has been recovered and documented from another region in any of the target species, a collaboration may be advisable, both to use previous data as well as to use the resistant colony for future research. It should be noted, however, that the genetic composition of insect populations varies geographically. Thus, the genetic basis of resistance could change from one region to the next.

12.5. Resistance Management Options for Vietnam

Options to delay resistance

Four approaches can be used to delay resistance evolution. The approach used most widely is to reduce the selection pressure (exposure) on the pests to Bt cotton by maintaining refuge plants. Specific issues to be addressed include: size, placement, time of planting and management of refuges (further detail below).

A second approach is to reduce the selection differential between resistant and susceptible insects. The selection differential is the fitness advantage of resistant phenotypes over susceptible phenotypes when both are exposed to the transgenic plant. This can be accomplished by suppressing pests emerging from the transgenic crop with other control tactics, such as insecticides, cultural controls, or more effective biological control.

A third approach is to reduce *RS* heterozygote fitness. Heterozygotes may have a susceptible or a resistant phenotype. If they are phenotypically susceptible, then they have low fitness on the Bt plant (resistance is recessive) and the rate of resistance evolution is slow. It is possible that natural enemies can alter heterozygote fitness; however, little is known about potential selective feeding by natural enemies in cotton. If IPM compatible options for the management of aphids and leafhoppers can be implemented, there may be potential for egg parasitoids and egg predators to reduce *H. armigera* heterozygote fitness.

The fourth approach can be used only with high dose strategies, such as for Bollgard II cotton varieties with two effective Bt toxins. For some species, it may be possible to manage the sex-specific movement and mating frequencies to delay resistance evolution (Andow and Ives, 2002). By using chemical and environmental attractants, it may be possible to enhance the movement of males and reduce simultaneously the movement of females from refuges and to transgenic fields, limiting the impact of source–sink dynamics (Caprio, 2001). Certainly, the simplest approach by far is to reduce selection pressure by maintaining refuges.

Refuge crops

A refuge is a habitat in which the target pest can maintain a viable population in the presence of Bt cotton fields and where there is no additional selection for resistance to Bt toxins and insects occur at the same time as in the Bt fields (Ives and Andow, 2002). Refuges can be structured [deliberately planted in association with the Bt crop] or unstructured [naturally present as part of the cropping system]. The refuge can be managed to control pest damage, as long as the control methods do not reduce the population to such low levels that susceptible populations are driven to extirpation (Ives and Andow, 2002). The effectiveness of any refuge will depend on its size and spatial arrangement relative to the Bt crop, the behavioural characteristics [movement, mating] of the target pests and the additional management requirements of the refuge which, in Vietnam smallholder systems, may work against the successful deployment of a refuge-based system.

A 'seed mixture' is often considered as a possible resistance management tactic, particularly for smallholder systems. It involves mixing the seeds of Bt and non-Bt cotton in the seed bags or planters so that a mixture of Bt and non-Bt plants occurs in each field and decisions about deployment of refuge are taken out of the control of the farmer. While it is true that seed mixtures delay resistance evolution compared to having no refuge at all (Tabashnik, 1994a), they can be compromised seriously by the movement of larvae between plants (Mallet and Porter, 1992). The worst case would occur when resistant heterozygotes can survive on the Bt plant long enough to move to a neighbouring non-Bt plant, where they can complete development and, vice versa, where susceptible larvae and resistant heterozygotes feeding on non-Bt plants move to Bt plants, where susceptibles are killed and heterozygotes survive, so reducing the value of the refuge.

Our understanding of the behaviour of the larvae of *S. exigua* and *H. armigera* suggests that interplant movement of larvae is sufficiently extensive as to compromise the effectiveness of seed mixture refuges; consequently, we advise strongly against consideration of 'seed mixtures' in this environment. *S. exigua* deposits egg masses from which larvae feed communally and may disperse considerable distances within fields. *H. armigera* lays eggs singly on certain plant parts, but the larvae move from plant to plant as they mature (King, 1994). Larvae of *P. gossypiella*, on the other hand, are very sedentary and rarely move between bolls on a plant. If this species were the only pest of cotton in Vietnam, seed mixtures might be a feasible tactic. However, there is no region in Vietnam where this species is the only lepidopteran pest of cotton and it is not a major consideration in irrigated production. Consequently, we conclude that seed mixtures should not be used.

The maximum distance from Bt cotton to the refuge depends on the frequency of movement and distances that adults disperse. The critical dispersal distances are the distances moved by males before mating and the distances moved by virgin females and mated females (David Andow, Minnesota, 2007, personal communication). High male dispersal is favourable for resistance management, while under most scenarios, high female dispersal is generally detrimental for resistance management (Caprio, 2001). Although detailed dispersal data do not exist for any of the target lepidopteran species in Vietnam, it may be reasonable to assume that all species undertake sufficient movement at a local scale of 1–2 km area from their emergence site. For the three at-risk species, it may be necessary to structure a resistance management strategy to take account of the least mobile of the three (*P. gossypiella*), where adult movements would likely be < 1 km (Tabashnik *et al.*, 1999). Refuges integrated at this spatial scale would be quite compatible with the smallholder Vietnamese production system.

Binh Thuan Province, a case study

As described previously, the diversity and interconnectedness of Vietnamese cropping systems with irrigated cotton occupying only a small fraction of the cropped landscape suggest that *H. armigera* and *S. exigua* at least may have considerable natural refuge among other non-Bt crops and some natural hosts. Thus, the first question to be addressed in seeking to design a pre-emptive resistance management strategy for irrigated cotton in Vietnam is to ask whether a structured refuge is required at all. To explore this question in detail, we chose to focus on Binh Thuan Province in the coastal lowlands region of Vietnam, where irrigated dry season cotton production was less than 1000 ha in 2003/04, but may increase to 38,000 ha in 5–10 years (Table 12.4). A trebling of rainfed production is also anticipated from 2500 ha to 7000 ha.

Table 12.4 summarizes the range of crops grown in the rainy season and the dry season and roughly quantifies the relative productivity of adult *Helicoverpa* or *Spodoptera* emerging from each crop. This was achieved by ranking the relative productivity of each crop, where the ranking summarizes the attractiveness of the crop for oviposition, its suitability for larval development and the expected survival through to adult moths. The rankings differ between crops and between the two species. No quantitative information of this type is available for Vietnam and so the rankings were derived from the expertise of the authors. It should be recognized that these are best-guess estimates. Multiplying the ranking by crop area provided a ready means to estimate the productivity of one crop relative to the others.

Table 12.4a shows actual crop areas in the 2003/04 cropping season, while Table 12.4b shows the projected areas following a dramatic increase in cotton area. Much of this more than tenfold increase [3200 ha to 45,000 ha] is projected to occur through replacement of irrigated rice, which would decrease by 26,000 ha.

Presently in Binh Thuan Province, only ~5–6% of *H. armigera* are expected to be generated on cotton in both the rainy and dry seasons (Table 12.4a). This

Table 12.4. Crop areas and estimated production of *Helicoverpa armigera* and *Spodoptera exigua* for each crop in the rainy season and the dry season in Binh Thuan Province, central coastal region of Vietnam. (a) Actual crop areas for 2003/04. (b) Projected crop areas in 5–10 years with an increase in cotton and decrease in rice. Relative production of adult moths from each crop is calculated. Relative production is calculated by assigning a rank to each crop to indicate its 'suitability' for the pest and then multiplying rank by crop area. Sesame, sweet potato, sugarcane, cassava and rice also occur in these landscapes, but they are not hosts for either *H. armigera* or *S. exigua*.

(a) Actual crop areas (2003/04)

Crop	Area rainy season	Area dry season	Area total	Rank Helico.	Rank Spod.	Production (%) Rainy season Helicoverpa	Spodoptera	Production (%) Dry season Helicoverpa	Spodoptera
Cotton	2,600	600	3,200	2	1	5,000 (6.0)	2,500 (2.9)	1,400 (4.7)	700 (2.9)
Maize	10,000	5,563	15,563	3	1	30,000 (36.2)	10,000 (11.8)	16,689 (55.8)	5,563 (23.2)
Groundnut	3,640	3,500	7,140	2	3	7,280 (8.8)	10,920 (12.9)	7,000 (23.4)	10,500 (44.0)
Soybean	220	210	430	1	1	220 (0.3)	220 (0.3)	210 (0.7)	210 (0.9)
Tobacco	0	369	369	1	3	369 (0.4)	1,107 (1.3)	0 (0.0)	0 (0.0)
Mung bean	20,000	2,295	22,295	2	3	40,000 (48.3)	60,000 (70.8)	4,590 (15.4)	6,885 (28.9)

(b) Projected crop areas with expansion of cotton

Crop	Area rainy season	Area dry season	Area total	Rank Helico.	Rank Spod.	Productivity (%) Rainy season Helicoverpa	Spodoptera	Productivity (%) Dry season Helicoverpa	Spodoptera
Cotton	7,000	38,000	45,000	2	1	14,000 (15.2)	7,000 (7.8)	76,000 (72.7)	38,000 (62.1)
Maize	10,000	5,563	15,563	3	1	30,000 (32.7)	10,000 (11.2)	16,689 (16.0)	5,563 (9.1)
Groundnut	3,640	3,500	7,140	2	3	7,280 (7.9)	10,920 (12.2)	7,000 (6.7)	10,500 (17.2)
Soybean	220	210	430	1	1	220 (0.2)	220 (0.2)	210 (0.2)	210 (0.3)
Tobacco	0	369	369	1	3	369 (0.4)	1,107 (1.2)	0 (0.0)	0 (0.0)
Mung bean	20,000	2,295	22,295	2	3	40,000 (43.5)	60,000 (67.2)	4,590 (4.4)	6,885 (11.3)

suggests that unstructured refuges will be sufficient to maintain susceptibility to Bt cotton in this province under present conditions. Moreover, even if Bt maize were to occur without a structured refuge in this province, only 42.2% of the *H. armigera* would be produced from cotton and maize during the rainy season, suggesting that an unstructured cotton and maize refuge would be sufficient. During the dry season, however, cotton and maize account for 60.5% of *H. armigera* production and additional analysis would be needed to determine if an unstructured refuge would suffice. Finally, mung bean provides the bulk of the unstructured refuge, so if mung bean, maize and cotton were all to have a Cry1 toxin, then resistance might occur quickly and it would likely be necessary to put a structured refuge into practice. Groundnut and mung bean provide unstructured refuges for *S. exigua*, so this species appears to have a lower resistance risk than *H. armigera* in this province. In addition, deployment of a Bt soybean would have little influence on resistance risk in this province.

The projected increase in cotton area will see production of *H. armigera* from cotton increase to 15.2% in the rainy season and a massive 72.7% in the dry season (Table 12.4b). If all the increase in cotton area in the dry season was through Bt varieties, the proportion of *H. armigera* potentially exposed to selection by Bt toxins could be so great that the risk of resistance and the necessity for active resistance management becomes much higher. Moreover, if Bt maize were introduced, 47.9% and 88.7% of *H. armigera* might be exposed to selection during the rainy and dry seasons, respectively. Additional analysis will be necessary to obtain a more accurate estimation of resistance risk if cotton area were to increase as projected in the dry season, and especially if Bt maize were commercialized with an unstructured refuge. The resistance risk for *S. exigua* appears to be less than for *H. armigera* because groundnut and mung bean remain excellent unstructured refuges. Deployment of Bt soybean, again, would have little effect on resistance risk, even under the projected increase in cotton area.

It is clear that many of the other crops grown in association with cotton in both rainy and dry seasons are suitable hosts for *H. armigera* and *S. exigua* and can thus be considered as part of a natural unstructured refuge, provided Bt sprays are not used extensively on those crops, as is presently the case in Vietnam. Either the introduction of Bt maize or the large increase in irrigated cotton could result in much higher resistance risk for *H. armigera*. In addition, it is possible that some of the important refuge crops (maize, groundnut or mung bean) could decline if they were replaced by Bt cotton, exacerbating this resistance risk. Such scenarios require carefully designed monitoring and response plans. In particular, it will be important to monitor resistance in *H. armigera* associated both with cotton and maize, and it will be necessary to quantify resistance risk more accurately to determine if a structured refuge is necessary.

Resistance management risk in Vietnam

In this section, we use a deterministic simulation model (Caprio, 1998b) to build on the previous discussion about potential changes in cropping pattern to assess the likelihood of resistance evolution under a range of scenarios. For simplicity, the simulations assumed *H. armigera* as the main target species, assumed selection with an effective high dose Bt toxin, though we incorporated, through sensitivity analysis, some probability that the resistance may be low dose, with functional dominance approaching additivity.

A model of resistance evolution in Vietnam must consider the agricultural context predominated by small landholders with a diversity of choices in crop production. Individual regions have relatively well defined cropping seasons, but each grower has a variety of options with regard to planting specific crops. Exposure of a given insect to a transgenic event will depend on grower choices, as well as the biology of the target insects in the crop landscape. In the case of irrigated cotton in the coastal lowland region of Vietnam, four generations of *H. armigera* are likely to be exposed to the transgenic (e.g. VIP3A or Bollgard II) cotton. We assume that generations of this pest that occur out of this time window are not subject to selection with Bt proteins and are therefore unlikely to play a role in increasing resistance gene frequencies. These generations could, however, contribute to a decline in resistance frequency if there is a fitness cost of resistance. We assume that the response to selection during the four affected generations is independent of population size (but see Neuhauser *et al.*, 2003), so we focus on gene frequencies in each landscape component. We investigate the role of refuges [either structured or unstructured] to reduce the resistance risk.

Given what is known of the ecology of the species at risk in Vietnam and the uncertainties about dose and efficacy of Bt cotton in Vietnam, we examined four different scenarios. The first is a low dose scenario for one of the targeted species. This scenario clarifies the challenges associated with low dose events and illustrates why a high dose event is superior for resistance management. The second, third and fourth scenarios are high dose scenarios. In the second scenario, we assume that cotton producers must manage resistance using structured cotton refuges. This scenario illustrates the challenges associated with structured cotton refuges. In the third and fourth, more realistic scenarios, we examine the risk of resistance evolution using unstructured refuges in the entire landscape. We consider both the present cropping system, with little dry season cotton (third scenario), and the projected cropping system, with large increases in dry season cotton production (fourth scenario, Table 12.4b).

Because field size is very small, we assume that the mobility of the species is high and that there is likely to be a high degree of movement between fields in the landscape (e.g. refuges, transgenic and non-transgenic cotton, other crops). In addition, we assume that resistance is inherited as a monogenic, autosomal trait with susceptible and resistance alleles in a diploid organism and that the initial resistance allele frequency is 0.001. We chose 60 generations (15 years assuming four generations/year on transgenic cotton) as a standard target for the lifetime of any transgenic toxin and scored failure as occurring when the resistance allele frequency exceeded 0.5.

Scenario 1. Low dose transgenic cotton

For the low dose scenario, we assumed that the toxin killed 90% of susceptible homozygotes and 80% of heterozygous (*RS*) individuals. We assumed that there was no additional mortality (e.g. insecticides) on insects in the refuges. In the absence of a refuge, resistance evolved in ten generations (Fig. 12.2a). In order to maintain the effectiveness of the transgene for at least 60 generations, the refuge must exceed 50% (Fig. 12.2a). These simulations were similar to other published simulations of low dose scenarios (Roush, 1997). With a low dose event, extensive areas of refuge were required to delay resistance for an acceptable period of time. For this reason, low dose events are not a preferred option.

Scenario 2. High dose, structured cotton refuges

For the high dose simulations, we assumed that the transgenic plants killed 99.9% of homozygous susceptible insects (and 99.8% of *RS* heterozygotes).

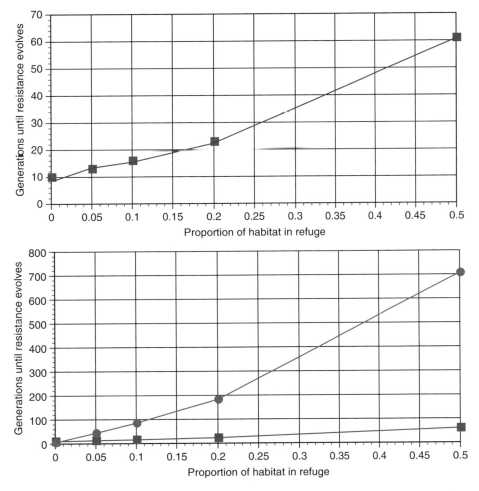

Fig. 12.2. Simulated time to resistance (in generations) with varying proportions of refuge crop. 12.2a. A low dose event as defined in the text. 12.2b. With a high dose event (circles) compared to a low dose event (squares). Note the difference in scale of the *x*-axis.

In this scenario, the structured refuge was conventional cotton which was untreated for *H. armigera*. In the absence of a refuge, resistance to the high dose transgene evolved in approximately half the time of the simulated low dose event (Fig. 12.2b). However, in the presence of a refuge as small as 6–7% of the total cotton habitat, resistance might be delayed longer than the target 60 generations. In this model, when the refuge area alone was treated to reduce survivorship of *H. armigera*, either by insecticidal sprays or other methods, its effectiveness could be reduced substantially. In this case, refuge size must be increased in order to compensate for the additional mortality in these areas. In other models with explicit population sizes, a larger refuge is needed to delay evolution for 60 generations, but evolution is not affected strongly by insecticidal sprays on the refuge (Ives and Andow, 2002).

Scenario 3. High dose, unstructured refuges in present-day landscape

In Scenario 3, we assumed that Vietnamese cotton growers were not required to plant a structured refuge of non-transgenic cotton, but could use an unstructured refuge comprised of fields of other crops and some wild hosts. We used the same model as in the previous scenarios, but incorporated uncertainty in the model parameters. For each uncertain model parameter, we specified the probability that it would take on certain values (Table 12.5). We selected a value for each uncertain parameter randomly and calculated the time to resistance failure. We repeated this several hundred times and graphed the distribution of time to failure. Three of the key uncertain parameters were:

Table 12.5. Parameter values and associated probabilities used to assess the risk of the evolution of resistance using unstructured refuges in Vietnam. For each value of the parameter (upper row), the probability of that value is given (lower row). '–' indicates no value and zero probability.

Parameter		Risk of resistance evolution				
		Very low	Low	Mid	High	Very high
Dominance	Value	–	0.001	0.01	0.1	0.5
	Probability	–	0.3	0.4	0.29	0.01
Initial resistance allele frequency	Value	0.01	0.001	0.0001	–	–
	Probability	0.05	0.65	0.3	–	–
F_{st}	Value	0.0	0.001	0.01	0.05	–
	Probability	0.2	0.4	0.3	0.1	–
Proportion of random mating between unstructured refuge and cotton	Value	1	0.8	–	–	–
	Probability	0.8	0.2	–	–	–
Proportion of unstructured refuge (Scenario 3 – present cropping system)	Value	–	0.4	0.6	0.8	0.9
	Probability	–	0.05	0.1	0.5	0.35
Proportion of unstructured refuge (Scenario 4 – expanded cotton area)	Value	0	0.1	0.2	0.3	–
	Probability	0.1	0.5	0.35	0.05	–

1. *Functional dominance*: while it was most likely that resistance alleles would be recessive (Tabashnik, 1994b), we included a 30% probability that resistance was low dominance and a 1% probability that resistance was additive. These values and probabilities were used because little was known about the dominance of resistance to the VIP3A protein and *S. exigua* might show low dose characteristics.

2. *Initial resistance allele frequency*: nothing was known about the initial frequencies of potential resistance alleles to the VIP3A protein, so we assumed that the initial frequency would be similar to that observed for Cry1A resistance to Bt crops. Estimates of initial allele frequency ranged from exceptionally high 0.16 in *P. gossypiella* (Tabashnik *et al.*, 2000) to 0.001 in *Heliothis virescens* (Gould *et al.*, 1997) and less in *Ostrinia nubilalis* (Bourguet *et al.*, 2003). We therefore gave 5% probability to an initial gene frequency of 0.01, 65% to 0.001 and 30% to 0.0001.

3. *Population substructure*: if movement and mating among populations in different regions of Vietnam were restricted, local variation in gene frequencies might exist and the frequency of rare recessive homozygotes might be somewhat more common in some local populations. Regional differentiation can increase the risk of the evolution of resistance. A standard measure of this differentiation among regions is the inbreeding coefficient Fst. When Fst = 0, there is no differentiation and when Fst = 1, there is complete differentiation. Relative to the scale of cropping in Vietnam, *H. armigera* is highly mobile, with probably little regional differentiation. Mobility of *S. exigua* is probably less than for *H. armigera* and we chose probabilities of regional differentiation to reflect uncertainty regarding this species (Table 12.5), although we weighted heavily the probability towards very low values of Fst (0.001–0.01).

Currently in Binh Thuan Province, the population of *H. armigera* emerging from dry season cotton probably constitutes only 5–6% of the total population during the dry season (Table 12.4a). In Scenario 3, we assumed that the current area of dry season cotton in Binh Thuan remained the same, but that all of this was replaced by Bt cotton. We conservatively used 10–20% emergence from dry season cotton (80–90% of the *H. armigera* population from the unstructured refuge) (Table 12.5).

The outcomes of the simulations show that the introduction of Bt cotton to the current cropping system has very little risk of generating resistance within 100 generations, with only a 0.6% likelihood of resistance arising before 60 generations (Fig. 12.3a).

Scenario 4. High dose, unstructured refuges in landscape with more cotton
Scenario 4 incorporated the projected increase in dry season cotton production (Table 12.4b). In this scenario, most of the *H. armigera* would emerge from cotton, so we changed the probability of emergence so that 80–90% came from cotton and 10–20% from unstructured refuges (Table 12.5). The proportion from unstructured refuges could be even lower if Bt maize and Bt mung bean were used. The values and probabilities for the parameters were unchanged.

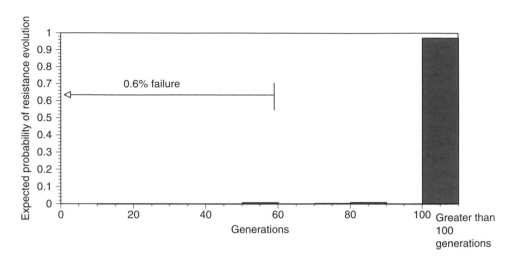

Fig. 12.3a. Expected probability of evolution of resistance to Bt toxins if Bt cotton replaced conventional cotton in the current Binh Thuan cropping system, with reliance on unstructured non-cotton refuges. Resistance does not emerge until at least 100 generations have passed. The probability of failure before 60 generations is 0.6%.

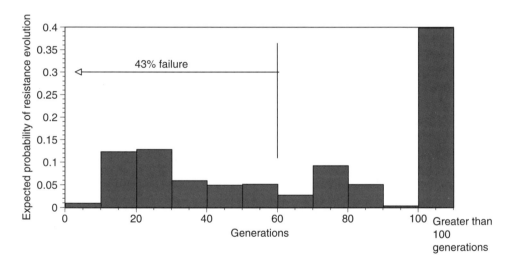

Fig. 12.3b. Expected probability of evolution of resistance to Bt toxins with adoption of Bt cotton and a 50-fold increase in dry season cotton area and reliance on unstructured non-cotton refuges. There are significant risks of resistance emerging quickly, with a 43% risk of failure by 60 generations.

The simulation outcomes indicate that reliance solely on unstructured refuge with a massively increased area of dry season Bt cotton will generate unacceptably high risks of resistance, with 43% risk of failure before 60 generations (Fig. 12.3b).

Finally, we examined the option of requiring a structured refuge of unsprayed non-Bt cotton during the dry season with increased cotton production. The

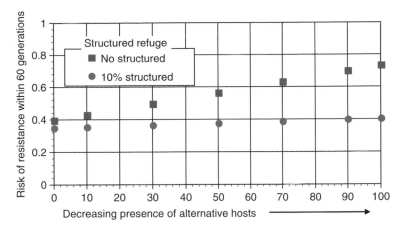

Fig. 12.4. Influence of the addition of 10% structured refuge with Bt cotton with varying proportions of alternative refuge crops on the risk of resistance within 60 generations in a simulated cropping system.

simulations included a 10% unsprayed cotton refuge and varied the probabilities for the size of the unstructured refuge (Table 12.5). Specifically, we changed the probability that there was no unstructured refuge from 0.1 to 1.0, adjusting the probability of larger unstructured refuges accordingly. Inclusion of a 10% structured refuge into the system can stabilize the risk of resistance (Fig. 12.4). As the probability of no unstructured refuge increases, the risk of resistance doubles.

12.6. Implementation of an Integrated Resistance Management Strategy

The above considerations suggest the following key components for a potentially practicable resistance management strategy for Vietnam, focusing primarily on *H. armigera* and secondarily on *S. exigua* and taking into consideration the likely introduction of Bt maize and Bt soybean into these cotton producing regions.

1. Use only high dose plants expressing two separate Bt proteins. These could include, for example, Cry1Ac/Cry2Ab, Cry1F/Cry1Ac, VIP3A/Cry1Ac or VIP3A/Cry1F.
2. Restrict use of Bt cotton to dry season, irrigated production systems. Rainfed production already has effective pest management and will not benefit as much from Bt cotton as dry season production. Cotton quality is much higher from dry season production, increasing even more the value of Bt cotton for dry season production.

By restricting Bt cotton to the dry season, the selection pressure is reduced greatly. If Bt cotton was grown widely in the rainy season, then eight generations of *H. armigera* would be exposed to selection and resistance would have to be delayed 120 generations to preserve the effectiveness of Bt cotton for 15 years. The probability of resistance failure increases substantially, especially if dry season cotton production is expanded (about 70% probability of failure). It should be practicable to restrict Bt cotton to the dry season, as farmers typically do not save seed from one crop to the next because the ginning facilities are centralized.

3. Incorporate Bt cotton as a component of an IPM approach. There is a critical need for non-disruptive, early season management of sucking pests, whether or not Bt cotton is used. If these pests can be controlled, natural enemy communities may be able to establish, which will provide additional control of *H. armigera* and *S. exigua* on Bt cotton and structured and unstructured refuges, which should further delay resistance. This is a critical research need.

4. Plant Bt cotton as early as possible (late November), avoid planting after December. Early planting will reduce attack from *H. armigera* and *S. exigua*, thereby reducing selection and prolonging the effectiveness of Bt cotton.

5. Maintain effective unstructured refuges. Reliance on unstructured refuges is more practicable.

- Maintain and monitor the proportion of effective unstructured refuge fields, such as maize, groundnut and mung bean for *H.* armigera (Table 12.6) and maize, mung bean, groundnut, tobacco and vegetables for *S. exigua* (Table 12.7).
- If the proportion of Bt cotton, Bt maize and Bt soybean is projected to increase to 50% of the total area of dry season host plants for either *H. armigera* or *S. exigua*, then (i) research should be initiated to estimate more accurately the risk of resistance for the species, and (ii) implementation of a structured 10% refuge should be initiated.

6. Avoid seed mixes of Bt and non-Bt seed in the same seed bag. Seed mixes will jeopardize resistance management. Intercrops of alternative crops (e.g. cotton–maize or cotton–mung bean) can be used.

7. Destroy Bt cotton crop residue immediately after harvest. Destruction of crop residues helps reduce pest carry-over to subsequent crops and delay

Table 12.6. Plants that can be used as refuges for *Helicoverpa armigera*.

Crop	Sensitive stage for insect (days after planting)	Growing time (days)	Proposed seeding dates
Bt cotton	35–120	140	1–15/12
Proposed refuge plants			
Non-Bt cotton	35–120	140	1–15/12
Maize	50–70	90–100	1–15/12
Mung bean	35–55	75	1–15/12; 15/2 – 5/3
Soybean	35–55	75	1–15/12; 15/2 – 5/3
Groundnut	35–60	90	1–15/12

Table 12.7. Plants that can be used as refuges for *Spodoptera exigua*.

Crop	Sensitive stage for insect (days after planting)	Growing time (days)	Proposed seeding dates
Bt cotton	7–120	140	1–15/12
Proposed refuge plants			
Non-Bt cotton	7–120	140	1–15/12
Maize	7–35	90–105	1–15/12
Mung bean	20–60	75	1–15/12; 15/2–5/3
Soybean	20–60	75	1–15/12; 15/2–5/3
Groundnut	7–80	90	1–15/12

resistance evolution by eliminating the possibility of pest propagation on ratooning cotton and by destroying potentially resistant insects in dormant stages associated with residues.

8. Implement an effective monitoring programme. This is discussed in greater detail below.

9. Develop and conduct educational programmes for farmers. Farmers need to be informed of the reasons for resistance management so that cooperation and compliance are enhanced.

Under present conditions, this resistance management plan is also likely to be effective for the other species (Table 12.1) that were considered to be a lower resistance risk. *Anomis flava*, *Earias vittella*, *S. litura* and *Sylepta derogata* are all polyphagous herbivores that feed on many of the same crops as *H. armigera* and *S. exigua*. This suggests that the unstructured refuges for *H. armigera* and *S. exigua* are likely to be effective for these other species as well. *P. gossypiella*, however, is a cotton specialist, so none of the other crops will act as an effective refuge for it. Presently, with cotton being a small part of the landscape, *P. gossypiella* is not a major pest, possibly because its cotton host is not common enough for it to produce large pestiferous populations. Thus, although the resistance risk may be high, under present conditions, the consequences of resistance may be small.

However, if cotton does increase substantially during the dry season, it should be expected that *P. gossypiella* will become more common and the consequences of resistance will increase. Thus, it will be essential to implement a 10% structured cotton refuge if Bt cotton increases substantially in area, irrespective of changes in the use of Bt maize or Bt soybean.

Principles for using the management plan

Stakeholders, such as farmers, farmer organizations, ginning facilities, seed companies and government research and outreach personnel may need incentives to implement resistance management. As long as resistance management can rely on unstructured refuges, there may be little need for additional incentives beyond educational efforts to keep growers informed of the need for continued vigilance. However, if a structured refuge becomes necessary, additional incentives may be important. These could range from awarding certificates of merit to participating farmers, providing financial incentives, insuring risks for structured refuges, fines against non-participants, withdrawal of distribution rights for transgenic cotton and so on. All of these issues will need to be considered prior to the development of an incentive programme for resistance management.

More critically, the responsible parties will need to be specified. Who is responsible for making sure that resistance management is followed? In the USA and Australia, individual farmers are required by contract with the technology provider to implement resistance management. Compliance is audited to a greater (Australia) or lesser (USA) degree. The government holds the technology provider responsible for individual growers. Such systems may not be necessary when an unstructured refuge is used, as we have suggested for the initial period in Vietnam.

12.7. Resistance Monitoring

The goal of resistance monitoring is to obtain timely information that can be used to avoid or lessen the adverse consequences of pest resistance (Andow and Ives, 2002). In the case of Bt cotton, this translates into using monitoring information to change the way that Bt cotton is deployed, prior to widespread control failures due to resistance, or to justifying continuation of the current use strategy. Necessary steps in achieving this goal with Bt cotton in Vietnam will include: (i) establishment of baseline susceptibility of target pests involving a dose/response assay; (ii) monitoring the frequency of resistance to determine if it is changing and when it might lead to control failures; (iii) investigation of putative field control failures; and (iv) documenting use of Bt cotton and compliance with the resistance management plan. Baseline susceptibility and resistance frequency monitoring should use the same methods. As discussed below, for low dose events, phenotypic methods may be appropriate and, for high dose events, genic methods are appropriate.

Although cotton is now a small part of the agricultural landscape, if the area expands as planned, or Bt maize is commercialized (which is likely), then the first two monitoring objectives necessitate the establishment and funding of centralized laboratory facilities in Vietnam to conduct resistance monitoring and to coordinate documentation for objective (iv). These laboratories could also be used for resistance monitoring associated with Bt maize. In addition to building the necessary infrastructure, personnel must be trained to conduct monitoring. These centralized facilities could be created at the Nhaho Research Institute for Cotton and Agricultural Development and/or the Plant Protection Research Institute. For Bt cotton, both *H. armigera* and *S. exigua* should be monitored but, in conjunction with Bt cotton, *H. armigera* is more important.

Investigating putative control failures, gathering data on Bt cotton (and Bt maize) use and recording information on compliance with resistance management could be conducted in coordination with the provincial pest protection personnel. Documentation of unstructured refuge could be coupled with GIS (geographic information systems), based on aerial digital imagery of the regional cropping areas. Additionally, Bt Cotton Resistance Working Groups should be convened annually in each region in order to evaluate the RM strategy in light of new findings, to disseminate new research information and to identify the most critical regional research and education needs.

Possible monitoring methods

It will be important to understand the geographic variability in the susceptibility to Bt toxin for the two target pests collected from the major cotton growing regions. This could be done using phenotypic screens for low dose events or genic screens for high dose events. Based on response of different populations to the Bt toxin, diagnostic concentrations could be defined for future resistance monitoring programmes. One possible diagnostic concentration is the concentration in the Bt plant itself. Huang *et al.* (2007b) used excised Bt maize leaf tissue as a diagnostic concentration.

Because of difficulties of finding larvae in Bt cotton fields, changes in susceptibility to Bt toxins can be estimated by collecting target pests in the refuge areas or alternative hosts (maize, groundnut, vegetables, etc.). When conducted in successive years using suitable genotypic methods, it allows detection of regional increases in resistance frequency. Collections of egg masses, larvae, pupae, or adults (using light traps) should be sent or delivered to the centralized regional laboratories for phenotypic screening (low dose events, Marçon *et al.*, 1999) or genic screening (high dose events, Gould *et al.*, 1997; Andow *et al.*, 2000). Both methods can be used to estimate the frequency of the resistance alleles.

Inevitably, some involved parties (farmers, consultants, provincial plant protection staff) will observe an increased damage or presence of insects in Bt cotton areas. Although such instances may indicate resistance and the potential for control failures, it is also common that plants are not really Bt plants, that the pests moved on to cotton from some other host, or many other possibilities having nothing to do with resistance evolution. It is critical that there is excellent coordination among the farmers, provincial plant protection personnel and the centralized laboratories so that surviving larvae can be collected from the field for bioassays in the centralized laboratories.

Monitoring for compliance to the resistance management strategy should focus first at the landscape level when unstructured refuges are the main defence. Aerial photographs of the region can be used to create GIS databases of the region and the crop types during the dry season can be interpreted from the photographs and entered into the database. Summary statistics, such as the proportion of area in each crop, the distance from cotton to a possible refuge field and potentially non-compliant areas can be obtained from the database.

If structured refuges are being used, compliance could focus first on the larger farms and fields. Non-compliance by large-scale producers would be a greater detriment to effective resistance management and will be easier to monitor. Such producers should be required to keep detailed maps of the placement of Bt and refuge cotton. Compliance could be estimated by statistical subsampling of cotton fields to evaluate the degree of concordance between maps provided by the producer and the observed size and location of Bt and non-Bt fields. Collection of leaf tissue or bolls could be made in Bt and non-Bt fields to corroborate designations, or antibody tests could be conducted for the same purpose.

12.8. Conclusions

For any insecticidal transgenic crop, there is no question that resistance is a potential risk to be managed as part of a comprehensive risk assessment process. Fortunately, there is accumulating evidence to show that resistance risks can be managed effectively (Tabashnik *et al.*, 2003). This is clearly the case for deployment of Bt cotton in Vietnam. Due to the diversity, small scale and seasonally dynamic nature of Vietnamese cropping systems, it is necessary to use appropriately innovative methods for evaluating resistance risks and their management. We evaluated a range of different scenarios to accommodate the possible deployment of Bt cotton into current cropping systems and also to evaluate possible future scenarios which reflect changes in the mix, spatial extent and spatial pattern of crops once Bt cotton is available. The pest Lepidoptera at risk of resistance evolution in Vietnam are *H. armigera*, *P. gossypiella* and *S. exigua*. These species exhibit differing sensitivities to Bt proteins and differing ecological traits. Although our analysis lacked specific data on the interaction of these species with the Vietnam cropping system, we concluded that the key species at risk is primarily *H. armigera* and secondarily *S. exigua*. This species is also a pest of maize and use of Bt maize would exacerbate the resistance risk.

In the present agricultural system in Vietnam, cotton is a minor crop. We expect that the resistance risk is small because there are many alternative host plants for *H. armigera* and *S. exigua*, which serve as an unstructured refuge. A practicable resistance management strategy relying on two-gene Bt cotton restricted to irrigated, dry season production can be developed and implemented. This strategy will provide opportunities to integrate the use of Bt cotton in newer, more effective IPM systems.

If cotton production increases substantially during the dry season, or Bt maize becomes common, the utility of the unstructured refuge will decline significantly. We suggest quantitative criteria to trigger revision of the resistance management strategy. If cotton and maize occupy more than 50% of the area of host plants of *H. armigera* or *S. exigua*, then additional research studies should be implemented to quantify more accurately the resistance risk and develop an alternative resistance management strategy. In addition, implementation of a 10% structured refuge should be initiated.

Monitoring should focus on estimating the frequency of resistance in *H. armigera*. This species will be at risk from both Bt cotton and Bt maize, both of which are expected to be commercialized in the near future (Chapter 1, this volume). A centralized laboratory facility should be established to conduct all the necessary monitoring for this pest.

References

Akhurst, R.J., James, W., Bird, L.J. and Beard, C. (2003) Resistance to the Cry1Ac delta-endotoxin of *Bacillus thuringiensis* in the cotton bollworm, *Helicoverpa armigera* (Lepidoptera: Noctuidae). *Journal of Economic Entomology* 96(4), 1290–1299.

Alstad, D.N. and Andow, D.A. (1995) Managing the evolution of insect resistance to transgenic plants. *Science* 268(5219), 1894–1896.

Andow, D.A. and Alstad, D.N. (1998) The F_2 screen for rare resistance alleles. *Journal of Economic Entomology* 91, 572–578.

Andow, D. and Hilbeck, A. (2004) Science-based risk assessment for non-target effects of transgenic crops. *BioScience* 54, 637–649.

Andow, D.A. and Ives, A.R. (2002) Monitoring and adaptive resistance management. *Ecological Applications* 12(5), 1378–1390.

Andow, D.A., Alstad, D.N., Pang, Y.-H., Bolin, P.C. and Hutchinson, W.D. (1998) Using an F2 screen to search for resistance alleles to *Bacillus thuringiensis* toxin in European corn borer (Lepidoptera: Crambidae). *Journal of Economic Entomology* 91(3), 579–584,

Andow, D.A., Olson, D.M., Hellmich, R.L., Alstad, D.N. and Hutchison, W.D. (2000) Frequency of resistance to *Bacillus thuringiensis* toxin Cry1Ab in an Iowa population of European corn borer (Lepidoptera: Crambidae). *Journal of Economic Entomology* 90(1), 26–30.

Bolin, P.C., Hutchison, W.D. and Andow, D.A. (1999) Long-term selection for resistance to *Bacillus thuringiensis* Cry1Ac endotoxin in a Minnesota population of European corn borer (Lepidoptera: Crambidae). *Journal of Economic Entomology* 92(5), 1021–1030.

Bourguet, D., Chaufaux, J., Séguin, M., Buisson, C., Hinton, J.L., Stodola, T.J., Porter, P., Cronholm, G., Buschman, L.L. and Andow, D.A. (2003) Frequency of alleles conferring resistance to Bt maize in French and US corn belt populations of the European corn borer, *Ostrinia nubilalis*. *Theoretical and Applied Genetics* 106(7), 1225–1233.

Caprio, M.A. (1998a) Evaluating resistance management strategies for multiple-toxins in the presence of external refuges. *Journal of Economic Entomology* 91, 1021–1031.

Caprio, M.A. (1998b) Non-random mating model. www.msstate.edu/Entomology/PGjava/ILSImodel.html (accessed April 2004).

Caprio, M.A. (2001) Source–sink dynamics between transgenic and non-transgenic habitats and their role in the evolution of resistance. *Journal of Economic Entomology* 94, 698–705.

Carrière, Y. and Tabashnik, B.E. (2001) Reversing insect adaptation to transgenic insecticidal plants. *Proceedings of the Royal Society of London Series B* 268(1475), 1475–1480.

Carrière, Y., Dennehy, T.J., Pedersen, B., Haller, S., Ellers-Kirk, C., Antilla, L., Liu, Y.-B., Willott, E. and Tabashnik, B.E. (2001) Large-scale management of insect resistance to transgenic cotton in Arizona: can transgenic insecticidal crops be sustained? *Journal of Economic Entomology* 94(2), 315–325.

Carrière, Y., Dutilleul, P., Ellers-Kirk, C., Pedersen, B., Haller, S., Antilla, L., Dennehy, T.J. and Tabashnik, B.E. (2004a) Sources, sinks, and the zone of influence of refuges for managing insect resistance to Bt crops. *Ecological Applications* 14(6), 1615–1623.

Carrière. Y., Sisterson, M.S. and Tabashnik, B.E. (2004b) Resistance management for sustainable use of *Bacillus thuringiensis* crops. In: Horowitz, A.R. and Ishaaya, I. (eds) *Insect Pest Management: Field and Protected Crops*. Springer, New York, pp. 65–95.

Chitkowski, R.L., Turnipseed, S.G., Sullivan, M.J. and Bridges, W.C. Jr (2003) Field and laboratory evaluations of transgenic cottons expressing one or two *Bacillus thuringiensis* var. *kurstaki* Berliner proteins for management of noctuid (Lepidoptera) pests. *Journal of Economic Entomology* 96(3), 755–762.

Cloud, G.L., Minton, B. and Grymes, C. (2004) Field evaluations of VipCotTM for armyworm and looper control. In: Dugger, P. and Richter D. (eds) *Proceedings of the 2004 Beltwide Cotton Conference*, 5–9 January, San Antonio, Texas. National Cotton Council of America, Memphis, Tennessee, p. 1353.

Cui, J.-J. and Xia, J.-Y. (2000) Effects of Bt (*Bacillus thuringiensis*) transgenic cotton on the dynamics of pest population and their enemies. *Acta Phytophylacica Sinica* 27(2), 141–145 (in Chinese with English abstract and tables).

Feng, H.-Q., Wu, K.-M., Ni, Y.-X., Cheng, D.-F. and Guo, Y.Y. (2005) High-altitude windborne transport of *Helicoverpa armigera* (Lepidoptera: Noctuidae) in midsummer in Northern China. *Journal of Insect Behavior* 18(3), 335–349.

Fitt, G.P. (1989) The ecology of *Heliothis* species in relation to agroecosystems. *Annual Review of Entomology* 34, 17–52.

Fitt, G.P. (2004) Implementation and impact of transgenic Bt cottons in Australia. In: *Cotton Production for the New Millennium. Proceedings of the Third World Cotton Research Conference*, 9–13 March, 2003, Cape Town, South Africa. Agricultural Research Council – Institute for Industrial Crops, Pretoria, South Africa, pp. 371–381.

Fitt, G.P., Andow, D.A., Carrière, Y., Moar, W.J., Schuler, T.H., Omoto, C., Kanya, J., Okech, M.A., Arama, P. and Maniania, N.K. (2004) Resistance risks and management associated with Bt maize in Kenya. In: Hilbeck, A. and Andow, D.A. (eds) *Environmental Risk Assessment of Genetically Modified Organisms, Volume 1: A Case Study of Bt Maize in Kenya*. CAB International, Wallingford, UK, pp. 209–250.

Fitt, G.P., Omoto, C., Maia, A.H., Waquil, J.M., Caprio, M., Okech, M.A., Cia, E., Huan, Nguyen Huu and Andow, D.A. (2006) Resistance risks of Bt cotton and their management in Brazil. In: Hilbeck, A., Andow, D.A. and Fontes, E.M.G. (eds) *Environmental Risk Assessment of Genetically Modified Organisms, Volume 2: Methodologies for Assessing Bt Cotton in Brazil*. CAB International, Wallingford, UK, pp. 300–345.

Génissel, A., Augustin, S., Courtin, C., Pilate, G., Lorme, P. and Bourguet, D. (2003) Initial frequency of alleles conferring resistance to *Bacillus thuringiensis* poplar in a field population of *Chrysomela tremulae*. *Proceedings of the Royal Society Biological Sciences Series B* 270(1517), 791–797.

Gould, F. (1998) Sustainability of transgenic insecticidal cultivars: integrating pest genetics and ecology. *Annual Review of Entomology* 43, 701–726.

Gould, F. and Tabashnik, B.E. (1998) Bt-cotton resistance management. In: Rissler, J. (ed.) *Now or Never: Serious New Plans to Save Natural Pest Control*. Union of Concerned Scientists Publications Department, Cambridge, Mississippi, pp. 67–105.

Gould, F., Anderson, A., Jones, A., Sumerford, D.V., Heckel, D.G., Lopez, J., Micinski, S., Leonard, R. and Laster, M. (1997) Initial frequency of alleles for resistance to *Bacillus thuringiensis* toxins in field populations of *Heliothis virescens*. *Proceedings of the National Academy of Sciences of the USA* 94(8), 3519–3523.

Guo, J.-Y., Dong, L. and Wan, F.-H. (2003) Influence of Bt transgenic cotton on larval survival of common cutworm *Spodoptera litura*. *Chinese Journal of Biological Control* 19(4), 145–148 (in Chinese with English abstract).

Haile, F.J., Braxton, L.B., Flora, E.A., Haygood, B., Huckaba, R.M., Pellow, J.W., Langston, V.B., Lassiter, R.B., Richardson, J.M. and Richburg, J.S. (2004) Efficacy of WideStrike cotton against non-heliothine Lepidopteran insects. In: Dugger, P. and Richter D. (eds) *Proceedings of the 2004 Beltwide Cotton Conference*, 5–9 January, San Antonio, Texas. National Cotton Council of America, Memphis, Tennessee, pp. 1339–1347.

Hilbeck, A., Baumgartner, M., Fried, P.M. and Bigler, F. (1998) Effects of transgenic Bt corn-fed prey on mortality and development time of immature *Chrysoperla carnea* (Neuroptera: Chrysopidae). *Environmental Entomology* 27, 480–487.

Huang, D. and Liu, H. (2005) Resistance of three transgenic cotton varieties to *Sylepta derogata*. *Jiangsu Journal of Agricultural Sciences* 21(2), 98–101 (in Chinese with English abstract and tables).

Huang, F., Buschman, L.L., Higgins, R.A. and McGaughey, W.H. (1999) Inheritance of resistance to *Bacillus thuringiensis* toxin (Dipel ES) in the European corn borer. *Science* 284, 965–967.

Huang, F., Leonard, B.R. and Andow, D.A. (2007a) Resistance to transgenic *Bacillus thuringiensis*-maize in sugarcane borer. *Journal of Economic Entomology* 100, 164–171.

Huang, F., Leonard, B.R. and Andow, D.A. (2007b) F_2 screen for resistance to a *Bacillus thuringiensis*-maize hybrid in sugarcane borer (Lepidoptera: Crambidae). *Bulletin of Entomological Research* 97(5), 437–444.

ILSI (International Life Sciences Institute) (1999) An evaluation of insect resistance management in Bt field corn: a science-based framework for risk assessment and risk management. Report of an Expert Panel. International Life Sciences Institute, Washington, DC.

Ives, A.R. and Andow, D.A. (2002) Evolution of resistance to Bt crops: directional selection in structured environments. *Ecology Letters* 5(6), 705–815.

King, A.B.S. (1994) Heliothis/Helicoverpa (Lepidoptera: Noctuidae). In: Matthews, G.A. and Tunstall, J.P. (eds) *Insect Pests of Cotton*. CAB International, Wallingford, UK, pp. 39–106.

King, A.B.S., Armes, N.J. and Pedgley, D.E. (1990) A mark-recapture study of *Helicoverpa armigera* dispersal from pigeonpea in southern India. *Entomologia Experimentalis et Applicata* 55, 257–266.

Kranthi, S., Kranthi, K.R., Siddhabhatti, P.M. and Dhepe, V.R. (2004) Baseline toxicity of Cry1Ac toxin against spotted bollworm, *Earias vittella* (Fab) using a diet-based bioassay. *Current Science* 87(11), 1593–1597.

Llewellyn, D.J., Mares, C.L. and Fitt, G.P. (2007) Field performance and seasonal changes in the efficacy against *Helicoverpa armigera* (Hübner) of transgenic cotton (VipCot) expressing the insecticidal protein Vip3A. *Agricultural and Forest Entomology* 9(2), 93–101.

Luo, S., Wu, K.M., Tian, Y., Liang, G., Feng, X., Zhang, J. and Quo, Y.Y. (2007) Cross-resistance studies of Cry1Ac-resistant strains of *Helicoverpa armigera* (Lepidoptera: Noctuidae) to Cry2Ab. *Journal of Economic Entomology* 100(3), 909–915.

Mallet, J. and Porter, P. (1992) Preventing insect adaptation to insect-resistant crops: are seed mixtures or refugia the best strategy? *Proceedings of the Royal Society of London Series B Biological Sciences* 250(1328), 165–169.

Marçon, P.C.R.G., Young, L.J., Steffey, K.L. and Siegfried, B.D. (1999) Baseline susceptibility of European corn borer (Lepidoptera: Crambidae) to *Bacillus thuringiensis* toxins. *Journal of Economic Entomology* 92(2), 270–285.

Morin, S., Biggs, R.W., Sisteron, M.S., Shriver, L., Ellers-Kirk, C., Higginson, D., Holley, D., Gahan, L.J., Heckel, D.G., Carrière, Y., Dennehy, T.J., Brown, J.K. and Tabashnik, B.E. (2003) Three cadherin alleles associated with resistance to *Bacillus thuringiensis* in pink bollworm. *Proceedings of the National Academy of Sciences of the USA* 100, 5004–5009.

Neuhauser, C., Andow, D.A., Heimpel, G., May, G., Shaw, R. and Wagenius, S. (2003) Community genetics – expanding the synthesis of ecology and genetics. *Ecology* 84(3), 545–558.

Olsen, K.M. and Daly, J.C. (2000) Plant–toxin interactions in transgenic Bt cotton and their effect on mortality of *Helicoverpa armigera* (Lepidoptera: Noctuidae). *Journal of Economic Entomology* 93(4), 1293–1299.

Onstad, D.W., Guse, C.A., Porter, P., Buschman, L.L., Higgins, R.A., Sloderbeck, P.E., Peairs, F.B. and Cronholm, G.B. (2002) Modeling the development of resistance by stalk-boring Lepidopteran insects (Crambidae) in areas with transgenic corn and frequent insecticide use. *Journal of Economic Entomology* 95(5), 1033–1043.

Roush, R.T. (1994) Managing pests and their resistance to *Bacillus thuringiensis:* can transgenics be better than sprays? *Biocontrol Science and Technology* 4, 501–516.

Roush, R.T. (1997) Managing resistance to transgenic crops. In: Carozzi, N. and Koziel, M.G. (eds) *Advances in Insect Control: the Role of Transgenic Plants*. Taylor & Francis, London, pp. 271–294.

Saito, O. (2000) Flight activity of three *Spodoptera* spp., *Spodoptera litura, S. exigua* and *S. depravata*, measured by flight actograph. *Physiological Entomology* 25(2), 112–119.

Selvapandiyan, A., Arora, N., Rajagopal, R., Jalali, S.K., Venkatesan, T., Singh, S.P. and Bhatnagar, R.K. (2001) Toxicity analysis of N- and C-terminus-deleted vegetative insecticidal protein from *Bacillus thuringiensis. Applied and Environmental Microbiology* 67(12), 5855–5858.

Sims, S.B., Greenplate, J.T., Stone, T.B., Caprio, M.A. and Gould, F. (1996) Monitoring strategies for early detection of Lepidoptera resistance to *Bacillus thuringiensis* insecticidal proteins. In: Brown, T.M. (ed.) *Molecular Genetics and Evolution of Pesticide Resistance*. ACS Symposium Series No 645, Washington, DC, pp. 229–242.

Storer, N.P., Peck, S.I., Gould, F., Van Duyn, J.W. and Kennedy, G.G. (2003) Spatial processes in the evolution of resistance in *Helicoverpa zea* (Lepidoptera: Noctuidae) to Bt transgenic corn and cotton in a mixed agroecosystem: a biology-rich stochastic simulation model. *Journal of Economic Entomology* 96(1), 156–172.

Tabashnik, B.E. (1994a) Delaying insect adaptation to transgenic plants: seed mixtures and refugia reconsidered. *Proceedings of the Royal Society of London Series B* 255 (1342), 7–12.

Tabashnik, B.E. (1994b) Evolution of resistance to *Bacillus thuringiensis. Annual Review of Entomology* 39, 47–79.

Tabashnik, B.E., Patin, A.L., Dennehy, T.J., Liu, Y.-B., Miller, E. and Staten, R.T. (1999) Dispersal of pink bollworm (Lepidoptera: Gelechiidae) males in transgenic cotton that produces a *Bacillus thuringiensis* toxin. *Journal of Economic Entomology* 92(4), 772–780.

Tabashnik, B.E., Patin, A.L., Dennehy, T.J., Liu, Y.-B., Carrière, Y., Sims, M.A. and Antilla, L. (2000) Frequency of resistance to *Bacillus thuringiensis* in field populations of pink bollworm. *Proceedings of the National Academy of Sciences of the USA* 97, 12980–12984.

Tabashnik, B.E., Liu, Y.-B., Dennehy, T.J., Sims, M.A., Sisterson, M.S., Biggs, R.W. and Carrière, Y. (2002) Inheritance of resistance to Bt toxin Cry1Ac in a field-derived strain of pink bollworm (Lepidoptera: Gelechiidae). *Journal of Economic Entomology* 95(5), 1018–1026.

Tabashnik, B.E., Carrière, Y., Dennehy, T.J., Morin, S., Sisterson, M.S., Roush, R.T., Shelton, A.M. and Zhao, J.-Z. (2003) Insect resistance to transgenic Bt crops: lessons from the laboratory and field. *Journal of Economic Entomology* 96(4), 1031–1038.

Venette, R.C., Hutchison, W.D. and Andow, D.A. (2000) An in-field screen for early detection and monitoring of insect resistance to *Bacillus thuringiensis* in transgenic crops. *Journal of Economic Entomology* 93(4), 1055–1064.

Wan, P., Zhang, Y.-J., Wu, K.M. and Huang, M.-S. (2005) Seasonal expression profiles of insecticidal protein and control efficacy against *Helicoverpa armigera* for Bt cotton in the Yangtze River valley of China. *Journal of Economic Entomology* 98(1), 195–201.

Wu, K.M. and Guo, Y.Y. (2005) The evolution of pest management practices in China. *Annual Review of Entomology* 50, 31–52.

Yu, C.G., Mullins, M.A., Warren, G.W., Koziel, M.G. and Estruch, J.J. (1997) The *Bacillus thuringiensis* vegetative insecticidal protein VIP3A lyses midgut epithelium cells of susceptible insects. *Applied and Environmental Microbiology* 63(2), 532–536.

Zhang, G.-F., Wan, F.-H., Liu, W.-X. and Guo, J.-Y. (2006) Early instar response to plant-delivered Bt-toxin in a herbivore (*Spodoptera litura*) and a predator (*Propylaea japonica*). *Crop Protection* 25(6), 527–533.

Zhao, J.-Z., Li, Y.-X., Collins, H.L. and Shelton, A.M. (2002) Examination of the F2 screen for rare resistance alleles to *Bacillus thuringiensis* toxins in the diamondback moth (Lepidoptera: Plutellidae). *Journal of Economic Entomology* 95(1), 14–21.

13 Challenges and Opportunities with Bt Cotton in Vietnam: Synthesis and Recommendations

DAVID A. ANDOW, NGUYỄN VĂN TUẤT, ANGELIKA HILBECK, EVELYN UNDERWOOD, A. NICHOLAS E. BIRCH, ĐINH QUYẾT TÂM, GARY P. FITT, MARC GIBAND, JILL JOHNSTON WEST, ANDREAS LANG, LÊ QUANG QUYẾN, LÊ THỊ THU HỒNG, GABOR L. LÖVEI, KRISTEN C. NELSON, NGUYỄN HỒNG SƠN, NGUYỄN HỮU HUÂN, NGUYỄN THỊ HAI, NGUYỄN THỊ THU CÚC, NGUYỄN VĂN HUỲNH, NGUYỄN VĂN UYỂN, PHĂM VẤN LẠM, PHAM VAN TOAN, EDISON R. SUJII, TRẦN ANH HÀO, TRẦN THỊ CÚC HOÀ, VŨ ĐỨC QUANG, RON E. WHEATLEY AND LEWIS J. WILSON

Vietnam is now a member of the World Trade Organization and its economy is being integrated increasingly into the global economy. The economy relies on agricultural products for more than half of its exports and the government is committed to make biotechnology a leading industry in Vietnam, with transgenic crops being one centre of activity. Active biosafety management of transgenic crops is deemed by the government to be essential to this commitment.

Vietnam is a signatory to the Cartagena Protocol and thus must develop policies and regulations on biological safety consistent with the Protocol to regulate import and export of transgenic plants and products. Vietnam has a biosafety decree (Decision 212/2005/QD-TTg)[1] which mandates that the Ministry of Natural Resources and Environment is responsible for coordination of this policy and regulation, and the Ministries of Health, Aquaculture,[2] Industry

[1] Decision No. 212/2005/QD-TTg promulgating the Regulation on Management of Biological Safety of Genetically Modified Organisms, Products and Goods Originating from Genetically Modified Organisms, signed by Prime Minister Phan Van Khai, effective date 26 August 2005 (National Biosafety Regulations).
[2] The Ministry of Aquaculture recently was combined with the Ministry of Agriculture and Rural Development. The new ministry retains the name of the Ministry of Agriculture and Rural Development.

and Trade, Science and Technology, and Agriculture and Rural Development are responsible for implementing different aspects of the regulation of agricultural biotechnology. Transgenic crops from field research to commercialization are the responsibility of the Ministry of Agriculture and Rural Development (MARD), which is now developing regulations that specify procedures for environmental risk assessment (ERA) of field trials and commercialization.

Vietnam has established a long-term plan for the development of agricultural biotechnology.[3] Under this plan, the country has decided to focus its efforts up to 2011 on the commercialization of transgenic cotton, soybean and maize. These crops are not exported (currently, considerable quantities must be imported to support the textile and animal production industries), so the commercialization of transgenic varieties is not expected to jeopardize international trade. The main desired trait is transgenic insect resistance, targeted against lepidopteran leaf- and boll-feeders in cotton, pod-borers in soybean and stem-borers and ear-feeders in maize. Vietnam is focusing on Bt cotton first and intends to use the experience gained from environmental risk assessment and commercialization of this crop for the development and commercialization of other transgenic crops.

This book is the first effort to synthesize scientific information relevant to the biosafety of transgenic crops in Vietnam, taking Bt cotton as an example. It provides biosafety tools that can be applied readily to assess Bt cotton and forms a foundation for future application to other transgenic plants. Nearly 50 Vietnamese scientists have used and evaluated these methods. The authors hope that this book will be useful as a technical tool that Vietnam can use in developing its biosafety regulations. Following these methods will ensure that the crops are assessed, managed and monitored in a safe, efficient and effective manner.

Cotton is a minor crop in Vietnam, currently cultivated on less than 15,000 ha (Chapter 2, this volume). Modern improved hybrid varieties of *Gossypium hirsutum* race *latifolium* are now grown in Vietnam, although small areas of *G. arboreum* continue to be cultivated. Currently, over 90% of the cotton area is rainfed, planted on small plots of less than 1 ha in mixed cropping systems and harvested by hand, with an average yield of around 1 t/ha. In recent years, cotton production has been declining, in part due to the falling price of cotton in relation to other crops, but also due to production difficulties caused by pests and diseases. Dry season irrigated cropping now produces about 20% of the total cotton production and the average yield has increased to 2 t/ha. Vietnam is looking for ways to increase the quantity and quality of cotton production in order to meet partially the large demand of the Vietnamese textile industry. Because fibre quality from irrigated cotton produced during the dry season is better than that during the rainy season, the focus of the Vietnam Cotton Company is to increase the area devoted to dry season cotton from the present ~1500 ha.

---

[3] Decision No. 11/2006/QD-TTg approving the key programme on development and application of biotechnology in the domain of agriculture and rural development up to 2020, signed by Prime Minister Phan Van Khai, effective date 12 January 2006.

For rainy season production, an integrated pest management (IPM) system has been developed and supported by educational efforts. This system is based on use of hairy (pilous) varieties resistant to leafhoppers (*Amrasca devastans*), control of wild and weedy hosts of pests and diseases, seed treatment with neonicotinoid insecticides to control aphids early in the season and the use of selective insecticides against lepidopteran boll pests (mainly *Helicoverpa armigera*) and leafhoppers later in the season. The limited use of foliar insecticide and the diverse small-scale cropping systems usually ensure effective natural biological control by predators, parasitoids and insect pathogens. However, this system has not been effective on dry season, irrigated cotton because pest pressure is much higher. The successful IPM system has also been disrupted by the recent introduction and outbreak of cotton blue disease (CBD) and a reduction in funding for farmers' education. CBD is transmitted by cotton aphids and has led to increased early season foliar insecticide use in attempts to control aphid abundance, and therefore disease spread. These sprays have disrupted natural enemy populations, leading to poor control of *H. armigera* and unacceptable levels of insect damage later in the season. CBD is now a significant challenge to cotton production in Vietnam.

Bt cotton is being evaluated in the hope that it will improve the marginal profitability of cotton in Vietnam by reducing late season insecticide application. Based on experiences in other countries (Chapter 1, this volume), Bt cotton potentially could increase yields and decrease pesticide use. However, effective protection against CBD is a major priority.

In the following, we summarize the most salient points from the individual chapters of the book and provide synthesis and recommendations concerning the biosafety of introducing Bt cotton into Vietnam.

13.1. Problem Formulation and Options Assessment for Environmental Risk Assessment

Problem formulation and options assessment (PFOA, Chapter 3, this volume) frames the environmental risk assessment process as one that informs decision making and societal consideration of the technology from the start. The goals of this process are to involve multiple stakeholders to consider how a transgenic crop may address their needs, identify the adverse effects related to using the transgenic crop in relation to other technological options and to provide information for decision makers about the societal risks and anticipated system changes to mitigate risk associated with the transgenic crop (Nelson and Banker, 2007).

PFOA may help Vietnamese decision making by providing a science-based, multi-stakeholder process to formulate problems and assess options when it is considering the introduction of a genetically modified organism. Public awareness and multi-stakeholder participation in the whole process are the main benefits of a PFOA approach. By following such an approach, government authorities would be sure that potential environmental risks have been assessed clearly on a scientific basis with the full participation of all stakeholders that would be affected, including companies, scientists and farmers.

Three main kinds of potential environmental risks of transgenic crops have been identified (Snow and Moran-Palma, 1997; Snow *et al.*, 2005). These involve possible adverse effects on non-target species and biodiversity, adverse effects due to gene flow and the evolution of resistance in the target pests to insecticidal transgene products such as Cry toxins. We have identified critical aspects of these kinds of risks for the case of Bt cotton in Vietnam and developed assessment strategies.

13.2. Bt Cotton Transgene Locus Structure and Expression

Vietnam plans to develop its own transgenic varieties in cooperation with the owners of useful transgenes (Chapter 4, this volume). Because undesired features of transgene locus structure can pose possible risks to the environment and human or animal health, the design and screening of transgene loci can be used to reduce risk. When possible, a transgenic locus should (i) have no unnecessary transgene DNA, including marker genes, or repeated copies of the transgene or transgene fragments, (ii) not disrupt any functioning plant genes, (iii) not create spurious open reading frames, and (iv) have minimal rearrangements of flanking genomic DNA. By eliminating these elements, the work necessary to characterize a transgenic event for risk assessment purposes can be reduced greatly. In addition, it will be essential to develop DNA primers for use in real time PCR to allow detection, identification, verification and quantification of the transgene, either in commercial products, exports or monitoring programmes. Development of these methods will prove crucial for the implementation of GMO management, including regulatory provisions such as labelling and tracing.

To support environmental risk assessment, it is essential to characterize transgene locus structure, inheritance and expression. Transgene locus structure should be characterized to enable prediction of the likely gene products and phenotypic effects of the transgene locus. The stability of transgene inheritance should be characterized to confirm that the transgene segregates as a normal Mendelian trait by segregation analysis to the F_4 or F_5 generation. Data on the expression of all transgenes is needed to support risk assessment. For non-target risks, expression should be characterized on at least monthly intervals on all relevant plant parts, including floral tissues. For gene flow risks, expression should be characterized in hybrids with relevant recipient populations and backcrosses of these hybrids to the recipient populations under representative environmental conditions. For resistance risks, transgene expression should be studied on the plant parts used by the target pests at times when the target pests are in early development stages and later development stages. The stability of transgene expression in different genetic backgrounds, under edaphic conditions and biotic and abiotic stresses should be characterized.

We have provided a review of information on transgene locus structure and expression of most Bt cotton transgenes that could be or are available to Vietnam (VIP3A and Cry1Fa/Cry1Ac in Chapter 4, this volume; Cry1Ac and Cry1Ac/Cry2Ab in Grossi-de-Sa *et al.*, 2006). This information may be helpful in considering which potential adverse effects merit further investigation prior to commercializing Bt cotton in Vietnam.

13.3. Non-target and Biodiversity Risk Assessment

Transgenic crops could affect biodiversity adversely in the environment where they will be grown or in areas to which they might spread. Adverse effects could include effects on the productivity or sustainability of the agricultural system, or negative effects on economic, conservation, cultural, spiritual or aesthetic values associated with biodiversity. Because there are thousands of species in the agroecosystem that could be affected, it is essential to select a few species that might be most associated with a possible risk. This will focus the risk assessment and make it practicable. In Chapters 6–10, we identify the most likely potential non-target and biodiversity risks (with the greatest potential adverse consequences) that a widely used Bt cotton variety might pose for Vietnam. To accomplish this, it is necessary to use prioritization and selection tools to identify the most significant adverse effects, the most important species or ecosystem processes and the most important mechanisms by which a risk might occur (Hilbeck *et al.*, 2006). A four-step method is described in Chapter 5, this volume.

First, it is essential to identify the potential adverse effects of greatest concern and the ecological functional groups associated with these effects. We focused on potential adverse effects to the production of cotton and nearby crops (damage to pollinators and natural pest control), agricultural sustainability associated with soils, honey production, wild pollinator genetic diversity and species of possible conservation concern. Five functional groups were considered: non-target herbivores on cotton, predators and parasitoids of these herbivores (natural biological control), cotton flower visitors and soil ecosystem processes.

Three of the functional groups of non-target insects (cotton herbivores, predators and parasitoids) are associated with the potential adverse effect of lowered profits from cotton or other nearby crops, either by reduced quantity or quality of yield, or increased pest management costs. Herbivore pest populations could be stimulated if Bt cotton was a better food resource or otherwise more attractive, if the pest was released from competition with the target pest, if the pest was favoured by reduced or more selective insecticide use on Bt cotton, or if it was released from control by predators or parasitoids (if they themselves were affected adversely by Bt cotton). CBD might increase if the aphid vector becomes more abundant or more effective at transmitting the disease.

The flower visitors group is associated with a number of possible adverse effects. An effect on pollinators, a subgroup of the flower visitors group, could lead to: (i) reduced pollination of crops that affects crop production negatively; (ii) adverse effects on the genetic or taxonomic diversity of wild pollinators; (iii) reductions in the production of bee products, leading to economic losses; or (iv) adverse effects on other protected species that feed on wild pollinators and/or their products. Other flower-visiting species that feed on flower resources without necessarily contributing to pollination could also be affected; for example, (v) an adverse effect on flower-visiting natural enemies could lead to a reduction in their biological control capacity; or there could be (vi) a negative effect on flower-visiting species of conservation concern or cultural or aesthetic value, such as butterflies.

Soils contain a substantial proportion of the biodiversity in agroecosystems and healthy soils are essential to maintain the long-term sustainability of production. Soil ecosystem processes can be prioritized by their role as indicators of soil quality or soil health (Arshad and Martin, 2002). In agricultural ecosystems, soil processes are driven largely by the types and amounts of carbon-containing materials entering the soils from plants, so risk assessment can focus on carbon input effects, from both living and dead plant material, on soil ecosystem processes in the soil.

Within each functional group, species and ecosystem processes associated with cotton in Vietnam were listed. A total of 152 species and 14 ecosystem processes were listed in the five functional groups (Tables 6.1, 7.1, 8.1, 9.1 and 10.1 in Chapters 6–10, this volume). We used a Selection Matrix (described in Chapter 5, this volume) to prioritize 25 non-target taxa and five ecosystem processes (Table 13.1), representing only 18% of the initial number.

In the third step, available information was used to focus on a subset of 19 of these selected species and processes to develop risk hypotheses. A risk hypothesis is a hypothetical causal chain leading from the transgenic plant (Bt cotton) via an exposure pathway to the selected species or process and continuing to a potential adverse effect on the environment. A risk hypothesis must be possible to test experimentally, in order to confirm or refute the possible risk (US EPA, 1998).

These risk hypotheses were prioritized by considering the relative likelihood of occurrence and the relative magnitude and irreversibility of the potential adverse effects. Of the 17 herbivore risk hypotheses, the six associated with *Aphis gossypii* and *A. devastans* were judged to be important to investigate more thoroughly. Of the 19 predator and parasitoid risk hypotheses, eight were recommended for additional investigation. Of the nine risk hypotheses associated with flower pollinators, perhaps only one should be considered for additional assessment, because most were considered to have relatively low likelihood or consequences (Table 9.2 in Chapter 9, this volume). Neither of the two soil process hypotheses was recommended for pre-release assessment because any significant adverse effect was considered subtle and unlikely to be detected in short-term experiments. Instead, it was suggested that soil organic content be monitored after the commercialization of Bt cotton. This prioritization process resulted in a focus on 15 risk hypotheses associated with seven non-target species, listed in Table 13.2. Because cotton is grown on small areas in Vietnam, most of the risk hypotheses are unlikely to have serious, irreversible effects on the environment and only the most serious concerns are likely to have any measurable effect in the landscapes where cotton is grown. Vietnam would be advised to consider which of these risk hypotheses merit investigation for the commercialization of Bt cotton, as some are not well connected to serious adverse effects. If we had been considering maize or soybean, which are grown much more widely in Vietnam, it would be justified to consider more species and processes for risk assessment.

Table 13.1. List of high priority non-target species and ecosystem processes identified using the Selection Matrix (Chapters 6–10, this volume).

Functional group and chapter	Species or taxon or ecosystem process	Order and family/taxonomic group
Herbivore pests (Chapter 6)	*Aphis gossypii* Glover	Homoptera: Aphididae
	Amrasca devastans (Distant)	Homoptera: Cicadellidae
	Bemisia tabaci (Gennadius)	Homoptera: Aleyrodidae
	Thrips palmi Kar.	Thysanoptera: Thripidae
	Scirtothrips dorsalis Hood	Thysanoptera: Thripidae
	Tetranychus urticae Koch	Acarina: Tetranychidae
	Spodoptera litura (Fabr.)	Lepidoptera: Noctuidae
	Spodoptera exigua (Hübner)	Lepidoptera: Noctuidae
Predators[a] (Chapter 7)	*Menochilus sexmaculatus* (Fabr.)	Coleoptera: Coccinellidae
	Eocanthecona furcellata Wolff.	Heteroptera: Pentatomidae
	Pardosa pseudoannulata (Boes. et Str.)	Araneida: Lycosidae
	Oxyopes javanus Thorell	Araneida: Oxyopidae
	Paederus fuscipes Curtis	Coleoptera: Staphylinidae
	Ischiodon scutellaris Fabr.	Diptera: Syrphidae
	Ophionea indica (Thunb.)	Coleoptera: Carabidae
Parasitoids[b] (Chapter 8)	*Apanteles* sp.	Hymenoptera: Braconidae
	Trichogramma chilonis Ishii	Hymenoptera: Trichogrammatidae
	Aphelinus sp.	Hymenoptera: Aphelinidae
Flower visitors (Chapter 9)	*Apis cerana* Fabr.	Hymenoptera: Apidae
	Apis mellifera L.	Hymenoptera: Apidae
	Bombus spp.	Hymenoptera: Apidae
	Megachile sp.	Hymenoptera: Megachilidae
	Precis atlites John.	Lepidoptera: Nymphalidae
	Phalanta sp.	Lepidoptera: Nymphalidae
	Didea fasciata Macquart	Diptera: Syrphidae
Soil ecosystem processes (Chapter 10)	Biomass decomposition	Soil macroorganisms, fungi, bacteria
	Cellulose and lignin breakdown	Fungi, bacteria
	Phosphorus and micronutrient uptake	Cotton mycorrhizae
	Soil particle aggregation	
	Water holding capacity	

Notes: [a]The predator group selection process used additional selection criteria to narrow down the selection from 14 taxa (Table 7.2 in Chapter 7, this volume) to seven taxa. [b]The parasitoid group selection process was incomplete due to knowledge gaps and these three species were chosen as case examples to represent a broad taxonomic and ecological range. They are not necessarily the highest priority species and the selection process should be repeated once more information is available.

Table 13.2. List of highest priority risk hypotheses identified for assessment of non-target and biodiversity risks of Bt cotton in Vietnam.

| Species | Risk hypotheses | | |
	Exposure pathway:	Leads to effect:	Adverse effect(s) pathway:[a]
Herbivore pests *Aphis gossypii* and *Amrasca devastans* (Chapter 6)	1/2. Improved food quality of Bt cotton increases survival of sucking pests 3/4. Increased attractiveness of Bt cotton increases immigration and/or oviposition of sucking pests on Bt cotton	Increased population of sucking pests either: (a) Early season, during vegetative growth and flowering, causing damage to seedling or leaf area of growing plant; (b) During boll maturation period, causing damage to and loss of squares and bolls	(a) Increased pest management costs;[b] (b) Cotton crop losses due to damage caused by sucking pests (delayed maturity leading to yield loss; yield loss due to damage to and loss of squares and bolls); (c) Yield loss of intercrop or nearby crops
Herbivore pest *A. devastans* (Chapter 6)	5. Reduction of late season insecticide sprays on Bt cotton in dry season results in lower mortality of *A. devastans* in the late season		
Herbivore pest *A. gossypii* + cotton blue disease (CBD) pathogen (Chapter 6)[c]	6. Improved food quality of Bt cotton and/or increased attractiveness of Bt cotton cause increased population density of *A. gossypii* and increased transmission of CBD pathogen	Increased prevalence of CBD	(a) Increased cotton crop losses due to CBD; (b) Increased aphid pest management costs[b]
Predator *Menochilus sexmaculatus* (Chapter 7)	7. Adult beetles feed on (a) Bt cotton pollen and/or (b) other prey on Bt cotton, and this leads to lethal or sublethal effect(s) on larvae or adults	Reduction in population density of *M. sexmaculatus* in cotton early in the growing season and subsequent reduction in biological control of *A. gossypii*	(a) Increased cotton crop losses due to CBD because of increased *A. gossypii* populations and higher transmission of CBD; (b) Increased intercrop losses due to increased aphid populations on intercrops and/or increased aphid transmitted diseases on intercrops; *Continued*

Table 13.2. *Continued*

Species	Exposure pathway:	Leads to effect:	Adverse effect(s) pathway:[a]
		Risk hypotheses	
	8. Bt cotton reduces the density of other prey early in the season (e.g. thrips), which leads to lower survival, oviposition and/or higher emigration prior to the appearance of large populations of *A. gossypii*		(c) Increased pest management costs[b]
Predator *Pardosa pseudoannulata* (Chapter 7)	9. Feeding on prey on Bt cotton and this leads to lethal or sublethal effects on the spider 10. Bt cotton reduces density of prey, such as small lepidopteran larvae, which leads to lower survival, oviposition and/or higher emigration of *P. pseudoannulata*	Reduction in population density of *P. pseudoannulata* in cotton and subsequent reduction in biological control of main prey species	(a) Increased cotton crop losses from increases in populations of some main prey species, such as possibly *A. devastans*; (b) Increased intercrop losses from increased populations of some main prey species, such as lepidopteran larvae on intercrops; (c) Increased pest management costs[b]
Predator *Eocanthecona furcellata* (Chapter 7)	11. Feeding on lepidopteran prey on Bt cotton and this leads to lethal or sublethal effects on *E. furcellata*	Reduction in population density of *E. furcellata* and subsequent reduction in biological control of other lepidopteran larvae	(a) Increased intercrop losses from increased populations of lepidopteran larvae associated with the intercrops; (b) Increased pest management costs[b]
Parasitoid *Apanteles* sp.[d] (Chapter 8)	12. Adult feeding on (a) nectar, (b) pollen, and/or (c) honeydew from sucking pests of Bt cotton and this leads to lethal or sublethal effects on adults	Reduction in population density of *Apanteles* sp. in cotton and subsequent reduction in biological control of main lepidopteran larvae on other crops (e.g. *H. armigera*, *A. flava*, *E. vitella*)	(a) Increased intercrop losses from increased populations of lepidopteran larvae (e.g. *H. armigera*, *A. flava*, *E. vitella*) on intercrops or neighbouring crops; (b) Increased pest management costs[b]

Continued

Table 13.2. *Continued*

| | Risk hypotheses | | |
Species	Exposure pathway:	Leads to effect:	Adverse effect(s) pathway:[a]
	13. Larvae *Apanteles* sp. develop inside *Spodoptera exigua* or *S. litura* on Bt cotton and this leads to lethal or sublethal effects on *Apanteles* sp		
	14. Bt cotton reduces the density of important prey on cotton, which reduces *Apanteles* larval survival, adult oviposition and/or increased adult emigration		
Flower visitor *Apis cerana* (Chapter 9)	15. Feeding on Bt cotton pollen or nectar reduces bee survival or reproduction, or Bt cotton replaces other important sources of pollen and nectar, resulting in less food for the bees	Reduction in bee colony density and/ or colony quality	(a) Reduced production of bee products (honey etc.); (b) Reduced honey production affects dependent species, such as civets, adversely

Notes: [a]In addition, the introduction of Bt cotton into cotton cropping systems could (i) increase cotton monoculture, reducing crop and habitat diversity; (ii) alter use of pesticide or fertilizer use, which could in turn affect non-target species, increase crop losses and/or pest management costs. [b]Any of these hypotheses could result subsequently in further adverse effect(s): increased pest abundances may spur farmers to increase insecticide spraying, which could (i) increase financial costs, (ii) increase labour requirements, (iii) cause negative effects on environmental and human health, (iv) stimulate resurgence of pests, which could result in additional damage to cotton and/or intercrops. In addition, the introduction of Bt cotton into cotton cropping systems could (i) increase cotton monoculture, reducing crop and habitat diversity; (ii) alter use of pesticide or fertilizer use; which could in turn affect non-target species, increase crop losses and/or pest management costs. [c]If cotton diseases transmitted by *Bemisia tabaci* are discovered in Vietnam, this hypothesis would apply to that *B. tabaci* + pathogen system. [d]The *Apanteles* species need to be identified before experiments can be conducted.

Fourth, for each of these priority risk hypotheses, the respective chapter develops an analysis plan and proposes experiments to begin the assessment process. For example, Risk Hypothesis 1 for *A. gossypii* (Table 13.2) could be tested initially in a laboratory trial if feeding on Bt cotton results in lower aphid survival. Risk Hypothesis 4 for *A. devastans* (Table 13.2) could be tested initially in a laboratory trial measuring settling, movement and oviposition of adults. Experimental details are provided in Chapter 6 (this volume). Other examples can be found in Chapters 6–10 (this volume).

Several additional research activities were identified which will support non-target and biodiversity risk assessment and post-commercialization monitoring of the Bt cotton in Vietnam. There is a need to collate, synthesize and make available existing knowledge on cotton pests and beneficial species in the different regions of Vietnam (Chapter 6, this volume), to conduct additional work on species identification (Chapters 6 and 8, this volume) and to use field experiments to measure population level and large-scale effects (Chapter 7, this volume).

It is also important to develop a long-term research and monitoring plan, including detection and monitoring of potential secondary pests after commercial release (Chapter 6, this volume). If Bt cotton is adopted widely in Vietnam, potentially it could change the cotton cropping systems to reduce habitat diversity for beneficial and valued species such as parasitoids, predators, pollinators, species of conservation concern and other biodiversity (Chapters 6–10, this volume), resulting in the adverse effects mentioned previously. To become aware of such potential effects, it was recommended to focus initial monitoring on the agronomic practices used for cotton production, quantifying the area of various cotton cropping systems (e.g. relay cropping, intercropping), use of fertilizer and insecticides and any other significant changes in crop management (Chapters 8 and 10, this volume).

13.4. Evaluating the Possibility and Consequences of Gene Flow from Bt Cotton

Movement of transgenes from transgenic plants into non-transgenic crop or wild recipient populations could lead to adverse ecological, economic or social effects and, once transgenes have moved into these recipient populations, they may be impossible to remove from the environment. Possible adverse economic and social consequences could include adverse effects on crop management, such as increased weediness from weeds receiving the transgene, threats to transgene-free production such as organic or conventional farming, or loss of revenue and export markets due to difficulties in segregating and labelling conventional and transgenic products. Adverse ecological consequences could include loss of valuable crop genetic diversity, or adverse effects on natural biodiversity.

The analysis of gene flow risks was considerably simpler in Vietnam (Chapter 11, this volume) compared to our previous analysis in Brazil (Johnston *et al.*, 2006), because of several factors. First, Vietnam is not a centre of origin for the genus *Gossypium* and there are no wild species of *Gossypium* in the country. Second, *G. hirsutum* race *latifolium* has been grown in Vietnam for a short period of time and, in the past several decades, much of the originally introduced germplasm has been replaced by modern varieties. Third, cotton is a minor crop in Vietnam, so there is limited potential for large-scale releases of Bt cotton.

We concluded that gene flow from Bt cotton (*G. hirsutum* race *latifolium*) had a very low likelihood of leading to any adverse ecological effect in Vietnam. The only potential recipient populations for transgenes from *G. hirsutum* race *latifolium* in Vietnam are *G. arboreum*, *G. herbaceum*, *G. barbadense* and other *G. hirsutum* race *latifolium*, all of which are cultivated crops. None of these are presently weedy and feral populations of the tetraploid crops are uncommon because the landscape is managed heavily. Thus, the possibility of increased weediness is remote. Two of these species, *G. arboreum* and *G. herbaceum*, are not sexually compatible with the tetraploid *G. hirsutum* race *latifolium* because they are diploids, so the likelihood of gene flow to these two species is very low. Both *G. barbadense* and *G. hirsutum* race *latifolium* have limited feral populations in Vietnam, so there is little probability that a transgene could persist in a naturally reproducing reservoir. Both species are also unlikely to harbour any significant genetic variation, as the original introductions into Vietnam were from a relatively narrow genetic base and most of the original germplasm has been replaced recently by modern varieties. Thus, there are no landraces and little possibility for unique and important genetic variation to exist in the present crop landscape (unlike that present in *G. arboreum*). However, minor exports of cottonseed meal might be affected economically for some markets by the need to test for transgene presence and segregate products.

13.5. Resistance Risk Assessment and Management for Bt Cotton in Vietnam

Experience with insecticides and basic consideration of evolutionary theory indicate that if a Bt crop is used extensively without appropriate management, resistance in target pests is a likely inevitable consequence. Even in Vietnam, where cotton is a minor crop, the potential evolution of resistance is a risk worth careful consideration. The main potential adverse consequences of resistance are control failures, yield loss and economic hardship when the pest is otherwise difficult to control. Increased use of pest management tactics back to levels similar to those required for conventional cotton could be the consequence. This might mean a return to the use of insecticides, which have significant adverse effects on human health and the environment and increased production costs for growers. It is noteworthy, however, that integration of Bt cotton into a background of IPM practices, which have been adopted widely in Vietnam, could assist with the preservation of Bt technology and help establish newer, more effective IPM systems.

Resistance risk assessment of Bt cotton in Vietnam was conducted by identifying the species potentially at risk of developing resistance and determining which species is at greatest risk by considering the history of resistance evolution, the likely 'dose' of the transgenic toxin for each species and the potential exposure of each species to the dose that may lead to selection in favour of resistance (Chapter 12, this volume). Seven lepidopteran pests were identified as potentially at risk; three of these were considered at higher risk; and

H. armigera was considered the species most likely to evolve resistance to Bt cotton in Vietnam. Moreover, *H. armigera* is an important pest of maize, so if Bt maize is commercialized in Vietnam, the resistance risk will be even greater.

To mitigate this risk, we considered several management options and focused on the use of refuges because this management strategy was likely to be more practicable for Vietnamese farmers than the other possibilities. A refuge is a habitat where the pest (in this case, *H. armigera*) can live and reproduce but is not subject to selection from a Bt crop.

To evaluate this proposed insect resistance management (IRM) strategy and assess its workability, we examined several scenarios for the use of Bt cotton in the central coastal lowland provinces of Ninh Thuan and Binh Thuan, where a large proportion of Vietnam cotton is grown and where increased dry season, irrigated cotton production is planned. We assumed that the present, effective, inexpensive IPM system would be the main control strategy used during the rainy season and that Bt cotton would be restricted to the dry season, where the need is greatest and effective control options are most limited. We examined four scenarios: (i) low dose Bt cotton; (ii) high dose Bt cotton with a structured refuge under present cotton production; (iii) high dose Bt cotton with an unstructured refuge under present cotton production; and (iv) high dose Bt cotton with a large expansion of dry season, irrigated cotton production. A structured refuge is one that is planted near Bt cotton deliberately and an unstructured refuge relies on the other crops already grown as part of the local cropping system and where Bt is not used. Unstructured refuges can be reliable when they are widespread in the landscape.

These scenarios demonstrated some important points for the use of Bt cotton in Vietnam: (i) low dose Bt cotton is likely to lead rapidly to resistance, even under present conditions, unless large refuges are planned and used. Vietnam should strive to use only high dose Bt cotton; (ii) under present production systems and with the low proportion of cotton, unstructured refuges provide adequate IRM for high dose Bt cotton. Thus, no special changes in production would be needed, except that growers should be informed of the need for continued vigilance; (iii) if dry season production is expanded up to or beyond the present expansion goals, the effectiveness of the unstructured refuge will decline substantially and the refuge requirements must be re-examined, when it may become necessary to implement a structured refuge; and (iv) this would be even more important if Bt maize or Bt mung bean start to be used widely in these provinces, because these two crops are the main unstructured refuges for Bt cotton.

Based on our assessment of resistance risk and our evaluation of workable IRM strategies, we recommend for Vietnam:

1. Use only high dose plants expressing two Bt proteins that do not share the same receptor in the target insect.
2. Incorporate Bt cotton in an IPM approach.
3. Plant Bt cotton as early as possible for dry season production.
4. Maintain effective unstructured refuges.
5. Monitor the proportion of the landscape in effective unstructured refuge fields.

6. Avoid seed mixes of Bt and non-Bt seed in the same seed bag.
7. Destroy Bt cotton crop residue by removal of stalks and cultivation immediately after harvest.
8. Implement an effective resistance monitoring programme.
9. Develop and conduct educational programmes for farmers.
10. Initiate research to find resistance genes in the target pests, especially *H. armigera*, which will be needed when Bt maize is evaluated.

13.6. Conclusion

The environmental risks associated with Bt cotton in Vietnam need to be assessed on a case-by-case basis, taking into account the Vietnamese environment (e.g. cropping systems and biodiversity) as it relates to a Bt gene in cotton. We have identified some issues that could be assessed for non-target and biodiversity risks for any Bt cotton transgene that might be introduced into Vietnam (Table 13.2). No significant ecological risks resulting from transgene movement from Bt cotton to other plants were identified. Resistance risks are significant, especially if the planned increases in dry season, irrigated cotton production are achieved. We conclude that these risks can be managed as long as similar Cry toxins do not become common in maize and mung bean.

Improving cotton production in Vietnam presents significant challenges, including the improvement of irrigation systems for dry season, irrigated production, the development and implementation of new IPM practices against CBD and its aphid vector, which is presently the major limiting factor of cotton production. Although lepidopteran boll-feeders, such as *H. armigera*, can be important factors reducing cotton yield, without effective, inexpensive control of CBD at present, Bt cotton may help improve production efficiency marginally so that cotton production can be maintained.

This book can be used as a technical manual to enable Vietnamese scientists to evaluate the potential environmental impacts of Bt cotton varieties prior to commercialization. With appropriate modification, this book can provide guidance for environmental risk assessment of any transgenic crop, especially those that Vietnam has targeted for commercialization in the next 5 years, Bt maize and Bt soybeans. We would expect that the risk hypotheses for non-target risk assessment will be different for these crops, but in Chapters 5–10 we have provided an overall model and illustrated its applicability to focus non-target risk assessment. In Chapter 11, we provide a structured procedure for analysis of the risks posed by transgene flow (Box 11.1 in Chapter 11, this volume), which can be applied for a screening level assessment and identification of key knowledge gaps, such as the genetic diversity in local maize varieties (NMRI, 2005), or the distribution of wild relatives of soybean (Lu, 2004). In addition, we suggested that cumulative effects of multiple Bt crops relying on similar Cry toxins might result in a much greater resistance risk, which may in turn require more active resistance management. Such cumulative risks should be examined carefully as each additional transgenic crop is assessed.

This book is part of a larger effort in Vietnam to increase investment in studying the environmental risks of transgenic plants, managing transgenic products safely and building modern facilities to assess their safety. Vietnam is committed to the development of biotechnology and transgenic crops and products to increase economic value for the country. This effort should be accompanied by scientific risk assessment in the agricultural ecological system, which is a foundation for creating a safe, sustainable and effective agriculture.

References

Arshad, M.A. and Martin, S. (2002). Identifying critical limits for soil quality indicators in agro-ecosystems. *Agriculture, Ecosystems and Environment* 88, 153–160.

Grossi-de-Sa, M.F., Lucena, W., Souza, M.L., Nepomuceno, A.L., Osir, E.O., Amugune, N., Tran Thi Cuc Hoa, Hai Truong Nam, Somers, D.A. and Romano, E. (2006) Transgene expression and locus structure of Bt cotton. In: Hilbeck, A., Andow, D.A. and Fontes, E. M.G. (eds) *Environmental Risk Assessment of Genetically Modified Organisms, Volume 2: Methodologies for Assessing Bt Cotton in Brazil*. CAB International, Wallingford, UK, pp. 93–107.

Hilbeck, A., Andow, D.A., Arpaia, S., Birch, A.N.E., Fontes, E.M.G., Lövei, G.L., Sujii, E., Wheatley, R.E. and Underwood, E. (2006) Methodology to support non-target and biodiversity risk assessment. In: Hilbeck, A., Andow, D.A. and Fontes, E.M.G. (eds) *Environmental Risk Assessment of Genetically Modified Organisms, Volume 2: Methodologies for Assessing Bt Cotton in Brazil*. CAB International, Wallingford, UK, pp. 108–132.

Johnston, J.A., Mallory-Smith, C., Brubaker, C.L., Gandara, F., Aragão, F.J.L., Barroso, P.A.V., Vu Duc Quang., Carvalho, L.P. de, Kageyama, P., Ciampi, A.Y., Fuzatto, M., Cirino, V. and Freire, E. (2006) Assessing gene flow from Bt cotton in Brazil and its possible consequences. In: Hilbeck, A., Andow, D.A. and Fontes, E.M.G. (eds) *Environmental Risk Assessment of Genetically Modified Organisms, Volume 2: Methodologies for Assessing Bt Cotton in Brazil*. CAB International, Wallingford, UK, pp. 261–299.

Lu, B.-R. (2004) Conserving biodiversity of soybean gene pool in the biotechnology era. *Plant Species Biology* 19(2), 115–125.

Nelson, K.C. and Banker (2007) *Problem Formulation and Options Assessment Handbook: A Guide to the PFOA Process and How to Integrate it into Environmental Risk Assessment (ERA) of Genetically Modified Organisms (GMOs)*. University of Minnesota, St Paul, Minnesota and GMO ERA Project. http://www.gmoera.umn.edu (accessed 11 February 2008).

NMRI (National Maize Research Institute) (2005) Result of 31 collected local maize populations in Vietnam in 2005. http://www.generationcp.org/capcorner/pop_diversity_wksp/presentations/presentation_1 (accessed 17 September 2007).

Snow, A.A. and Moran-Palma, P. (1997) Commercial cultivation of transgenic plants: potential ecological risks. *BioScience* 47(2), 86–97.

Snow, A., Andow, D.A., Gepts, P., Hallerman, E.M., Power, A., Tiedje, J.M. and Wolfenbarger, L.L. (2005) Genetically engineered organisms and the environment: current status and recommendations. *Ecological Applications* 15(2), 377–404.

US EPA (United States Environmental Protection Agency) (1998) *Guidelines for Ecological Risk Assessment*. EPA/630/R095/002F. United States Environmental Protection Agency, Risk Assessment Forum, Washington, DC.

Index

Note: Page numbers in italic refer to tables or figures

345